ISNM 74:
International Series of Numerical Mathematics
Internationale Schriftenreihe zur Numerischen Mathematik
Série internationale d'Analyse numérique
Vol. 74

Edited by
Ch. Blanc, Lausanne; R. Glowinski, Paris;
G. Golub, Stanford; P. Henrici, Zürich;
H. O. Kreiss, Pasadena; A. Ostrowski, Montagnola;
J. Todd, Pasadena

Birkhäuser Verlag
Basel · Boston · Stuttgart

Delay Equations, Approximation and Application

International Symposium at the
University of Mannheim,
October 8–11, 1984

Edited by

G. Meinardus
G. Nürnberger

1985

Birkhäuser Verlag
Basel · Boston · Stuttgart

Editors

G. Meinardus
Universität Mannheim
Lehrstuhl für Mathematik IV
D–6800 Mannheim 1

G. Nürnberger
Universität Mannheim
Fakultät für Mathematik
und Informatik
D–6800 Mannheim

Supported by »Stiftung Volkswagenwerk«

CIP-Kurztitelaufnahme der Deutschen Bibliothek

Delay equations, approximation and application:
internat. symposium at the Univ. of Mannheim,
October 8–11, 1984 / ed. by G. Meinardus ;
G. Nürnberger. – Basel ; Boston ; Stuttgart : Birkhäuser,
1985.
  (International series of numerical mathematics ;
  Vol. 74)
  ISBN 978-3-0348-7378-9   ISBN 978-3-0348-7376-5 (eBook)
  DOI 10.1007/978-3-0348-7376-5

NE: Meinardus, Günter [Hrsg.]; Universität ⟨Mannheim⟩;
GT

5

CONTENTS

PREFACE

The international symposium held in October 1984 at the University of Mannheim was the first with the special aim to expose the connection of the Theory of Delay Eauations and Approximation Theory with the emphasis on constructive methods and applications. Although the separate character of both domains is reflected by their historical development, the latest research shows that the numerical treatment of Delay Equations leads to various approximation and optimization problems. An introductory survey of this circle of problems written by the editors is included at the beginning of the book.

Delay Equations have their origin in domains of applications, such as physics, engineering, biology, medicine and economics. They appear in connection with the fundamental problem to analyse a retarded process from the real world, to develop a corresponding mathematical model and to determine the future behavior.

Thirty mathematicians attended the conference coming from Germany, West- and Eastern Europe and the United States- more than twenty of them presented a research talk.

The lectures about Delay Equations were mainly oriented on the following subjects: single-step, multi-step and spline methods; monotonicity methods for error estimations; asymptotic behavior and periodicity of solutions. The topics of the talks on Approximation Theory covered different aspects of approximation by polynomials, splines and rational functions and their numerical realization. Additionally included in the scientific program was a special session on Open Problems, where several suggestions were made for further research concerning both fields.

The free and harmonic atmosphere which lasted during the whole conference led to intensive and fruitful scientific discussions.

Our special thank is due to the "STIFTUNG VOLKSWAGENWERK" for their generous financial support of the symposium. Moreover, we would like to thank the "BIRKHÄUSER-VERLAG" for the kind agreement to publish this volume.

Our hope is that this meeting will stimulate the experts from Delay Equations and Approximation Theory to utilize the methods from both domains for solving at least some of the open problems.

Günter Meinardus   Günther Nürnberger

PARTICIPANTS

Priv.-Doz. Dr. H. Arndt, Bonn

Prof. Dr. A. Bellen, Triest, Italien

Prof. Dr. H.-P. Blatt, Eichstätt

Prof. Dr. D. Braess, Bochum

Prof. Dr. L. Collatz, Hamburg

Priv.-Doz. Dr. F. J. Delvos, Siegen

Prof. Dr. F. Deutsch, University Park, PA, USA

Prof. Dr. B. Dreseler, Siegen

Priv.-Doz. Dr. M.H. Gutknecht, Zürich, Schweiz

Priv.-Doz. Dr. U. an der Heiden, Bremen

Dr. M. Hollenhorst, Gießen

Frau Dipl.-Math. G. Kaiser, Mannheim

Dipl.-Math. U. Kaiser, Mannheim

Prof. Dr. G. Meinardus, Mannheim

Prof. Dr. G. Nürnberger, Mannheim

Prof. Dr. V. A. Popov, Sofia, Bulgarien

Prof. Dr. Q. I. Rahman, Montréal, Canada

Dipl.-Math. R. Scharlach, Eichstätt

Prof. Dr. W. Schempp, Siegen

Prof. Dr. G. Schmeißer, Erlangen

Prof. Dr. L.L. Schumaker, College Station, Texas, USA

Prof. Dr. W. Sippel, Kassel

Prof. Dr. M. Sommer, Eichstätt

Dr. G. Still, Mannheim

Prof. Dr. H. Strauß, Erlangen

Prof. Dr. G. D. Taylor, Fort Collins, Colorado, USA

Prof. Dr. R. S. Varga, Kent, Ohio, USA

Prof. Dr. H. O. Walther, München

Prof. Dr. H. Werner, Bonn

Prof. Dr. M. Zennaro, Triest, Italien

International Series of
Numerical Mathematics, Vol. 74
© 1985 Birkhäuser Verlag Basel

APPROXIMATION THEORY AND NUMERICAL METHODS FOR

DELAY DIFFERENTIAL EQUATIONS

Günter Meinardus and Günther Nürnberger

Fakultät für Mathematik und Informatik
Universität Mannheim
6800 Mannheim, Federal Republic of Germany

ABSTRACT

The latest research shows that the numerical treatment of delay differential equations leads to various approximation and optimization problems. In this manuscript of introductory character several aspects of this circle of problems are described. Moreover, some problems are stated and a few new observations are added.

INTRODUCTION. According to their historical development the theory of delay equations and approximation theory can be considered as separated domains in mathematics. On the other hand, the research in the very last years shows that various approximation (and also optimization) problems, concerning in particular spline functions, appear in the numerical treatment of delay equations.

In this paper of introductory character a few aspects of this circle of problems are shortly described to inspire the experts to look from one field to the other. Moreover, some questions, partially of standard type, are raised and some new observations are added. According to our limited experience with delay equations we have subjectively selected certain topics which seem to be of interest.

Section 1 deals with linear initial value problems (asymptotic behavior of solutions, approximation methods, examples, computational realization). In section 2 nonlinear initial value problems are discussed (adjustment of standard methods for ordinary differential equation, spline collocation methods, examples, theoretical aspects on periodic, oscillating and "chaotic" solutions). Section 3 is devoted to boundary value problems with arbitrary functional argument (method of collocation and Galerkin's method for splines, examples). In section 4 numerical methods for delay equations of monotonic type in the sense of L. Collatz which yield quantitative error bounds are considered. On the other hand, we do not consider shooting methods (see e.g. De Nevers & Schmitt [1971] and Bellen [1984b]), iterative methods (see e.g. Chocholaty & Slahor [1979], Hutson [1980] and Bellen [1979], [1983]) and the transformation of delay differential equations into ordinary differential equations (see e.g. Banks & Burns [1978], Banks, Burns & Cliff [1981] and Ascher & Russel [1981]).

The reader interested in more detailed information is refered to the books of Bellman & Cooke [1963], El'sgol'ts & Norkin [1973], Hale [1977], Driver [1977] and Cushing [1977] on the theory of delay equations and to the survey articles of Cryer [1972] and

Bellen [1985] on numerical methods for delay equations. A selection of monographs on numerical methods in approximation theory includes Meinardus [1964], Collatz & Krabs [1973] and Watson [1980] and on spline functions Schoenberg [1973], Böhmer [1974], de Boor [1978] and Schumaker [1981].

The fundamental problem to analyse a retarded process from the real world, to give a description by a mathemerical model and to determine the subsequent behavior leads to delay equations. Such type of equations appear in many fields of application:

PHYSICS AND ENGINEERING:

nuclear reactions

electrodynamic processes

elasticity theory

control systems

diffusion

rocket engines

transistor design

BIOLOGY:

population growth

ecological models

spread of epidemics

host-parasite-models

two-species competition

MEDICINE:

pharmacokinetic models

spread of infection diseases

production of red blood cells

ECONOMICS:

    business cycles

    economic growth.

The reader should consult the above mentioned books on delay equations, where references to many other examples can be found.

We briefly describe a standard example taken from Driver [1977]. Let $y(t)$ be the number of animals in an isolated region at the time $t > 0$. The simplest model for the growth of population is

$$y'(t) = cy(t) \quad , \quad t \in [0, \infty) \quad ,$$
$$y(0) = y_0 \quad ,$$

where c is a positive constant. The solution $y(t) = y_0 e^{ct}$ reflects an exponential growth. Being more realistic, a biological self-regulatory factor $1 - \frac{1}{d}y(t)$ (resulting possibly from overcrowding or shortage of food) can be involved which yields the ordinary differential equation

$$y'(t) = c\left(1 - \frac{1}{d}y(t)\right)y(t) \quad , \quad t \in [0, \infty) \quad ,$$
$$y(0) = y_0 \quad ,$$

where d is a positive constant. Since it is reasonable to assume that the self-regulation reacts with some delay $\omega > 0$, we are led to the delay differential equation

$$y'(t) = c\left(1 - \frac{1}{d}y(t-\omega)\right)y(t) \quad , \quad t \in [0, \infty) \quad .$$

If, in addition, the development

$$y(t) = \phi(t) \quad , \quad t \in [-\omega, 0] \quad ,$$

is taken into consideration, we finally obtain an initial value problem.

1. LINEAR INITIAL VALUE PROBLEMS. We first discuss initial value
problems for <u>linear delay differential equations</u> with constant
coefficients and constant delay:

$$y'(t) + ay(t) + by(t-\omega) = f(t) \quad , \quad t \in [0,\infty) \quad , \tag{1.1}$$

$$y(t) = \phi(t) \quad , \quad t \in [-\omega,0] \quad , \tag{1.2}$$

where a,b and $\omega > 0$ are real numbers.

A <u>solution</u> of this initial value problem is a continuous function $y : [-\omega,\infty) \to \mathbb{R}$ satisfying (1.1) and (1.2) (where $y'(0)$ means
the right-hand derivative).

If f and ϕ are continuous functions, then there exists a
unique solution of (1.1) and (1.2) (see Hale [1977]).

The first simple observation in trying to solve equation (1.1)
and (1.2) is that the problem can be reduced to a sequence of
ordinary differential equation. We first solve (numerically) the
ordinary initial value problem

$$y'(t) + ay(t) = f(t) - b\phi(t-\omega) \quad , \quad t \in [0,\omega] \quad , \tag{1.3}$$

$$y(0) = \phi(0) \quad .$$

Then we continue with this method on $[\omega,2\omega]$, $[2\omega,3\omega]$ and so on.

Before describing numerical methods which make use of this
observation, we consider the question how to determine the asymptotic behavior of the solution $y(t)$ for $t \to \infty$, one of the fundamental problems for delay equations.

Suppose that $f = 0$, then the solution y of (1.1) and (1.2)
satisfies $\lim_{t \to \infty} y(t) = 0$ for every delay ω and every initial function ϕ if and only if $-a < b \leq a$ (Zennaro [1985b], see also Barwell [1975]).

Moreover, another approach can be used. The equation

$$z + a + be^{-\omega z} = 0 \tag{1.5}$$

is called the <u>characteristic equation</u> of

$$y'(t) + ay(t) + by(t-\omega) = 0 \quad . \tag{1.6}$$

The zeros of (1.5) are called <u>characteristic roots</u>.

If $z_1, \ldots, z_n$ are characteristic roots of (1.6) and p_i is any polynomial of degree less than the multiplicity of z_i, $i=1, \ldots, n$, then

$$y(t) = \sum_{i=1}^{n} p_i(t) e^{z_i t} \tag{1.7}$$

is a solution of (1.6) (see Bellman & Cooke [1963]).

Every solution of (1.6) satisfies $\lim_{t \to \infty} y(t) = 0$ if and only if every characteristic root of (1.5) has negative real part (see Bellman & Cooke [1963]).

<u>PROBLEM 1.1</u>. How can the characteristic roots of (1.5) be calculated (for theoretical aspects see Bellman & Cooke [1963, chapter 12 and 13])?

The above results do not give a complete solution of the following

<u>PROBLEM 1.2</u>. Under which conditions does the solution of (1.1) and (1.2) satisfy $\lim_{t \to \infty} y(t) = 0$?

A numerical approach to determine the asymptotic behavior of solutions (at least on bounded intervals) can be described as follows.

Let $G = \text{span}\{g_1, \ldots, g_n\}$ be an n-dimensional subspace of $C(\mathbb{R})$, the space of continuous real-valued functions on $\mathbb{R}$, with the property that $g_1 = 1$ and $g_i(0) = 0$, $i=2,\ldots,n$. We set

$$G_0 = \{g \in G : g(0) = \phi(0)\} =$$

$$= \{\phi(0) + \sum_{i=2}^{n} a_i g_i : a_2,\ldots,a_n \text{ real}\}$$

and

$$L(y)(t) = y'(t) + ay(t) \quad .$$

Then we solve the following approximation problem. Determine a function $g \in G_0$ such that

$$L(g)(t) \approx f(t) - b\phi(t-\omega) \quad , \quad t \in [0,\omega] \quad . \tag{1.8}$$

(By the unusual symbol "$\approx$" we mean that the error

$$\sup_{t \in [0,\omega]} |f(t) - b\phi(t-\omega) - L(g)(t)|$$

shall be small.)

Obviously (1.8) is equivalent to determine real numbers $a_2,\ldots,a_n$ such that

$$\sum_{i=2}^{n} a_i L(g_i)(t) \approx f(t) - b\phi(t-\omega) - a\phi(0) \quad , \quad t \in [0,\omega] \quad . \tag{1.9}$$

A standard approach to solve the approximation problem (1.9) is the method of <u>collocation</u>. This means, that we choose points $0 \le t_1 < \ldots < t_{n-1} \le \omega$ and solve the linear system of equations

$$\sum_{i=2}^{n} a_i L(g_i)(t_j) = f(t_j) - b\phi(t_j-\omega) - a\phi(0) \quad , \quad j=1,\ldots,n-1 \quad . \tag{1.10}$$

Then we continue with this method on $[\omega,2\omega]$, $[2\omega,3\omega]$ and so on.

<u>PROBLEM 1.3.</u> Under which conditions does the above system (1.20) of linear equations have a unique solution?

Prototypes of subspaces G which are used in the above approach are spaces of spline functions, defined as follows. The space of <u>polynomials</u> of degree at most m is denoted by Π_m. Now, let points $0 = x_0 < x_1 < \ldots < x_k < x_{k+1} = \omega$ be given. We call

$$S_m^r(x_1,\ldots,x_k) = \left\{ s \in C^r[0,\omega] : s\big|_{[x_i,x_{i+1}]} \in \Pi_m \, , \, i=0,\ldots,k \right\} \qquad (1.11)$$

the space of r-times continuously differentiable <u>spline functions</u> of degree m with fixed knots $x_1,\ldots,x_k$. Such type of functions are frequently used in the method of collocation.

On the other hand, since certain exponential functions (see (1.7)) are solutions of equation (1.6), an appropriate method could be to use <u>exponential splines</u>. Spaces of such functions are defined analogously as $S_m^r(x_1,\ldots,x_k)$, where Π_m is replaced by

$$E_m = \text{span}\left\{1, e^{\gamma_1 t}, \ldots, e^{\gamma_m t}\right\} \, ,$$

the space of <u>exponential functions</u> with frequences fixed $\gamma_1,\ldots,\gamma_m$.

Moreover, in certain cases the following approach to solve (1.1) and (1.2) numerically could be used. Suppose that the function f has the property that its values on $[\omega,2\omega]$, $[2\omega,3\omega]$ and so on can be obtained by a simple relation from its values on $[0,\omega]$ (e.g. if $f = 0$, f is periodic or f is an exponential function). Then we approximate the functions f,ϕ respectively e^{-at} by spline functions from $P_m^r(x_1,\ldots,x_k)$ or by piecewise polynomials with free knots on $[0,\omega]$, $[-\omega,0]$ respectively $[0,\omega]$ with high accuracy. Algorithms to compute such spline approximations are available (see Nürnberger & Sommer [1983] and Strauß [1984] for fixed knots and Nürnberger, Sommer & Strauß [1984] for free knots). If we replace the functions by the corresponding approximations we can solve the arising ordinary initial value problems exactly proceeding knot by knot on $[0,\omega]$, $[\omega,2\omega]$ and so on (compare (1.3) and (1.4)).

At this point we note that in domains of applications also general equations like

$$a_0 y'(t) + a_1 y'(t-\omega) + b_0 y(t) + b_1 y(t-\omega) = f(t) \tag{1.12}$$

appear. Equation (1.22) is said to be of <u>neutral type</u>, if $a_0 \neq 0$ and $a_1 \neq 0$, and it is said to be of <u>advanced type</u>, if $a_0 = 0$ and $a_1 \neq 0$.

Of course, also such type of equations are of interest in the case that the coefficients are not constant, and possibly $t-\omega$ is replaced by an arbitrary function.

We close this section by giving several typical examples of delay equations for which numerical results were obtained:

$$\begin{aligned} y'(t) + y(t-1) &= 0 \quad , \quad t \in [1,\infty) \quad , \\ y(1) &= 1 \quad , \quad t \in [0,1] \quad . \end{aligned} \tag{1.13}$$

$$\begin{aligned} y'(t) + y(t) + y(t-1) &= 0 \quad , \quad t \in [1,\infty) \quad , \\ y(t) &= 1 \quad , \quad t \in [0,1] \quad . \end{aligned} \tag{1.14}$$

$$\begin{aligned} y'(t) + 5y(t) + y(t-1) &= 0 \quad , \quad t \in [1,\infty) \quad , \\ y(t) &= e^{-5t} \quad , \quad t \in [0,1] \quad . \end{aligned} \tag{1.15}$$

$$\begin{aligned} y'(t) - y(t-1) &= 100(t+1)^2 \quad , \quad t \in [0,\infty) \quad , \\ y(t) &= 0 \quad , \quad t \in [-1,0] \quad . \end{aligned} \tag{1.16}$$

$$\begin{aligned} y'(t) + ay(t) + by(t-1) &= ce^{-dt} \quad , \quad t \in [0,\infty) \quad , \\ y(t) &= 1 \quad , \quad t \in [-1,0] \quad . \end{aligned} \tag{1.17}$$

For the above examples see Cryer [1972].

$$y'(t) + by(t-1) = 0 \quad , \quad t \in [0,\infty) \quad ,$$
$$y(t) = 1 \quad , \quad t \in [-1,0] \quad . \tag{1.18}$$

For (1.18) see Oppelstrup [1978].

$$y'(t) + 50y(t) - 40y(t-1) = 0 \quad , \quad t \in [0,\infty) \quad ,$$
$$y(t) = e^{-t} \quad , \quad t \in [-1,0] \quad . \tag{1.19}$$

For (1.19) see Zennaro [1985a].

Numerical studies were also made for delay equations with non-constant coefficients:

$$y'(t) + \frac{1}{t}y(t-1) = 0 \quad , \quad t \in [1,\infty) \quad ,$$
$$y(t) = 1 \quad , \quad t \in [0,1] \quad . \tag{1.20}$$

$$y'(t) + a(t)y(t) + a(t)y(t-\omega) = 0 \quad , \quad t \in [0,\infty) \quad . \tag{1.21}$$

Moreover, equations of the following type were tested numerically:

$$y'(t) + y(t-1-e^{-t}) - 1 = 0 \quad , \quad t \in [0,\infty) \quad ,$$
$$y(t) = 0 \quad , \quad t \in (-\infty,0] \quad . \tag{1.22}$$

$$y'(t) - y(t) + 2ty(t - \frac{t}{1+t}) = 0 \quad , \quad t \in [1,4] \quad ,$$
$$y(t) = t \quad , \quad t \in [\tfrac{1}{2},1] \quad . \tag{1.23}$$

$$y'(t) + y(0.8t) + y(t) = 0 \quad , \quad t \in [0,1] \quad ,$$
$$y(0) = 1 \quad . \tag{1.24}$$

$$y'(t) + y(g(t)) = 0 \quad , \quad t \in [0,\infty) \quad ,$$
$$y(0) = 1 \quad . \tag{1.25}$$

For these examples see Cryer's survey [1972].

Finally, we make some remarks on the computational realization. Numerical results which can be found in the literature are obtained on intervals up to a length of 40. (Does this give enough information about the asymptotic behavior of the solution?) The total computing time - concerning also difficult non-linear problems - varies from less than one second up to several minutes. The relative error at the end point of the considered interval usually lies between 10^{-10} and 10^{-2} (see e.g. Oberle & Pesch [1981]). All this, of course, depends on the complexity of the problem and the efficiency of the available method.

2. NONLINEAR INITIAL VALUE PROBLEMS. We now turn to a much more general class of equations, namely initial value problems for nonlinear delay differential equations. In these considerations we restrict ourselves to the case of constant delays:

$$y'(t) = F(t,y(t),y(t-\omega)) \quad , \quad t \in [0,\infty) \quad , \tag{2.1}$$

$$y(t) = \phi(t) \quad , \quad t \in [-\omega,0] \quad , \tag{2.2}$$

where ω is a positive number.

If F is continuous and Lipschitzian in the last two variables and ϕ is continuous, then there exists a unique solution of (2.1) and (2.2) (for details see Hale [1977] and Driver [1977]).

We now briefly describe two different methods which were frequently used in the last years to solve (2.1) and (2.2) numerically.

The first approach is the adjustment of standard methods for ordinary differential equations described as follows (see Stetter

[1965], Neves [1975], Oppelstrup [1978], Bock & Schlöder [1981], McKee [1981], Oberle & Pesch [1981], Arndt [1983], [1984], van der Houwen & Sommeijer [1983]).

A single-step or multi-step method or a Runge-Kutta-method is applied to the ordinary initial value problem

$$y'(t) = F(t,y(t),\phi(t-\omega)) \quad , \quad t \in [0,\omega] \quad , \qquad (2.3)$$
$$y(0) = \phi(0) \quad . \qquad (2.4)$$

In this way an approximate solution on a finite set $\{t_1,\ldots,t_n\}$ with $0 \le t_1 < \ldots < t_n \le \omega$ (together with approximate values for the derivatives from (2.3)) is obtained. In order to be able to proceed with this method on $[\omega,2\omega]$ and so on, an approximation problem appears, because one needs an approximate solution on the whole interval $[0,\omega]$. Such an approximation is usually obtained by Hermite interpolation with piecewise polynomials using the given values at $t_1,\ldots,t_n$.

Roughly speaking, if the method for ordinary differential equations has the order of consistency p and the interpolating polynomials are of order q, then the global order of convergence is min$\{p,q\}$ (for details see Neves [1975], Oppelstrup [1978] and Oberle & Pesch [1981]). Thus, to preserve the order we must have $q \ge p$.

In order to avoid the above second procedure by which approximate solutions on intervals are obtained, one may use a different global approach which yields an approximation on each considered interval in one step. This can be done by the method of collocation with spline functions, described as follows.

Let knots $0 = x_0 < x_1 < \ldots < x_k < x_{k+1} = \omega$ be given and let $N = \dim S_m^r(x_1,\ldots,x_k)$. Then we choose points $0 < t_1 < \ldots < t_{N-1} < \omega$ and consider the approximation problem to determine a spline function $s \in S_m^r(x_1,\ldots,x_k)$ such that

$$s'(t_j) = F(t_j, s(t_j), \phi(t_j - \omega)) \quad , \quad j=1,\ldots,N-1 \quad , \tag{2.5}$$

$$s(0) = \phi(0) \quad . \tag{2.6}$$

We then continue with this method on $[\omega, 2\omega]$ and so on.

Under certain assumptions and for a suitable choice of collocation points (e.g. the zeros of Legendre polynomials, called Gaussian points, relative to the intervals $[x_i, x_{i+1}]$, $i=0,\ldots,k$) problem (2.5) and (2.6) has a unique solution, if $h = \max_{i=0,\ldots,k} (x_{i+1} - x_i)$ is sufficiently small (for details see Bellen [1984a]). In order to solve (2.5) and (2.6) numerically, iterative methods have to be used (e.g. Newton's method).

PROBLEM 2.1. Suppose that the equation (2.1) is linear. Under which conditions does a unique spline function exist which satisfies (2.5) and (2.6), if h is not required to be sufficiently small?

The phenomena that the order of convergence at the collocation points is, roughly speaking, of double precision compared with the global error is called superconvergence (for details see Bellen [1984a]; also Hulme [1972] and de Boor & Swartz [1973] for ordinary differential equations and Brunner [1984]).

PROBLEM 2.2. What can be said about the error of the described methods (see the above mentioned papers)? Which other type of functions could be used instead of piecewise polynomials and splines (e.g. exponential splines)? Which other approximation methods could be used to solve problem (2.1) and (2.2) numerically?

We now give a few typical examples by which the above or other methods were tested.

$$y'(t) = -y(t-1) -y^2(t) \quad , \quad t \in [0,\infty) \quad ,$$
$$y(t) = 0.1 \quad , \quad t \in [-1,0] \quad .$$

$$(2.6)$$

For (2.6) see Cryer's survey [1972].

$$y'(t) = -ay(t-1)(1+y(t)) \quad , \quad t \in [0,\infty) \quad ,$$
$$y(t) = \phi(t) \quad , \quad t \in [-1,0] \quad .$$

$$(2.7)$$

For (2.7) see Jones [1962a], Oberle & Pesch [1981] and Bellen [1984a].

More generally, also equations with variable delay have to be considered, e.g.

$$y'(t) = a\frac{t-1}{t}y(t-(\ln t)-1) \, y(t) \quad , \quad t \in [1,\infty)$$
$$y(t) = 1 \quad , \quad t \in [0,1] \quad .$$

$$(2.8)$$

For (2.8) see Bellen [1984a].

We close this section by discussing some theoretical aspects. A fundamental and deep problem is to describe the asymptotic behavior of solutions for delay equations, such as periodicity, oscillation, bifurcation and convergence to zero for $t \to \infty$.

Although much progress was gained in the last years, the description is not complete at present (see e.g. Hale [1977] and

Nussbaum's survey [1979]). In particular, certain phenomena of "chaotic" behavior of solutions, discovered by numerical experiments (see e.g. Grafton [1972] and Mackey & Glass [1977]), could not be explained mathematically up to now.

We now give a few known results in this area.

A continuous function $y : [0,\infty) \to \mathbb{R}$ is called a _slowly oscillating periodic function_, if there exist real numbers p_0, p with $p_0 > 1$, $p - p_0 > 1$ such that $y(0) = 0$, $y(t) > 0$, $t \in (0, p_0)$, $y(t) < 0$, $t \in (p_0, p)$ and $y(t+p) = y(t)$, $t \in [0, \infty)$.

By using fixed point principles the following theorems could be proved.

If $\alpha > \frac{\pi}{2}$, then there exists a slowly oscillating periodic solution of

$$y'(t) = -\alpha y(t-1)(1+y(t)) \tag{2.9}$$

(see Jones [1962a], [1962b], where also numerical results are given).

If $\alpha > \frac{\pi}{2}$, then there exists a _unique_ slowly oscillating solution of

$$y'(t) = -\alpha y(t-1)(1-y^2(t)) \tag{2.10}$$

(see Nussbaum's survey [1979]).

In particular, it is not known, if the special solution of (2.9) is unique.

It is easy to see that $y(t) = e^{z(t)} - 1$ transforms equation (2.9) into $z'(t) = -\alpha f(z(t-1))$, where $f(z) = e^z - 1$. This leads to the study of the more general _autonomous delay equations_

$$y'(t) = -\alpha f(y(t-1)) \tag{2.11}$$

(see e.g. Nussbaum's survey [1979] and Walther [1985]).

3. <u>BOUNDARY VALUE PROBLEMS</u>. In this section we consider boundary value problems of a very general form. Let the following boundary value problem for <u>nonlinear differential equations with arbitrary functional argument</u> be given:

$$y''(t) = F(t,y'(t),y'(h_1(t)),y(t),y(h_0(t))) \quad , \quad t \in [a,b], \quad (3.1)$$

$$y(t) = \begin{cases} \phi_1(t) & , \quad t \in (-\infty,a] \\ \\ \phi_2(t) & , \quad t \in [b,\infty) \end{cases} \quad (3.2)$$

Without loss of generality we may assume that $\phi_1(a) = \phi_2(b) = 0$ (see Reddien & Travis [1974]).

Only very few papers on numerical methods for boundary value problems of this type do exist. For special cases of problem (3.1) and (3.2) shooting methods were used by De Nevers & Schmitt [1971] and iterative methods were used by Bellen [1979], [1983], Chocholaty & Slahor [1979] and Hutson [1980].

In the following we briefly describe approximation methods which use spline functions. In order to solve problem (3.1) and (3.2) numerically we consider the following approximation problem.

Let knots $a = x_0 < x_1 < \ldots < x_k < x_{k+1} = b$ be given. Determine a function $s \in C(\mathbb{R})$ such that $s\big|_{[a,b]} \in S_m^r(x_1,\ldots,x_k)$ and

$$s''(t) \approx F(t,s'(t),s'(h_1(t)),s(t),s(h_0(t))) \quad , \quad t \in [a,b] \quad , \quad (3.3)$$

$$s(t) = \begin{cases} \phi_1(t) & , \quad t \in (-\infty,a] \\ \\ \phi_2(t) & , \quad t \in [b,\infty) \end{cases} \quad (3.4)$$

This problem can be solved by the method of collocation with spline functions from $S_m^r(x_1,\ldots,x_k)$, similarly as described in section 2.

Under certain conditions and for a suitable choice of collocation points (e.g. Gaussian points), this collocation problem has a unique solution, if $h = \max\limits_{i=0,\ldots,k} (x_{i+1}-x_i)$ is sufficiently small (for details see Reddien & Travis [1974] and Bellen & Zennaro [1984]; also de Boor & Swartz [1973] for ordinary boundary value problems).

<u>PROBLEMS 3.1</u>. Suppose that the equation (3.1) is linear. Under which conditions does a unique spline function exist which solves the above mentioned collocation problem, if h is not required to be sufficiently small?

A different approach for solving (3.3) and (3.4) is to apply Galerkin's method, described as follows.

Let $\{s_1,\ldots,s_{N-2}\}$ be a basis of the space

$$\{s \in S_m^r(x_1,\ldots,x_k) : s(a) = s(b) = 0\} \quad .$$

Determine a function $s \in C(\mathbb{R})$ such that $s\big|_{[a,b]} \in S_m^r(x_1,\ldots,x_k)$ and

$$\int_a^b s''(t)s_j(t)dt = \int_a^b F(t,s'(t),s'(h_1(t)),s(t),s(h_0(t)))s_j(t)dt \tag{3.5}$$

$$j=1,\ldots,N-2 \quad ,$$

$$s(t) = \begin{cases} \phi_1(t) & t \in (-\infty,a] \\ & \text{if} \\ \phi_2(t) & t \in [b,\infty) \end{cases} \tag{3.6}$$

Under certain conditions the approximation problem (3.5) and (3.6) has a unique solution, if h is sufficiently small (see Reddien & Travis [1974]).

PROBLEM 3.2. What can be said about the error of the described methods (see the above mentioned papers)? Which other approximation methods could be used to solve problem (3.3) and (3.4)?

Finally, some typical examples are given by which the above methods were tested.

$$y''(t) + 10y'(t) + 25y'(t-0.5) + 100y(t) - 0.04y'(t-0.5)^3 + 50 = 0 \ ,$$
$$t \in [0,1] \ ,$$
$$y(0) = y(1) = 0 \ ,$$
$$y'(t) = 2\pi\cos(2\pi t) \ , \quad t \in [-0.5,0] \ .$$

$$(3.7)$$

For (3.7) see Bellen & Zennaro [1984].

$$y''(t) + \frac{1}{16}\sin y(t) + (t+1)y(t-1) - t = 0 \ , \quad t \in [0,2] \ ,$$
$$y(t) = t - \frac{1}{2} \ , \quad t \in [-1,0] \ ,$$
$$y(2) = -\frac{1}{2} \ .$$

$$(3.8)$$

For (3.8) see De Nevers & Schmitt [1971], Reddien & Travis [1974] and Chocholaty & Slahor [1979].

$$y''(t) - 0.25(1 - y^2(t))y'(t - \frac{1}{2}) - y(t) + \sin(2\pi t) = 0 \ . \qquad (3.9)$$

For (3.9) see Bellen [1983].

We close this section with an observation about the existence and uniqueness of splines satisfying certain linear collocation conditions.

Let the following linear boundary value problem

$$L(y)(t) = y''(t) + f_1(t)y'(t) + g_1(t)y'(h_1(t)) + f_0(t)y(t) +$$

$$+ g_0(t)y(h_0(t)) = f(t) \quad , \quad t \in [a,b] \quad , \tag{3.10}$$

$$y(t) = \begin{cases} \phi_1(t) & , \quad t \in (-\infty, a] \quad , \\ \\ \phi_2(t) & , \quad t \in [b, \infty) \quad , \end{cases} \tag{3.11}$$

be given, where ϕ_1, ϕ_2 are continuously differentiable functions and $f_1, g_1, h_1, f_0, g_0, h_0$ are continuous functions. Moreover, let knots $a = x_0 < x_1 < \ldots < x_k < x_{k+1} = b$ and points $a < t_2 < \ldots < t_{N-1} < b$ be given, where $N = m + k(m-r) + 1 = \dim S_m^r(x_1, \ldots, x_k)$ (see Schumaker [1981]) and $r < m$.

We consider the collocation problem to determine a function $s \in C(\mathbb{R})$ such that $s\big|_{[a,b]} \in S_m^r(x_1, \ldots, x_k)$ and

$$L(s)(t_j) = f(t_j) \quad , \quad j = 2, \ldots, N-1 \quad , \tag{3.12}$$

$$s(t) = \begin{cases} \phi_1(t) & , \quad t \in (-\infty, a] \quad , \\ \\ \phi_2(t) & , \quad t \in [b, \infty) \quad , \end{cases} \tag{3.13}$$

where f is a continuous function. We recall that the appearing derivatives are meant to be right-hand derivatives. The reader should compare problem (3.3) and (3.4).

We now identify the set of knots $\{x_1, \ldots, x_k\}$ with $\{y_1, \ldots, y_{k(m-r)}\}$ where each knot x_i appears $(m-r)$-times in the last set, $i = 1, \ldots k$. Let points $a \leq w_1 < \ldots < w_N \leq b$ be given. The well-known Theorem of Schoenberg & Whitney (see Schumaker [1981]) says that for each function $f \in C[a,b]$ there exists a unique spline $s \in S_m^r(x_1, \ldots, x_k)$ with

$$s(w_j) = f(w_j) \quad , \quad j=1,\ldots,N \quad , \tag{3.14}$$

if and only if

$$w_j < y_j < w_{j+m+1} \quad , \quad j=1,\ldots,k(m-r) . \tag{3.15}$$

We now prove the following "density"-result on the choice of the set $\{t_j\}_{j=2}^{N-1}$ such that problem (3.12) and (3.13) has a unique solution.

THEOREM. Suppose that for every $s \in C(\mathbb{R})$ with $s(t) = 0$, $t \in (-\infty,a] \cup [b,\infty)$ and $s\big|_{[a,b]} \in S_m^r(x_1,\ldots,x_k)$ and every $i \in \{0,\ldots,k\}$ the function $L(s)$ does not vanish on a subinterval of $[x_i,x_{i+1}]$, if $s\big|_{[x_i,x_{i+1}]}$ is not the zero function. If points $a = w_1 < w_2 < \ldots < w_{N-1} < w_N = b$ are given such that

$$w_j < y_j < w_{j+m+1} \quad , \quad j=1,\ldots,k(m-r) , \tag{3.16}$$

then for each $\varepsilon < 0$ there exist points $a < t_2 < \ldots < t_{N-1} < b$ with $|t_j - w_j| < \varepsilon$, $j=2,\ldots,N-1$, such that the collocation problem (3.12) and (3.13) has a unique solution.

REMARK. (1) The requirement on L in the above theorem is a natural condition and is satisfied for most of the standard examples.

(2) A completely analogous result to the above theorem holds for initial value problems (see section 2) and of course for ordinary differential equations.

Proof of the theorem. We first define the following set

$$S = \{s \in C(\mathbb{R}) \; : \; s\big|_{[a,b]} \in S_m^r(x_1,\ldots,x_k), \; s(t) = \phi_1(t) \cdot ,$$
$$t \in (-\infty,a], \; s(t) = \phi_2(t), \; t \in [b,\infty)\} \tag{3.17}$$

and

$$\tilde{S} = \{s \in C(\mathbb{R}) : s\big|_{[a,b]} \in S_m^r(x_1,\ldots,x_k), \; s(t) = 0,$$
$$t \in (-\infty,a] \cup [b,\infty)\} \quad . \tag{3.18}$$

By the above stated Theorem of Schoenberg & Whitney there exist

points $a = v_1 < v_2 < \ldots < v_{N-1} < v_N = b$ and a basis $\{s_1,\ldots,s_N\}$ of

$S_m^r(x_1,\ldots,x_k)$ such that

$$s_i(v_j) = \begin{cases} 1 & j = i \\ & \text{if} \\ 0 & j \neq i \end{cases} \quad , \quad i,j=1,\ldots,N \; . \tag{3.19}$$

Then for each $s \in S$ there exist uniquely determined real numbers

$a_2,\ldots,a_{N-1}$ such that

$$s(t) = \phi_1(a)s_1(t) + \sum_{i=2}^{N-1} a_i s_i(t) + \phi_2(b)s_N(t) \quad , \quad t \in [a,b]. \tag{3.20}$$

Let the function $s_0 \in S$ be defined by

$$s_0(t) = \begin{cases} \phi_1(a)s_1(t) + \phi_2(b)s_N(t) & , \quad t \in [a,b] \\ \phi_1(t) & , \quad t \in (-\infty,a] \\ \phi_2(t) & , \quad t \in [b,\infty) \quad . \end{cases} \tag{3.21}$$

Then by (3.20) and (3.21) for each $s \in S$ there exists a unique

function $\tilde{s} \in \tilde{S}$ such that $s = s_0 + \tilde{s}$. Therefore, the collocation pro-

blem (3.12) and (3.13) is equivalent to determine a function

$\tilde{s} \in \tilde{S}$ such that

$$L(\tilde{s})(t_j) = f(t_j) - L(s_0)(t_j) \quad , \quad j=2,\ldots,N-1 \; . \tag{3.22}$$

The set $\{\tilde{s}_2,\ldots,\tilde{s}_{N-1}\}$ is a basis of the $(N-2)$-dimensional sub-

space $\tilde{S}$, where

$$\tilde{s}_i(t) = \begin{cases} s_i(t) & , \quad t \in [a,b] \\ & \\ 0 & , \quad t \in (-\infty,a] \cup [b,\infty) \end{cases} \quad , \quad i=2,\ldots,N-1. \tag{3.23}$$

Thus, in order to prove the theorem we have to show that for each $\varepsilon > 0$ there exist points $a < t_2 < \ldots < t_N < b$ with $|t_j - w_j| < \varepsilon$, $j=2,\ldots,N-1$, such that

$$\det(L(\tilde{s}_i)(t_j))_{i,j=2}^{N-1} \neq 0 \; .$$

Suppose that this is false. Then there exists a real number $\varepsilon > 0$ such that the functions $L(\tilde{s}_2),\ldots,L(\tilde{s}_{N-1})$ are linearly independent on $I = \bigcup_{j=2}^{N-1} [w_j-\varepsilon, w_j+\varepsilon]$, i.e. there exists a nontrivial function $\tilde{s} \in \tilde{S}$ such that

$$L(\tilde{s})(t) = 0 \quad \text{for all} \quad t \in I \; . \tag{3.24}$$

Let $j \in \{2,\ldots,N-1\}$ be given. We will show that $\tilde{s}(w_j) = 0$. Suppose to the contrary that $\tilde{s}(w_j) \neq 0$. Then by continuity of $\tilde{s}$ we have $\tilde{s}(t) \neq 0$ for all $t \in [w_j-\tilde{\varepsilon}, w_j+\tilde{\varepsilon}]$ and all sufficiently small $\tilde{\varepsilon} > 0$. It follows from (3.24) and the assumption on L that

$$\tilde{s}(t) = 0 \quad , \quad t \in [w_j-\tilde{\varepsilon}, w_j+\tilde{\varepsilon}] \quad ,$$

a contradiction to $\tilde{s}(w_j) \neq 0$.

This shows that

$$\tilde{s}(w_j) = 0 \quad , \quad j=1,\ldots,N \; .$$

Then it follows from (3.16) and the Theorem of Schoenberg & Whitney that $\tilde{s} = 0$ which is a contradiction. This proves the theorem.

4. QUANTITATIVE ERROR ESTIMATION AND MONOTONICITY. Most of the results on the error arising in the numerical treatment of delay differential equations (even in the classical case of ordinary differential equations) are of qualitative nature. In these cases the existence of a constant $K > 0$ is guaranteed such that the error is less than Kh^q, where the step size h and the exponent q are known. On the other hand, in general no special information

about the size of K is available. But, in practice when calculations are made on a computer it is desirable to have real bounds for the error. Such quantitative error estimations can be obtained for a certain class of differential operators described as follows.

Let Y and Z be spaces of real-valued functions, defined on a subset of the real line. An operator $L : Y \to Z$ is called of <u>monotonic type</u>, if $L(y_1) \leq L(y_2)$ implies $y_1 \leq y_2$ for all $y_1, y_2 \in Y$ (see Collatz [1981], [1983]).

Now, if an equation $L(y) = f$ is given, where L is of monotonic type, then inclusions for the solutions y can be obtained in the following way. The task is to determine approximations $y_1, y_2 \in Y$ with $L(y_1) \leq f \leq L(y_2)$. Then it follows that $y_1 \leq y \leq y_2$ which yields the quantitative error estimation

$$|y - \tilde{y}| \leq \frac{1}{2} |y_2 - y_1|$$

for $\tilde{y} = \frac{1}{2}(y_1 + y_2)$.

In the case of periodic functions conditions are known such that the operator

$$L(y)(t) = y'(t) + ay(t) + by(t-\omega) \tag{4.1}$$

is of monotonic type (see Bellen & Zennaro [1983] and Zennaro [1984]).

<u>PROBLEM 4.1.</u> What can be said about monotonicity for nonperiodic functions? Which statements can be derived, if only those functions are considered which satisfy certain initial or boundary conditions? What can be said about other types of delay differential operators?

In order to determine functions $y_1, y_2 \in Y$ with $L(y_1) \leq f \leq L(y_2)$ we can proceed as follows.

Let G be a finite-dimensional subspace of Y (e.g. a space of spline functions). Then we have to compute functions $g_1, g_2 \in G$ with $L(g_1) \approx f$, $L(g_2) \approx f$ and $L(g_1) \leq f \leq L(g_2)$, where various approximation methods and norms may be used. Thus we are confronted with one-sided approximation problems.

The following observation may be of some help. Suppose that the space $\{L(g) : g \in G\}$ contains a strictly positive function. Then we can solve an "ordinary" approximation problem to get a function $g_0 \in G$ with $L(g_0) \approx f$ and obtain the desired functions by shifting $L(g_0)$ with the existing positive function up and down.

A more direct approach is to discretize (i.e. to consider the problem on a finite subset of the domain) and to solve the resulting optimization problem on the discrete set:

$$\text{Minimize } d \text{, subject to} \tag{4.2}$$
$$0 \leq g_1 - g_2 \leq d \quad, \quad L(g_1) \leq f \leq L(g_2) \quad \text{and} \quad g_1, g_2 \in G \ .$$

This can be done by applying standard optimization algorithms (e.g. the Simplex-algorithm).

PROBLEM 4.2. Can we take advantage of the special structure of the space $\{L(g) : g \in G\}$ to compute approximate solutions on the whole domain?

We close this paper with a general remark. By applying the known numerical approaches approximate solutions of delay equations are computed on bounded intervals. Since a fundamental problem is to determine the asymptotic behavior of solutions for $t \to \infty$, it would be desirable to develop global numerical methods for unbounded intervals.

REFERENCES

Arndt H. (1983), The influence of interpolation on the global error in retarded differential equations, in Differential-difference equations, Collatz L., Meinardus G. and Wetterling W. eds., ISNM 62, Birkhäuser-Verlag, Basel, 9-17.

Arndt H. (1984), Numerical solution of retarded initial value problems: local and global error and stepsize control, Numer. Math. 43 (1984), 343-360.

Ascher U. and Russel R.D. (1981), Reformulation of boundary value problems into "standard" form, SIAM Rev. 23, 238-254.

Banks H.T. and Burns J.A. (1978), Heriditary control problems: Numerical methods based on averaging approxiamtions, SIAM J. Control Optim. 16, 169-208.

Banks H.T., Burns J.A. and Cliff E.M. (1981), Parameter estimation and identification for systems with delays, SIAM J. Control Optim. 19, 791-828.

Barwell V.K. (1975), Special stability problems for functional differential equations, BIT 15, 130-135.

Bellen A. (1979), Cohen's iteration process for boundary value problems for functional differential equations, Rend. Ist. Mat. Univ. Trieste 11, 32-46.

Bellen A. (1983), Monotone methods for periodic solutions of second order scalar functional differential equations, Numer. Math. 42, 15-30.

Bellen A. (1984a), One-step collocation for delay differential equations, J. Comp. Appl. Math. 10, 275-283.

Bellen A. (1984b), A Runge-Kutta-Nystrom method for delay differential equations, preprint.

Bellen A. (1985), Constrained mesh methods for functional differential equations, this volume.

Bellen A. and Zennaro M. (1983), Maximum principles for periodic solutions of linear delay differential equations, in Differential-difference equations, Collatz L., Meinardus G. and Wetterling W. eds., ISNM 62, Birkhäuser-Verlag, Basel, 19-24.

Bellen A. and Zennaro M. (1984), A collocation method for boundary value problems of differential equations with functional arguments, Computing 32, 307-318.

Bellman R. and Cooke K. (1963), Differential difference equations, Academic Press, New York.

Bock H.G. and Schlöder J. (1981), Numerical solution of retarded differential equations with statedependent time lags, ZAMM 61, 269-271.

Böhmer K. (1974), Spline-Funktionen, Teubner-Verlag, Stuttgart.

de Boor C. (1978), A practical guide to splines, Springer-Verlag, New York.

de Boor C. and Swartz B. (1973), Collocation at Gaussian points, SIAM J. Numer. Anal. 10, 582-606.

Brunner H. (1984), Implicit Runge-Kutta methods of optimal order for Volterra integro-differential equations, Math. Comp. 42, 95-109.

Chocholaty P. and Slahor L. (1979), A method to boundary value problems for delay equations, Numer. Math. 33, 69-75.

Collatz L. (1981), Anwendung von Monotoniesätzen zur Einschließung der Lösungen von Gleichungen, Jahrbuch Überblicke Mathematik 1981, Bibliographisches Institut, Mannheim, 189-225.

Collatz L. (1983), Einschließung periodischer Lösungen bei einer Klasse von Differenzen-Differentialgleichungen, in Differential-difference equations, Collatz L., Meinardus G. and Wetterling W. eds., ISNM 62 Birkhäuser-Verlag, Basel, 49-54.

Collatz L. and Krabs W. (1973), Approximationstheorie, Teubner-Verlag, Stuttgart.

Cryer C.W. (1972), Numerical methods for functional differential equations, in Delay and functional differential equations and their applications, Schmitt K. ed., Academic Press, New York, 17-101.

Cushing J.M. (1977), Integrodifferential equations and delay models in population dynamics. Lecture Notes in Biomathematics 20, Springer-Verlag, Berlin.

De Never K. and Schmitt A. (1971), An application of the shooting method to boundary value problems for second order delay equations, J. Math. Anal. Appl. 36, 588-597.

Driver R.D. (1977), Ordinary and delay differential equations, Springer-Verlag, Berlin.

El'sgol'ts L.E. and Norkin S.B. (1973), Introduction to the theory and application of differential equations with deviating arguments, Academic Press, New York.

Grafton R. (1972), Periodic solutions of Lienard equations with delay: some theoretical and experimental results, in Delay and functional differential equations and their applications, Schmitt K. ed., Academic Press, New York, 321-334.

Hale J. (1977), Theory of functional differential equations, Springer-Verlag, Berlin.

van der Houwen P.J. and Sommeijer B.P. (1983), Improved absolute stability of predictor corrector methods for retarded differential equations, in Differential-difference equations, Collatz L., Meinardus G. and Wetterling W. eds., ISNM 62, Birkhäuser-Verlag, Basel, 137-148.

Hulme B.L. (1972), One-step piecewise polynomial Galerkin methods for initial value problems, Math. Comp. 26, 415-426.

Hutson V.C.L. (1980), Boundary-value problems for differential difference equations, J. Differential equations 36, 363-373.

Jones G.S. (1962a), On the nonlinear differential difference equation $f'(x) = -\alpha f(x-1)[1+f(x)]$, J. Math. Anal. Appl. 4, 440-469.

Jones G.S. (1962b), The existence of periodic solutions of $f'(x) = -\alpha f(x-1)[1+f(x)]$, J. Math. Anal. Appl. 5, 435-450.

Mackey M.C. and Glass L. (1977), Oscillation and chaos in physiological control systems, Science 197, 287-295.

McKee S. (1981), Fixed step discretisation methods for delay differential equations, Comp. Maths. Appl. 7, 413-423.

Meinardus G. (1964), Approximation von Funktionen und ihre numerische Behandlung, English edition (1967), Springer-Verlag, Berlin.

Neves K.W. (1975), Automatic integration of functional differential equations: an approach, ACM Trans. Math. Software 1, 357-368.

Nürnberger G. and Sommer M. (1983), A Remez type algorithm for spline functions, Numer. Math. 41, 117-146.

Nürnberger G., Sommer M. and Strauß H. (1984), An algorithm for segment approximation, preprint.

Nussbaum R.D. (1979), Periodic solutions of nonlinear autonomous functional differential equations, in Functional differential equations and approximation of fixed points, Peitgen H.-O. and Walther H.-O. eds., Lecture Notes in Mathematics 730, Springer-Verlag, Berlin, 283-325.

Oberle H.J. and Pesch H.J. (1981), Numerical treatment of delay differential equations by Hermite interpolation, Numer. Math. 37, 235-255.

Oppelstrup J. (1978), The RKFHB4 method for delay differential equations, in Numerical treatment of differential equations, Bulirsch R., Grigorieff R.D. and Schröder J. eds., Lecture Notes in Mathematics 631, Springer-Verlag, Berlin, 133-146.

Reddien G. and Travis C.C. (1974), Approximation methods for boundary value problems of differential equations with functional arguments, J. Math. Anal. Appl. 46, 62-74.

Schoenberg I.J. (1973), Cardinal spline interpolation, CBMS 12, SIAM, Philadelphia.

Schumaker L.L. (1981), Spline functions: Basic theory, John Wiley & Sons, New York.

Stetter H.J. (1965), Numerische Lösung von Differentialgleichungen mit nacheilendem Argument, ZAMM 45, 79-80.

Strauß H. (1984), An algorithm for the computation of strict approximations in subspaces of spline functions, J. Approximation Theory 41, 329-344.

Walther H.-O. (1985), Homoclinic solution and chaos in $\dot{x}(t) = f(x(t-1))$, J. Nonlinear Analysis, to appear.

Watson G.A. (1980), Approximation Theory and Numerical Methods, John Wiley & Sons, Chichester.

Zennaro M. (1984), Maximum principles for linear difference-differential operators in periodic function spaces, Numer. Math. 43, 121-139.

Zennaro M. (1985a), On the P-stability of one-step collocation for delay differential equations, this volume.

Zennaro M. (1985b), Private communication.

International Series of
Numerical Mathematics, Vol. 74
© 1985 Birkhäuser Verlag Basel

Numerical Integration of Retarded Differential
Equations with Periodic Solutions

H. ARNDT, P.J. VAN DER HOUWEN, B.P. SOMMEIJER

ABSTRACT: It is the purpose of this paper to show that the minimax versions of linear multistep methods, originally derived for <u>ordinary</u> differential equations with a periodic solution, are also suitable for the integration of <u>retarded</u> differential equations possessing a periodic solution. Especially for this type of equations it is extremely useful to have methods yielding highly accurate results for relatively large time steps h. We consider several examples of first-order and second-order equations with constant and state-dependent delay and compare the numerical results with that of the conventional methods.

1. INTRODUCTION

We consider the following initial value problem for retarded differential equations

$$(1.1) \qquad \begin{aligned} y'(t) &= f(t, y(t), z(t)) \quad \text{for} \quad t \in [0, b] \\ y(t) &= \psi(t) \quad \text{for} \quad t \in [-s, 0] \\ \text{with} \quad & z(t) = y(t - \tau(t, y(t))). \end{aligned}$$

Here ψ is the initial function, $\psi : [-s, 0] \to \mathbb{R}^m$, the function f is defined in an open subset $\Omega \subset \mathbb{R} \times \mathbb{R}^m \times \mathbb{R}^m$, such that

$$(0, \psi(0), \psi(t)) \in \Omega \qquad \text{for all} \qquad t \in [-s, 0],$$

and the (state-dependent) delay τ is a real function defined on an open subset Ω^* of $\mathbb{R} \times \mathbb{R}^m$, where Ω^* is the projection of Ω on its first $m + 1$ components, and τ is nonnegative and bounded on Ω^*, $0 \le \tau(t, y) \le \tau_0$.

If f, ψ, τ are continuous there exists a solution of the initial value problem (1.1) on an interval $[0, b]$; if, in addition, $f(t, y, z)$ is Lipschitz-continuous with respect to y and z and $\tau(t, y)$ is Lipschitz-continuous with respect to y we have uniqueness and continuous dependence on the data; see Driver [3] and Hale [5].

Typeset by $\mathcal{A}_{\mathcal{M}}\mathcal{S}$-TEX

We will study linear multistep methods for the numerical integration of (1.1) in the special case where it is known that (1.1) possesses a <u>periodic</u> solution. We will consider Adams-Moulton and Milne-Simpson methods. Both families of methods will be applied in conventional form and in the so-called <u>minimax</u> form as described in [6]. The minimax versions take into account the periodicity of the solution and are rather effective in the case of ordinary differential equations.

Since a rich source of periodic problems is formed by initial value problems involving retarded differential equations of <u>second</u> order (cf. El'sgol'ts and Norkin[4, p.187]), we also consider linear multistep methods for second-order equations; in particular, we will investigate the conventional and the minimax version of the symmetric method of Lambert and Watson [7] (this family of methods was specifically designed for the integration of periodic problems).

Summarizing, it is the purpose of this paper to show that the minimax versions of linear multistep methods, originally derived for <u>ordinary</u> differential equations with a periodic solution, are also suitable for the integration of <u>retarded</u> differential equations possessing a periodic solution. Especially for this type of equations it is extremely useful to have methods yielding highly accurate results for relatively large time steps h; this is because delay equations require the storage of $y-$ (or $f-$) vectors, the number of which is roughly equal to $\max_{t,y(t)} \tau(t, y(t))/h \leq \tau_0/h$.

2. LINEAR MULTISTEP METHODS

We consider the application of linear multistep (LM) methods known from ordinary differential equations $(ODEs)$ to retarded problems. We will discuss the case of first-order $ODEs$, but the case of second-order $ODEs$ can be dealt with in an analogous manner. Generally, the application of an LM method can be accomplished by the following algorithm (for convenience let us first consider the case of a constant delay $\tau > 0$):

Choose a grid $0 = t_0 < t_1 < t_2 < ... < t_N = b$ with constant stepsize $h = t_j - t_{j-1} < \tau$. Let (ρ, σ) be a linear multistep method of order p for ordinary differential equations with characteristic polynomials

$$\varrho(\varsigma) = \sum_{j=0}^{k} a_j \varsigma^j, \qquad \sigma(\varsigma) = \sum_{j=0}^{k} b_j \varsigma^j.$$

Assume that we have starting values $y_0, y_1, ..., y_{k-1}$ as approximations for $y(t_0), y(t_1), ..., y(t_{k-1})$. Then if $y_0, y_1, ..., y_{n+k-1}$ are known compute y_{n+k} from

$$\rho(E)y_n = h\sigma(E)f_n$$

$$\text{with} \qquad f_n := f(t_n, y_n, z_n)$$

$$\text{and} \quad z_n := \begin{cases} \psi(t_n - \tau), & \text{if } t_n - \tau \le 0, \\ u(t_n - \tau), & \text{if } t_n - \tau > 0. \end{cases}$$

Here the function u is a suitable interpolate of the known values, e.g.

$$u(t_j) = y_j, \quad u'(t_j) = f_j, \quad j = 0, 1, ..., n + k - 1;$$

most often piecewise polynomial interpolation of fixed order q (degree $q - 1$) for suitable q is used.

The same arguments apply for nonconstant delay τ; for state dependent delay we use $\tau(t_n, y_n)$ as an approximation of $\tau(t_n, y(t_n))$.

Some inherent difficulties may occur: In general the solution y of (1.1) is differentiable only once on $[0, b]$ even if the data are analytic as can be seen from the simple example

$$\begin{aligned} y'(t) &= y(t - 1) \quad \text{for } t \ge 0, \\ y(t) &= 1 \qquad\qquad \text{for } t \le 0, \end{aligned}$$

the solution of which possesses so called jump discontinuities at the natural numbers. In this example the solution gets smoother with increasing t because of the constant delay but this is not the case in general. If $y^{(j)}(t_-) \ne y^{(j)}(t_+)$ and $y^{(j-1)}(t_-) = y^{(j-1)}(t_+)$ then y has a jump discontinuity of order j at t. If all jump discontinuities of order p or less belong to the grid and the method is started again after each of these discontinuities the order of the resulting method is $\min(p, q)$. Observe that in this case one generally has to use a nonequidistant grid.

A second difficulty arises if τ is very small. On the one hand the jump discontinuities in this case lie near together which may reduce the stepsize drastically, on the other hand if $\tau < h$ then the interpolation formula is implicit with respect to the unknown value y_{n+k}.

Last but not least note that in general the exact solution y is only continuous at t_0. Therefore one cannot use the known values $y_{-k+1}, y_{-k+2}, ..., y_0$ of the initial function as starting values of the multistep method, that is $y_{-j} = y(-jh) = \psi(-jh), \quad j = 0, 1, ..., k - 1$. The starting values can be computed e.g. with a onestep method or with LM methods that increase the order.

3. Periodic Solutions

We call a solution y of (1.1) periodic with period $T > 0$ and frequency $\omega = 2\pi/T$ on the interval $[0, b]$ if

$$y(t + T) = y(t) \quad \text{for all} \quad t \in [0, b - T].$$

In general the solution is not periodic on the interval $[-s, b]$. For example consider the problem

$$y'(t) = (y(t) - \sin t) \cdot g(t, y(t), y(t - \tau)) + \cos t, \quad t \geq 0$$
$$y(t) = \psi(t), \quad t \leq 0$$

where g is arbitrary. For all initial functions ψ with $\psi(0) = 0$ the function $y(t) = \sin t$ is a periodic solution of the problem on $[0, \infty)$ but in general y is not periodic on $[-s, \infty)$.

If f, ψ and τ are p-times differentiable then the solution y of (1.1) is at least $(p+1)$-times differentiable between jump discontinuities. Because of the bounded delay the solution will be globally $(p+1)$-times differentiable for large enough t. If in addition y is periodic on $[0, b]$ then for sufficiently large b we can conclude that y is $(p+1)$-times differentiable on the whole interval $[0, b]$. Consequently we need not obey the jump discontinuities for periodic solutions of the initial value problem and may use a constant stepsize - at least for tests.

4. Predictor - Corrector Methods

Suppose one decides to solve the implicit relations, arising in the application of an LM method, by a predictor-corrector method. Then we are faced with the problem of choosing the order of the predictor. In the case of non state-dependent delays one could use a predictor of order $p - 1$, when the corrector is of order p, because the problem can be handled similar to ordinary differential equations. In problems with state-dependent delays τ it is better to use a predictor with the same order as the corrector because a good approximation y_{n+k}^* is needed for the computation of the retarded argument

$$t_{n+k} - \tau(t_{n+k}, y_{n+k}^*)$$

which is used in the corrector step.

5. INTERPOLATION

We want to give some more detailed comments to the type of interpolation procedure. Assume that an Adams method for ordinary differential equations is used,

$$y_{n+k} - y_{n+k-1} = h \sum_{j=0}^{k} b_j f_{n+j}.$$

Such formulas are constructed by approximating the integrand in the Volterra equation

$$y(t_{n+k}) - y(t_{n+k-1}) = \int_{t_{n+k-1}}^{t_{n+k}} f(s, y(s)) ds$$

by a polynomial P_{n+k} that interpolates at t_{n+j} the value $f_{n+j}, j = 0, 1, ..., k$, that is

$$y_{n+k} - y_{n+k-1} = \int_{t_{n+k-1}}^{t_{n+k}} P_{n+k}(s) ds = h \sum_{j=0}^{k} b_j f_{n+j}.$$

If in the case of retarded differential equations an approximation for the value $y(t^*)$ with $t^* = t_{n+k} - \tau(t_{n+k}, y(t_{n+k}))$ is needed and we have $t_{l-1} \leq t^* < t_l$ for some $1 \leq l < n + k$, it is very natural to take

$$y(t^*) \approx y_{l-1} + \int_{t_{l-1}}^{t^*} P_l(s) ds$$

where P_l interpolates the values f_{l-j} at $t_{l-j}, j = 0, 1, ..., k$, see Bock, Schlöder [2]. One can show that with these formulas not only the local integration error can be controlled but the local interpolation error as well, cf. Arndt [1].

Another possibility for the interpolation procedure is given by the above mentioned Hermite-interpolation at points $t_{l\pm j}$ for certain j such that t^* lies nearly in the middle of these points. These formulas come along with fewer grid points and therefore theoretically lead to a smaller error.

6. Minimax Methods

The minimax modification of a linear multistep method (ρ, σ) for an $m - th$ order ODE is defined by the equations (cf. [6], [8])

$$\varphi_m(i\nu^{(l)}) = 0, \quad l = 1, 2, ..., r,$$

(6.1)
$$\varphi_m(z) := \rho(e^z) - z^m \sigma(e^z),$$

$$\nu^{(l)} := \left[\frac{1}{2}\left(\overline{\nu}^2 + \underline{\nu}^2\right) + \frac{1}{2}\left(\overline{\nu}^2 - \underline{\nu}^2\right)\cos\left(\frac{2l-1}{2r}\pi\right)\right]^{\frac{1}{2}}.$$

Here, $[\underline{\omega}, \overline{\omega}] =: h^{-1}[\underline{\nu}, \overline{\nu}]$ is an estimate of the inverval of dominant frequencies in the exact solution of the ODE. The value of r is determined by the number of free coefficients in the polynomials ρ and σ. Generally, the system (6.1) represents a (linear) system with complex coefficients so that, in order to obtain real-valued coefficients, we should have $2r$ free coefficients in (ρ, σ). In the special case of symmetric methods (i.e., $\rho(\eta) = \eta^k \rho(\eta^{-1})$ and $\sigma(\eta) = \eta^k \sigma(\eta^{-1})$), the system (6.1) has a real coefficient matrix, so that we need only r free parameters in (ρ, σ).

We conclude this section by deriving a relation for the truncation error in the case of a retarded differential equation with periodic solution. Assuming the localizing assumption to be satisfied ($y(t_j) = y_j, j = 0, ..., n$), we may write

$$y(t) = u(t) + I(t, h),$$

where $u(t)$ is the interpolating function introduced in Section 2, and $I(t, h)$ denotes the interpolation error. The truncation error at t_{n+k} is given by

$$T_{n+k} := \rho(E)y(t_n) - h^m \sigma(E) f(t_n, y(t_n), u(t_n - \tau))$$
$$\approx \rho(E)y(t_n) - h^m \sigma(E)\left[f(t_n, y(t_n), y(t_n - \tau)) - \frac{\partial f}{\partial z} I(t_n, h)\right].$$

Recalling the definition of $\varphi_m(z)$ we find

(6.2)
$$T_{n+k} \approx \left[\varphi_m(h\frac{d}{dt})y(t) + h^m \sigma(\exp(h\frac{d}{dt}))\frac{\partial f}{\partial z} I(t, h)\right]_{t=t_n},$$

where m is the order of the differential equation. If φ_m corresponds to a minimax method adapted to the periodic solution $y(t)$, than the truncation error is dominated by the second term containing the interpolation error $I(t, h)$. From this we conclude that high accuracies can be expected provided that we use interpolation of sufficiently high order.

7. NUMERICAL EXPERIMENTS

In this section we want to demonstrate the performance of the minimax modification of linear multistep methods both in PECE mode and when using Newton iteration. All methods tested are of order $p = 6$. When applied in PECE mode we used (for first-order equations) the Adams-Bashforth method of order 6 (AB_6) as predictor and the Adams-Moulton method of order 6 (AM_6) or the Milne-Simpson method of order 6 (MS_6) as corrector. In the case of the minimax-modification of the PECE method, both the predictor and corrector were modified. In the case of second-order equations, we applied the 4-step symmetric method of Lambert and Watson of order 6 (LW_6) (cf.[7, p.198]). In all experiments, interpolation polynomials of degree 9 (i.e. of order 10) were employed. The abscissas used are: $t_{l-9}, t_{l-8}, ..., t_l$, where l is determined by $t_{l-1} \le t^* < t_l$ and t^* is the retarded argument (cf. Section 5).

In the tables of results given in the following subsections, the accuracy is measured by the number of correct digits in the numerical solution at the end point t_N, i.e., by

$$sd := -\log_{10}(|\, y_N - y(t_N)\, |).$$

7.1 DELAY EQUATIONS OF FIRST ORDER

First we consider an example possessing a <u>constant delay</u>:

$$(7.1) \qquad \begin{aligned} y'(t) &= y(t) + y(t - \pi) + 3\cos t + 5\sin t, \quad t \in [0, 10], \\ y(t) &= 3\sin t - 5\cos t, \quad t \le 0, \end{aligned}$$

with exact solution $y(t) = 3\sin t - 5\cos t$.

In Tables 7.1 and 7.2 we list respectively the accuracies of the conventional and minimax methods, obtained for several values of the step length h.

h	$AB_6 - AM_6$ (PECE)	AM_6 (Newton)	$AB_6 - MS_6$ (PECE)	MS_6 (Newton)
2/5	-0.1	0.1	0.1	0.4
1/5	1.8	1.4	1.9	1.8
1/10	3.6	3.0	5.7	3.4
1/20	4.9	4.7	5.5	5.1
1/40	6.6	6.5	7.0	6.9

Table 7.1

sd-values for problem (7.1) using conventional methods

h	$AB_6 - AM_6$ (PECE)	AM_6 (Newton)	$AB_6 - MS_6$ (PECE)	MS_6 (Newton)
2/5	4.6	3.0	4.8	3.0
1/5	6.5	6.0	6.6	5.8
1/10	8.3	7.7	11.3	8.3

Table 7.2

sd-values for problem (7.1) using minimax methods with $[\underline{\omega}, \overline{\omega}] = [0.95, 1.05]$

In this example, the choice of the corrector (AM or MS) is of minor importance whereas the way in which the corrector has been solved (either PECE-mode or Newton iteration) is more crucial. However, the improvement obtained by the minimax versions is easily recognized. It should be noted that the additional effort required by the minimax methods is almost negligible.

Mention should be made of the fact that, for this example, $\lambda := \partial f(t, y, z)/\partial y$ ist <u>positive</u>. As the principle root of the characteristic equation approximates $e^{\lambda h}$ for $h \to 0$ we must reckon with amplification of roundoff errors. For this example, in which $\lambda = 1$ and the endpoint of integration equals 10, the accumulated amplification can be as bad as $(e^h)^{\frac{10}{h}} = e^{10} \simeq 2 \cdot 10^4$ (for small h). Hence, in requiring a result which is accurate in say n digits, we should use a machine which performs the calculations in at least $n + 4$ digits.

In our second example we consider a state-dependent delay term:

$$(7.2) \quad y'(t) = \omega \cdot \cot(g(t)) \cdot y(t) - \frac{\omega}{\sin(g(t))} y(t - \tau(t, y(t))), \quad t \in [0, 10]$$

$$y(t) = \sin(\omega t), \quad t \leq 0$$

$$\text{with} \quad \tau(t, y) := \frac{1}{\omega}(2 + \frac{1}{5}e^y)$$
$$\text{and} \quad g(t) := \omega \tau(t, \sin(\omega t))$$

which has the exact solution $y(t) = \sin(\omega t)$.

We applied the various methods for different values of the frequency ω. In the minimax versions we employed the frequency interval $[\underline{\omega}, \overline{\omega}] = [0.95\omega, 1.05\omega]$. The results can be found in Tables 7.3 an 7.4.

ω	h	$AB_6 - AM_6$ (PECE)	AM_6 (Newton)	$AB_6 - MS_6$ (PECE)	MS_6 (Newton)
	2/5	3.7	4.3	3.9	3.3
1	1/5	5.4	5.7	6.7	6.1
	1/10	7.2	7.4	8.0	7.2
	1/20	9.1	9.2	9.2	9.8
	1/10	3.5	3.8	4.6	0.2
3	1/20	5.5	5.9	2.6	1.2
	1/40	7.9	7.7	3.0	1.1

Table 7.3

sd-values for problem (7.2) using conventional methods

ω	h	$AB_6 - AM_6$ (PECE)	AM_6 (Newton)	$AB_6 - MS_6$ (PECE)	MS_6 (Newton)
	2/5	8.4	5.7	8.6	6.2
1	1/5	10.1	9.3	11.5	10.9
	1/10	11.9	11.8	12.6	11.7
3	1/10	8.2	6.0	8.6	2.7
	1/20	10.2	9.4	7.3	5.0

Table 7.4

sd-values for problem (7.2) using minimax methods with $[\underline{\omega}, \overline{\omega}] = [0.95\omega, 1.05\omega]$

The results for $\omega = 1$ give rise to the same conclusions as in the previous example. However, both in the conventional as well as in the minimax version, the Adams-Moulton method is superior to the Milne-Simpson method as the frequency ω increases. This is due to the better stability properties of the Adams-type methods.

Finally, we consider the influence of an inaccurate estimate of the frequency. For $\omega = 3$ we obtain in case of the AM_6-minimax method the following results

$\underline{\omega}$	$\overline{\omega}$	sd-value for h=1/10
2.85	3.15	6.0 (see Table 7.4)
2.5	3.5	5.9
2.0	4.0	5.2
2.0	2.5	4.9
2.5	3.0	6.0
3.0	3.5	6.1
3.5	4.0	4.6

7.2 DELAY EQUATIONS OF SECOND ORDER

Our first example is the second-order equivalent of problem (7.1):

$$\begin{aligned}
y''(t) &= -y(t) - y(t - \frac{3\pi}{2}) + 3\cos t + 5\sin t, \quad t \in [0, 10], \\
(7.3) \qquad y(t) &= 3\sin t - 5\cos t, \qquad t \leq 0, \\
y'(t) &= 3\cos t + 5\sin t, \qquad t \leq 0
\end{aligned}$$

with exact solution $y(t) = 3\sin t - 5\cos t$. Table 7.5 shows the results for the Lambert-Watson method and for its minimax variant using the frequency interval $[0.95, 1.05]$. In these tests the implicit relations were solved using Newton's method. Again, a substantial gain in accuracy is obtained.

h	LW_6 (conventional)	LW_6 (minimax)
2/5	4.1	5.3
1/5	6.0	7.9
1/10	7.8	11.1

Table 7.5

51

As a second example we consider a Bessel-type equation involving a state-dependent delay:

$$
\begin{aligned}
& y''(t) + (100 + \tfrac{1}{4t^2})y(t) + y(t - 1 - y^2(t)) = g(t), && t \in [3, 10] \\
& \qquad\qquad y(t) = t^{\frac{1}{2}} J_0(10t), && t \le 3 \\
& \qquad\qquad y'(t) = \tfrac{1}{2}t^{-\frac{1}{2}}\left[J_0(10t) - 20t J_1(10t)\right], && t \le 3
\end{aligned}
$$

(7.4)

where J_0 and J_1 are the Bessel functions of first and second kind, respectively. The inhomogeneous term $g(t)$ is chosen in such a way that we have the almost periodic solution

$$
y(t) = t^{\frac{1}{2}} J_0(10t).
$$

The results can be found in Table 7.6. Obviously, the frequency is approximately equal to 10; hence, the minimax method was applied using the frequency interval $[9.9, 10.1]$.

h	LW_6 (conventional)	LW_6 (minimax)
1/10	1.7	4.0
1/20	3.7	6.4
1/40	5.5	10.1

Table 7.6

REFERENCES

1. Arndt, H., *Numerical Solution of Retarded Initial Value Problems: Local and Global Error and Stepsize Control*, Numer. Math. **43** (1984), 343 - 360.
2. Bock and H.G., Schlöder, J., *Numerical Solution of Retarded Differential Equations with State-dependent Time Lags*, Z. Angew. Math. Mech. **61** (1981), T269 - 271.
3. Driver, R.D., *Existence Theory for a Delay-differential System*, Contribution to Diff. Equations **1** (1963), 317 - 336.
4. El'sgol'ts, L.E. and Norkin, S.B., *Introduction to the Theory and Application of Differential Equations with Deviating Arguments*, Academic Press, New York - London (1973).
5. Hale, J., *Theory of Functional Differential Equations*, Springer, Berlin - Heidelberg - New York (1977).
6. van der Houwen, P.J. and Sommeijer, B.P., *Linear Multistep Methods with Reduced Truncation Error for Periodic Initial Value Problems*, IMA J. Numer. Anal. **4** (1984), 479 - 489.
7. Lambert, J.D. and Watson, I.A., *Symmetric Multistep Methods for Periodic Initial Value Problems*, JIMA **18** (1976), 189 - 202.
8. Sommeijer, B.P., van der Houwen, P.J. and Neta, B., *Symmetric Linear Multistep Methods for Second-Order Differential Equations with Periodic Solutions*, Report NM-R8501, Centre for Math. and Comp. Sc. , Amsterdam (1985), (preprint).

H. Arndt, Institut für Angewandte Mathematik, Wegelerstr. 6, D-5300 Bonn
P.J. van der Houwen and B.P. Sommeijer, Centre for Mathematics and Computer Science, Kruislaan 413, NL-1098 SJ Amsterdam

International Series of
Numerical Mathematics, Vol. 74
© 1985 Birkhäuser Verlag Basel

CONSTRAINED MESH METHODS FOR FUNCTIONAL DIFFERENTIAL EQUATIONS

by A. Bellen

Istituto di Matematica

Università di Trieste

1. AN INTRODUCTORY SURVEY.

Consider the following Volterra Delay Integro Differential Equation (VDIDE):

$$(1) \quad y'(t)=f(t,y(t), \int_{t_o}^{t} K(t,s,y(s))ds, \ y(t-\tau(t)), \ \int_{t_o}^{t-\sigma(t)} K'(t,s,y(s))ds)$$

with initial conditions:

$$y(t_o)=y_o$$

and $\quad y(t):=g(t) \quad$ for $t<t_o$

where $y: [t_o,b] \to \mathbb{R}^n$, $f \in [t_o,b] \times \mathbb{R}^{4n} \to \mathbb{R}^n$ and both K,K' map: $\{(t,s): t_o \leq s \leq t \leq b\} \times \mathbb{R}^n \to \mathbb{R}^n$. Moreover assume the delays τ and σ are continuous and strictly positive. For $K=K'=0$ (1) reduces to a Delay Differential Equation (DDE), and for $\tau=\sigma=0$ to a Volterra Integro Differential Equation (VIDE).

Besides (1), consider the Neutral Differential Equation (NDE):

$$(2) \quad y'(t)=f(t, y(t), y(t-\tau(t)), y'(t-\sigma(t)))$$

and the second order DDE:

$$(3) \quad y''(t)=f(t, y(t), y'(t), y(t-\tau(t)), y'(t-\sigma(t)))$$

where $y(t):=g(t)$; $y'(t):=h(t)$ for $t<t_o$. We can consider initial condition $y(t_o)=y_o$ for (2) and either initial conditions $y(t_o)=y_o$ $y'(t_o)=y'_o$ or boundary conditions $y(t_o)=y_o$, $y(b)=y_b$ for (3).

It is of great interest in the applications to find also periodic solutions to DDE's and NDE's, i.e. functions $y: \mathbb{R} \to$

53

$\mathbb{R}^n$ which are periodic and satisfy the equation for all t of $\mathbb{R}$.

Equations (1-3) belong to a larger class of equations, called Functional Differential Equations (FDE's), since the right hand side terms depend, for each t, on certain functionals acting on y(s),also other than the point functional y(t) leading to ODE's.

The solution y belongs to a suitable function space which depends on the order of the equation, on the initial, boundary or periodicity conditions, and on the smoothness of f,K,K', τ,σ as well as of the assigned functions g and h,if any.

For the numerical approximation of y, it is worth to remark that,if g and h are assigned on the left of t_0, in general they don't join smoothly y and y' at t_0. Therefore discontinuities on the higher derivatives of y (called "primary discontinuities") spread forward because of the delays. Other discontinuities can rise from those of f,K,K',τ,σ,g and h, and they are spread by the delays,too. All these discontinuity points are called "breaking points". Since we have assumed the delays are continuous and strictly positive, there is a finite number of breaking points in any bounded interval (t_0,b). For this reason we are allowed to assume the solution piecewise as smooth as necessary on suitable subintervals.

For a theoretical approach to the problems above, and to more general classes of FDE's, we refer the interested reader to the books of Bellman and Cooke [20], Driver [34], Elsgolts and Norkin [35], Hale [36],Lakshmikantham and Leela [44].

About applications, FDE's provide a flexible tool for modelling a wide range of biological, physical, engineering, economical and environmental processes. Although the first applications of functional equations go back to the last century and the famous book of Volterra [59] on "the struggle for life" to 1931, the interest of applied mathematicians in FDE's has blown up in

the last 15 years, especially in delay equations. Among the huge
literature which is concerned with particular FDE's related to
real-life problems, let me quote the contribution papers in
Schmitt [55], Collatz-Meinardus-Wetterling [29],Kappel-Schappacher
[42] and the monography of Cushing [32].

Also from the numerical standpoint, the interest in
delay equations has considerably risen. Until 1972, when Cryer's
survey [30] appeared, most of the methods where concerned with initial value problems and almost all of them consisted in suitable
adaptations of techniques developped for ODE's. An exhaustive analytical presentation of such methods is available in [30] together with a description of some sparse papers for periodic solutions and boundary value problems. A different approach to DDE's,
quoted in [30] too, was given by the method of steps, due to
Bellman , where the initial value problem for DDE's is transformed
into a larger system of ODE's [18-19]. More recent is the book of
Hall and Watt [37] (1976) including chapters on i.v.p.'s for DDE's
and VIDE's due to Baker.

In the last 15 years, the interest of applied mathematicians in more general problems for delay equations (such as b.v.p.'s
periodic solutions, identification problems, optimal control, partial differential equations etc. with finite, infinite and state-
dependent delays) lead to a great deal of work on numerical analysis of FDE's. Nevertheless no specific DDE-routine is included to
date in any of the most popular and qualified libraries of mathematical software, not even for the simple i.v.p.. The only, not private, available code come to my knowledge is that of Neves [51]
appeared in 1975 in the "Collected Algorithms" of ACM under the
number 497.

Confining our interest to equations (1,2,3) with initial,
boundary and periodicity conditions, we can approximately arrange

the available methods in the following classes. What follows, afar from pretending to be exhaustive, would like to give a short list of methods developped in the last few years for the problems we are interested in, avoiding overlaps to [30].

Transformation of DDE's into ODE's.

Similarly to the method of steps for i.v.p.'s, also b.v.p.'s can be transformed, by a change of variable, into a larger system of ODE's with suitable boundary conditions [4] , so as to take advantage of the existence of several efficient ODE-codes. The aim is gained at the cost of a severe growth of the system-size, as it was for the i.v.p.'s.

In a quite different theoretical approach, the delay initial value problem is regarded as an infinite dimensional system, i.e. as an abstract Cauchy problem. The idea goes back to Krasowski [43] and has been successfully developped by Banks and Burns for several kinds of problems (mainly control problems-see [7 - 8]). The numerical approach consists in approximating the infinite dimensional system by a sequence of finite dimensional ones, i.e. ordinary differential equations. In this line are the papers [9 - 41] the last of which concerned with infinite delay.

Iterative methods.

They are suitable for several problems and consist in transforming the original delay equation into a sequence of somewhat simpler equations. There are different ways to do this. In [26] the authors consider the two point b.v.p.: $y''(t)=f(t,y(t),y(t-r))$ $t\in[a,b]$, $r=const.,y(t)=g(t)$ for $t\in[a-r,a]$, $y(b)=y_b$, and the sequence of (nonlinear) ODE's:

$$y''_{n+1}(t)=f(t,y_{n+1}(t),y_n(t-r))$$
$$y_{n+1}(a)=g(a) \quad ; \quad y_{n+1}(b)=y_b \qquad n=0,1,\ldots$$

with a given starting function y_o. Under suitable conditions on f, each iteration can be solved monotonically by the method of

Newton-Kantorovich and the sequence y_i converges to the solution.

For periodic solutions of: $y^{(i)}(t)=f(t,y(t),y(t-r(t)))$ $i=1,2$ the following sequence of linear delay equations with constant coefficients:

$$(4)\quad y_{n+1}^{(i)}(t)+ay_{n+1}(t)+by_{n+1}(t-r(t))=f(t,y_n(t),y_n(t-r(t)))+$$
$$+ay_n(t)+by_n(t-r(t))\quad n=0,1,\ldots$$

is considered in [10-14-15-60].

By starting from lower and upper periodic solutions, (4) yields monotonically convergent sequences y_i, provided a and b are such that the left-hand differential operator satisfies certain maximum principles, and the right-hand function satisfies certain growth properties with respect to $y_n(t)$ and $y_n(t-r(t))$.

For only theoretical purposes, the same technique has been used in [39] in connection with the two points b.v.p. of second order DDE's.

<u>Collatz-type methods.</u>

The well known method of Collatz, giving inclusions for the wanted solution via optimization techniques, extends to equations depending on functional differential operators for which a minimum-maximum principle is available. See [27-28].

<u>Adaptations of ODE-methods.</u>

By this expression we mean to refer to discrete ODE-methods endowed with suitable extensions of the discrete solution at extra-nodal points. Since Cryer's survey, this is the class where the most deal of work has been done. Probably I'm nearly right in saying that any method for ODE's has been employed, with suitable adaptations, for DDE's.

The new trends of numerical analysis in the treatment of differential equations, with the demand for "robust" and "reliable" codes and the subsequent effort in studing stability, efficient error estimates and step control of the algorithms employed, was reflected also in delay differential equations.

Among the tens of papers appeared after 1972, let me quote some of them which are more in line with the mentioned trends.

As for i.v.p.'s of DDE's, the papers [21-22-31-38-49-57] deal with multistep methods, [52-53] with Runge-Kutta-Fehlberg method, [50] with Runge-Kutta-Merson, [12-13-17-58-61-62] with implicit Runge-Kutta methods (one-step collocation), [33] with extrapolation methods, [1-40-63] with stability of various methods while [2-3] contain an analysis of the influence of interpolation on global error estimates.

Again for i.v.p.'s, the papers [6-24-45-46-48] deal with VIDE's and [47] with VDIDE's.

Finally, [5-12-16-54] deal with b.v.p.'s and [11-23] with periodic solutions of DDE's.

2. A UNIFYING APPROACH TO FDE-METHODS.

In this section we want to investigate, in a unifying approach, some methods of the last above considered class for the model problem (VDIDE of neutral type):

$$(5) \quad y'(t)=f(t,y(t),\int_{t_o}^{t} K(t,s,y(s))ds, \ y(t-\tau(t)),\int_{t_o}^{t-\sigma(t)} K'(t,s,y(s))ds, \ y'(t-\rho(t)))$$

$$y(t)=g(t) \quad \text{for } t \leq t_o$$

including DDE's, VIDE's, VDIDE's and NDE's.

Suppose to have a "local discrete VIDE-method" (in short: "local method") available for (5), that is a method suited to give, for any t_k, an approximate solution at $t_{k+1}:=t_k+h$, provided an approximation y^* of y is known up to t_k. Furthermore assume the step-length h is so small to have $s-\tau(s)<t_k$, $s-\sigma(s)<t_k$ and $s-\rho(s)<t_k$ for all s in (t_k,t_{k+1}). In other words, for any t_k, the local method yields an approximation z_{k+1} of $z(t_{k+1})$, $z(t)$ being the exact solution of the VIDE:

$$(6) \quad z'(t) = f\!\left(t,\, z(t),\, \int_{t_k}^{t} K(t,s,z(s))\,ds + F_{t_k}(t,y^*),\, y^*(\phi(t)),\, F'_{\psi(t)}(t,y^*),\, y^{*\prime}(\xi(t))\right)$$

$$z(t_k) = y^*(t_k)$$

where: $\phi(t) := t-\tau(t)$; $\psi(t) := t-\sigma(t)$; $\xi(t) := t-\rho(t)$ and, for any

real r and t ($t_o \leq r \leq t \leq b$), and for any $u(s) : \mathbb{R} \to \mathbb{R}^n$,

$$F_r(t,u) := \int_{t_o}^{r} K(t,s,u(s))\,ds \quad ; \quad F'_r(t,u) := \int_{t_o}^{r} K'(t,s,u(s))\,ds.$$

Moreover assume the local method has order p, i.e.:

$$\left| z_{k+1} - z(t_{k+1}) \right| = O(h^{p+1})$$

uniformly with respect to t_k in (t_o,b).

In principle, any local method together with a continuous extension of the approximate solution to the whole interval
$[t_k, t_{k+1}]$, and with a quadrature rule for approximating the integrals $F_{t_k}(t,y^*)$ and $F'_{\psi(t)}(t,y^*)$ at any t of $[t_k, t_{k+1}]$, gives rise to
a FDE-method defined recursively in the following way:

- define $y^*(t) := g(t)$ for $t \leq t_o$;

- once the approximate solution y^* is given up to t_k,

  apply the local method to the equation:

$$w'(t) = f\!\left(t,w(t),\, \int_{t_k}^{t} K(t,s,w(s))\,ds + \tilde{F}_{t_k}(t,y^*),\, y^*(\phi(t)),\, \tilde{F}'_{\psi(t)}(t,y^*),\, y^{*\prime}(\xi(t))\right)$$

$$w(t_k) = y^*(t_k)$$

  where $\tilde{F}$ and $\tilde{F}'$ are quadrature rules for F and F', so as

  to obtain an approximation w_{k+1} of $w(t_{k+1})$;

- construct a local continuous extension between t_k and

  t_{k+1}, that is a function $w^*(t)$ such that $w^*(t_k) = w(t_k)$,

  $w^*(t_{k+1}) = w_{k+1}$ and $\max_t | w^*(t) - w(t) | = O(h^m)$ for $t \in (t_k,$

  $t_{k+1})$ independently on t_k (in this case the extension is

  said to have order m);

- prolong y^* up to t_{k+1} by $y^* = w^*$.

The local continuous extension w^* is possibly given by
the local method (when the method itself is continuous, like one-
step collocation, linear multistep Adams method etc.) or, in gene-

ral, by "a posteriori" interpolation (preferably Hermite interpolation - see [56]) over two or more nodal values $y^*(t_i)$.

Actually the previously quoted methods make use of extensions of both types. Nevertheless, in author's opinion, there are two reasons for avoiding "a posteriori" interpolation over more than two nodes:

-it gives rise to some trouble near the breaking points, where the solution is not smooth enough;

-it is not intrinsic to the underlying local method.

In any case one is faced with the following crucial question:

"Given a local method of order p, how accurate must the continuous extension be between the nodes t_i, and how accurate must the quadrature rule be on each interval (t_i, t_{i+1}), in order for the FDE-method to preserve the global order p at the nodes?"

Like for the ODE-methods, a FDE-method is said to have global order p if $\max_i |y(t_i) - y^*(t_i)| = O(h^p)$.

For DDE's it has been proved by Neves [50] and Oberle-Pesch [52], that an interpolation of accuracy order p is sufficient to this aim. For VIDE's it is important, for any t, the accuracy of the functionals $\tilde{F}_{t_k}(t, y^*)$ rather than that of $y^*(t)$. Lubich in [45] (and earlier Neves) proved that the accuracy order p for the mentioned functional, indepentently on t and k, is still sufficient for preserving the global order p of the VIDE-method. However, he noticed that, for an s-level VIDE-Runge-Kutta method, this condition can be achieved either by a continuous extension of order p together with an arbitrary quadrature rule of the same order, or by a particular quadrature rule of order p acting on suitable nodes $t_k + c_i h$, $0 \leq c_i \leq 1$, $i = 1, \ldots, s$ where the values $y^*(t_k + c_i h)$ are accurate of order p', possibly strictly less than p.

He called the two methods respectively: "mixed" and "extended"
VIDE-R-K-method.

      The result that in dealing with VIDE's it is not neces-
sary for y* to be p-order accurate, arises also from the paper
of Brunner [24]. He considers the collocation and the fully discretized collocation at Gaussian points (which are implicit R-K-
methods) where the approximate solution is given,in the whole interval,by means of piecewise polynomials. The global order is p=
2s (s being the number of collocation points per step, and hence
the number of levels of the corresponding R-K-method) despite the
uniform accuracy order in (t_o,b) is only s+1. In the optic of
Brunner's paper this discrepancy is viewed as the phenomenon of
superconvergence at the nodes for collocation at Gaussian points,
well-known for ODE's.

      With some constraints in the choice of the nodes t_i,
the same "superconvergence preservation law" has been proved by
the author in [12-13] for 1st and 2nd order DDE's and, in a quite
different way, by Zennaro [61] for NDE's.

      In this paper we shall extend this law to equations
such as (5),and all these superconvergence results for FDE's will
be viewed in a unifying approach based on the following perturbation analysis which draws inspiration f r o m Zennaro's paper
[61]. The sense of this approach consists in giving an answer to
the question stated above, i.e. in finding conditions on the extended solution y* in order to preserve, for the FDE-method, the
global order of the local method.

3. <u>PERTURBATION ANALYSIS</u>.

      Let $\Delta=\{t_o<t_1<\ldots<t_N\equiv b\}$ be a mesh,and h be the maximum
length between consecutive nodes.

DEFINITION: Δ is said to be a "constrained mesh" for (5) if each
    function $\phi(t),\psi(t)$ and $\xi(t)$ maps any subinterval of the mesh

either on the left of t_0, or one-to-one onto a previous one.

Remark that a constrained mesh automatically includes the primary discontinuities. Henceforward assume the mesh Δ is a constrained mesh including a number of breaking points of (5) so that the solution y is piecewise smooth of class C^{p+1} between two of them. Besides y, let u be a perturbed function of class $C_\Delta^{p+1}(t_0, t_k)$ (i.e. of class C^{p+1} on each subinterval of the mesh up to t_k) matching y up to t_0.

Assume the perturbation y-u satisfies the following conditions:

i_1) $\displaystyle\int_{t_{i-1}}^{t_i} G(s)(y^{(j)}(s)-u^{(j)}(s))ds=O(h^{p+1})$ $\qquad$ j=0,1

$\qquad$ for all i$\leq$k and for any sufficiently smooth matrix valued function G.

i_2) $\displaystyle\max_{t_0\leq t\leq t_k}|y^{(j)}(t)-u^{(j)}(t)|=O(h^{d+1-j})$ $\qquad$ j=0,1,...,d

i_3) the higher derivatives $u^{(j)}$ j=d+1,...,p+1, are uniformly bounded as h$\to$0.

Let w and z be the solutions of the VIDE's:

(7)
$$w'(t)=f(t,w(t),\int_{t_k}^t K(t,s,w(s))ds+F_{t_k}(t,u), u(\phi(t)), F'_{\psi(t)}(t,u), u'(\xi(t)))$$
$$w(t_k)=u(t_k)$$

and

(8)
$$z'(t)=f(t,z(t),\int_{t_k}^t K(t,s,z(s))ds+F_{t_k}(t,y), y(\phi(t)), F'_{\psi(t)}(t,y), y'(\xi(t)))$$
$$z(t_k)=u(t_k).$$

The following theorem shows how the perturbation y-u in $[t_0,t_k]$ is reflected on the difference w-z in $[t_k,t_{k+1}]$.

THEOREM 1. Assume the functions f,K and K' are sufficiently smooth, and y-u satisfy the conditions above for 2d$\geq$p. Then:

j_0) $w(t_{k+1})-z(t_{k+1})=O(h^{p+1})$

j_1) $\displaystyle\int_{t_k}^{t_{k+1}} G(s)(w^{(j)}(s)-z^{(j)}(s))ds=O(h^{p+1})$ $\qquad$ j=0,1

62

for any sufficiently smooth matrix valued function G.

$$j_2) \quad \max_{t_k \leq t \leq t_{k+1}} |w^{(j)}(t) - z^{(j)}(t)| = O(h^{d+1-j}) \qquad j = 0,1,\ldots,d$$

$j_3)$ the higher derivatives $w^{(j)}$ and $z^{(j)}$ are uniformly bounded as $h \to 0$.

Sketch of the proof. By defining the defect:

$$(9) \quad \delta(t) := w'(t) - f(t, w(t), \int_{t_k}^{t} K(t,s,w(s))\,ds + F_{t_k}(t,y), y(\phi(t)), F'_{\psi(t)}(t,y), y'(\xi(t)))$$

we get (see Brunner [25]):

$$(10) \qquad w(t) - z(t) = \int_{t_o}^{t} R(t,s)\,\delta(s)\,ds.$$

By the smoothness of f, (7) yields

$$|w(t) - z(t)| \leq \int_{t_k}^{t} L|R(t,s)| \{ |F_{t_k}(s,y) - F_{t_k}(s,u)| + |y(\phi(s)) - u(\phi(s))| +$$

$$+ |F'_{\psi(s)}(s,y) - F'_{\psi(s)}(s,u)| + |y'(\xi(s)) - u'(\xi(s))| \}\,ds$$

with the constant L independent on t and t_k. Condition $i_2)$, the smoothness of K and K', and the last inequality yield:

$$\max_{t_k \leq t \leq t_{k+1}} |w(t) - z(t)| = O(h^{d+1}).$$

By subtracting (8) to (7) and by differentiating, we get $j_2)$, while conditions on the higher derivatives of y and u, immediately yield $j_3)$.

In order to prove $j_1)$, differentiation of (10) yields

$$w'(t) - z'(t) = \int_{t_k}^{t} (d/dt) R(t,s)\,\delta(s)\,ds + R(t,t)\,\delta(t)$$

and hence

$$(11) \qquad \int_{t_k}^{t_{k+1}} G(t)(w'(t) - z'(t))\,dt = \int_{t_k}^{t_{k+1}} H(t_{k+1}, s)\,\delta(s)$$

where

$$H(t_{k+1}, s) = \int_{s}^{t_{k+1}} G(r)(d/dr) R(r,s)\,dr + G(s) R(s,s)$$

and the smoothness of G(t) and R(t,s) are inherited by $H(t_{k+1}, s)$.

By expanding the function f in (9) and substituting in (11), we obtain, by (7),

$$(12) \qquad \int_{t_k}^{t_{k+1}} G(t)(w'(t) - z'(t))\,dt = \int_{t_k}^{t_{k+1}} \{ G_1(s)(F_{t_k}(s,y) - F_{t_k}(s,u)) +$$

$$+G_2(s)(y(\phi(s))-u(\phi(s)))+G_3(s)(F'_{\psi(s)}(s,y)-F'_{\psi(s)}(s,u))+$$

$$+G_4(s)(y'(\xi(s))-u'(\xi(s)))\}ds + R$$

where the functions G_i, $i=1,2,3,4$, depend on the partial derivatives of f and belong to $C^{p+1}[t_k,t_{k+1}]$, while, by j_2), the remainder R is an $O(h^{2d+1})$, and hence at least $O(h^{p+1})$.

The crucial part of the proof consists in proving that the four terms of the sum at right-hand of (12) are all $O(h^{p+1})$. For $d\geq p$ ($d\geq p-1$ in non neutral case), the assertion is immediately proved, by i_2), for any t_{k+1}, i.e. without any restriction on the mesh. Assume $d\leq p-1$. By i_2), the first integral to be handled can be written:

$$(14) \qquad \int_{t_k}^{t_{k+1}} G_1(s)\int_{t_o}^{t_k}(\partial/\partial y)K(s,r,y(r))(y(r)-u(r))drds+O(h^{2d+3})$$

where, by i_1) (for $j=0$), the interior integral is an $O(h^p)$ and hence (14) is an $O(h^{p+1})$. Remark that, since delays don't occur in (14), the result is achieved independently on the location of t_{k+1}, i.e. for any mesh.

By the change of variable $s=\phi^{-1}(r)$, with ϕ^{-1} smooth, the second integral at right-hand of (12) can be written

$$\int_{t_i}^{t_{i+1}}(d/dr)(\phi^{-1}(r))G_2(\phi(r))(y(r)-u(r))dr \qquad \text{for some } i$$

and hence, by i_1) (for $j=0$), it is an $O(h^{p+1})$.

Similarly one can handle the fourth integral of (12). By combining the arguments used for the first and second integrals, the assertion can be proved for the third one too. Summarizing, (12) is an $O(h^{p+1})$ and j_1) is proved for $j=1$.

By integrating j_1) for $G(s)\equiv I$ and $j=1$, one obtain j_0). Finally, integration by parts of j_1)(for $j=1$) and j_0, yield j_1) for $j=0$. This completes the proof.

It is not difficult to realize that the same properties

$j_0)-j_3)$ hold if w is the exact solution of (7) modified by $\tilde{F}$ and $\tilde{F}'$ instead of F and F', provided the quadrature errors are both $O(h^p)$ indepentently on t and t_k. Furthermore, instead of w, consider an approximate solution w* given by a local method of order p together with a continuous extension satisfying:

$k_1)$ w* is a polynomial of degree d

$k_2)$ $\displaystyle\max_{t_k\le t\le t_{k+1}}|w^{*'}(t)-w'(t)|=O(h^d)$

$k_3)$ $\displaystyle\int_{t_k}^{t_{k+1}}G(t)(w^{*'}(t)-w'(t))dt=O(h^{p+1})$

for any sufficiently smooth matrix valued function G. One easily observes that the hypothesis on the order of the local method, together with conditions $k_1)-k_3)$ on w*, imply all the properties $j_0)-j_3)$ for the difference w*-w and hence for w*-z.

DEFINITION: Given a local method of order p for (7), we call

Natural Continuous Extension (NCE) of degree d any continuous

extension w* satisfying $k_1)-k_3)$.

By the previous perturbation analysis and standard arguments for i.v.p.'s, one can easily prove, by induction, the following theorem:

THEOREM 2. Any local method of order p gives rise to a FDE-method

for (5), which preserves the global order p at the nodes if:

1)   the continuous extension is a NCE of degree d, with $d\ge\left\lceil\dfrac{p+1}{2}\right\rceil$

2)   the quadrature rule has order p on any finite interval.

3)   the mesh is a "constrained mesh".

Condition 3) drops either for integral equations, or for delay equations if $d\ge p$ ($d\ge p-1$ in non-neutral case).

A question arises naturally: "Do NCE's exist for some VIDE-method ?". The answer is:"yes". In fact the polynomial given by the mentioned collocation and fully discretized collocation methods based on s Gaussian nodes are both NCE's of degree s.

So the superconvergence results of Brunner [24] is restated and extended to VDIDE's of neutral type. Furthermore, for ODE's, Zennaro has proved,in [61],that any Runge-Kutta method of order p has a NCE of (minimal) degree $q=\left\lceil\frac{p+1}{2}\right\rceil$.

We conjecture that the same statement holds for any VIDE-R-K-method. If so, we could perform FDE-methods which realize a compromise between Lubich's mixed and extended VIDE-R-K-methods, improving both of them.

4. <u>CONCLUSIONS</u>.

In the foregoing exposition we have seen how to get p-order FDE-methods from p-order VIDE-methods by using NCE's and avoiding interpolation at too many nodes. With the exception of VIDE's, the aim was attained by selecting constrained meshes only, unless p was small enough. In this case, Hermite interpolation at the end points could provide a sufficiently accurate continuous extension for any step-size.

Obviously the constrained mesh selection is more expensive, since it requires the precomputation of the nodes (unless the delays are constant), and avoids the step-size control. On the other hand, it seems to provide the best stability properties for some FDE-methods (for definitions of stability of FDE's see [62]). In particular, any A-stable one-step collocation method for ODE's gives rise to a DDE-method which inherits a similar property,provided the nodes form a constrained mesh (see [63]).

In conclusion, constrained mesh methods seem to be more advisable, especially in dealing with FDE's which reveal some stiffness.

REFERENCES

1. AL-MUTIB,A.N. Stability properties of numerical methods for solving delay differential equations.J.Comput.Appl.Math. 10,71-79 (1984).

2. ARNDT,H. The influence of interpolation on the global error
   in retarded differential equations.In [29],9-17 (1983)

3. ARNDT,H Numerical solution of retarded initial value problem:
   Local and global error and step-size control.Numer.Math.
   43, 343-360 (1984).

4. ASCHER,U.,RUSSEL,R.D. Reformulation of boundary value problems into "standard" form. SIAM Rew. 23,238-254(1981).

5. BADER,G. Solving boundary value problems for functional differential equations by collocation. To appear.

6. BAKER,C.T.H.,MAKROGLOU,A.,SHORT,E. Stability regions for Volterra integro-differential equations. SIAM J.Num.Anal. 16,
   890-910(1979).

7. BANKS,H.T.,BURNS,J.A. Hereditary control problems:Numerical
   methods based on averaging approximations. SIAM J.Control
   Optim. 16,169-208(1978).

8. BANKS,H.T.,BURNS,J.A.,CLIFF,E.M. Parameter estimation and
   identification with systems with delays. SIAM J.Control
   Optim. 19,791-828(1981).

9. BANKS,H.T.,KAPPEL,F. Spline approximation for functional differential equations.J.Differential Equations 34,496-522(1979).

10. BELLEN,A. Cohen's iteration process for boundary value pro-
    blems for functional differential equations. Rend.Istit.
    Matem.Univ.Trieste 11,32-46(1979).

11. BELLEN,A. The collocation method for the numerical approxima-
    tion of the periodic solutions of functional differential
    equations. Computing 23,55-66(1979).

12. BELLEN,A. A Runge-Kutta-Nystrom method for delay differential
    equations. To appear.

13. BELLEN,A. One-step collocation for delay differential equa-
    tions. J.Comput.Appl.Math. 10,275-283(1984).

14. BELLEN,A.,ZENNARO,M. Maximum principles for periodic solutions of linear delay differential equations. In [29],19-24(1983).

15. BELLEN,A.,ZENNARO,M. Sulla ricerca di soluzioni periodiche di equazioni e disequazioni differenziali ordinarie e con ritardo. Boll.UMI (6)2-B,803-817(1983).

16. BELLEN,A.,ZENNARO,M. A collocation method for boundary value problems of differential equations with functional arguments. Computing 32,307-318(1984).

17. BELLEN,A.,ZENNARO,M. Numerical solutions of delay differential equations by uniform corrections to an implicit Runge Kutta method. To appear.

18. BELLMAN,R. On the computational solution of Differential-Difference Equations.J.Math.Anal.Appl.2,108-110(1961).

19. BELLMAN,R.,COOKE,K.L.,On the computational solution of a class of functional differential equations. J.Math.Anal. Appl.12,495-500(1965).

20. BELLMAN,R.,COOKE,K.L.Differential-Difference Equations. New York-London Academic Press 1963.

21. BOCK,H.G.,SCHLÖDER,J. Numerical solutions of retarded differential equations with statedependent time lags. ZAMM 61, T269-T271(1981).

22. BOCK,H.G.,SCHLÖDER,J. A variable order variable step Adams method for general retarded differential equations. To appear.

23. BOUC,R. Equations différentielles et fonctionelles non linéaire. Equa-Diff-73, Paris Hermann 1974.

24. BRUNNER,H. Implicit Runge-Kutta methods of optimal order for Volterra integro-differential equations. Math.Comp.42,95-109(1984).

25. BRUNNER,H. The application of the variation of constants formulas in the numerical analysis of integral and integro-

differential equations. Utilitas Math. 19,255-280(1981).

26. CHOCHOLATY,P.,SLAHOR,L. A method to boundary value problems for delay equations. Numer.Math. 33,69-75(1979).

27. COLLATZ,L. Einschliessung periodischer Lösungen bei einer Klasse von Differenzen-Differentialgleichungen. In ⌊29⌋, 49-54(1983).

28. COLLATZ,L. Einschliessung periodischer Lösungen bei Delay-Gleichungen. Theese Proceedings.

29. COLLATZ,L.,MEINARDUS,G.,WETTERLING,W. Differential-Difference equations.ISNM 62, Birkhâuser 1983.

30. CRYER,C.W. Numerical methods for functional differential equations. In [55],17-101(1972).

31. CRYER,C.W. Highly stable multistep methods for retarded differential equations. SIAM J.Numer.Anal.11, 788-797(1974).

32. CUSHING,J.M. Integrodifferential equations and delay models in population Dynamics. Lect. Notes in Biomath.20. Springer Verlag 1977.

33. DE GEE,M. Extrapolated midpoint rule for delay differential equations. V.U.Amsterdam Rapp.nr.255(1984).

34. DRIVER,R.D. Ordinary and delay differential equations. Springer-Verlag -New York-Heidelberg-Berlin 1977.

35. ELSGOLC,L.E.,NORKIN,S.B. Introduction to the theory of differential equations with deviating arguments. Academic Press New York 1973.

36. HALE,J. Theory of functional differential equations. Springer-Verlag New York-Heidelberg-Berlin 1977.

37. HALL,G.,WATT,J.M. Modern numerical methods for ordinary differential equations. Clarendon Press Oxford 1976.

38. van der HOUVEN,P.J.,SOMMEIJER,B.P. Improved absolute stability of predictor-corrector methods for retarded differential equations. In[29],137-148(1983).

39. HUTSON,V.C.L. Boundary-value problems for differential diffe-
    rence equations.J.Differential Equations 36,363-373(1980).

40. JACKIEWICZ,Z. Asymptotic stability analysis of θ-methods for
    functional differential equations.Numer.Math. 43,389-396(1984).

41. KAPPEL,F.,KUNISCH,K. Approximation of the state of infinite
    delay and Volterra-type equations.In [29],149-168(&983).

42. KAPPEL,F.,SCHAPPACHER,W. Infinite dimensional Systems. Lect.
    Notes in Math. 1076. Springer-Verlag Berlin Heidelberg
    New York Tokyo 1984.

43. KRASOVSKIJ,N.N. The approximation of a problem of analytic
    design of controls in a system with time-lag. J.Appl.Math.
    mech. 28,876-885(1964).

44. LAKSHMIKANTHAM,V.,LEELA,S. Differential and integral inequa-
    lities. Vol.I-II Academic Press, New York 1969.

45. LUBICH,C. Runge-Kutta theory for Volterra integrodifferential
    equations. Numer.Math. 40,119-135(1982).

46. MAKROGLOU,A. Convergence of a block-by-block method for non
    linear Volterra integrodifferential equations. Math.Comp.
    35,783-796(1980).

47. MAKROGLOU,A. A block-by-block method for the numerical solu-
    tion of Volterra delay integro-differential equations.
    Computing 30,49-62(1983).

48. MATTHYS,J. A-stable linear multistep methods for Volterra
    integro-differential equations. Numer.Math.27,85-94(1976).

49. McKEE,S. Fixed step discretisation methods for delay differen
    tial equations. Comp.& Maths.with Appl.7,413-423(1981).

50. NEVES,K.W. Automatic integration of functional differential
    equations: An approach. ACM Trans.Math.Soft.1,357-368(1975).

51. NEVES,K.W. Automatic integration of functional differential
    equations. Collected Algorithms from ACM.Alg.497(1975).

52. OBERLE,H.J.,PESCH,H.J. Numerical treatment of delay differen-

tial equations by Hermite interpolation. Numer.Math.37, 235-255(1981).

53. OPPELSTRUP,J. The RKFHB4 method for delay differential equations. Lect. Notes in Math.631,133-146 Springer-Verlag(1978).

54. REDDIEN,G.W.,TRAVIS,C.C. Approximation methods for boundary value problems of differential equations with functional arguments. J.Math.Anal.Appl.46,62-74(1974).

55. SCHMITT,K. Delay and functional differential equations and their applications. Academic Press, New York, 1972.

56. STETTER,H.J. Numerische Lösung von Differentialgleichungen mit nacheilendem Argument.ZAMM 45,T79-T80 (1965).

57. TAVERNINI,L. Linear multistep methods for the numerical solution of Volterra functional differential equations.J.Appl. Anal.3,169-185(1973).

58. VERMIGLIO,R. A one-step subregion method for delay differential equations. To appear.

59. VOLTERRA,V. Leçons sur la théorie mathématique de la lutte pour la vie. Gauthier-Villars, Paris, 1931.

60. ZENNARO,M. Maximum principles for linear difference-differential operators in periodic function spaces. Numer.Math.43, 121-139(1984).

61. ZENNARO,M. Natural continuous extensions of Runge-Kutta methods. To appear.

62. ZENNARO,M. On the P-stability of one-step collocation for delay differential equations. Theese Proceedings.

63. ZENNARO,M. P-stability properties of Runge-Kutta methods for delay differential equations. To appear.

This work has been performed within the activity of CNR (Italian National Council of Research), Progetto Finalizzato INFORMATICA, sottoprogetto P1, Obiettivo SOFMAT.

International Series of
Numerical Mathematics, Vol. 74
© 1985 Birkhäuser Verlag Basel

DISTRIBUTION OF ZEROS OF POLYNOMIAL SEQUENCES,

ESPECIALLY BEST APPROXIMATIONS

H.-P. Blatt and E. B. Saff

Mathematisch-Geographische Fakultät, Katholische Universität Eichstätt,D-8078 Eichstätt,Federal Republik of Germany

Department of Mathematics, University of South Florida, Tampa, FL 33620, U.S.A.

In [2] the asymptotic behavior of the zeros of polynomials of near best approximation to functions f on a compact set E was studied in the case when f is not everywhere analytic on E. For example, suppose E is a finite union of compact intervals of the real line and f is continuous, but not analytic on E; then we have shown that every point of E is a limit point of zeros of the polynomials of best uniform approximation to f on E. Moreover, if the complement K of E is simply connected and the boundary of E consists of a finite number of analytic Jordan arcs, then the distribution of the zeros of the polynomials of best uniform approximation was analyzed. The purpose of this paper is to give a new interpretation of this distribution, namely to show that these zeros are uniformly distributed with respect to the normal derivative of $G(x,y)$ on ∂E, where $G(x,y)$ is Green's function of K; furthermore results are obtained for the general case when K is connected.

1. Introduction

Let E be a closed bounded set in the z-plane ($z = x+iy$)

whose complement K (with respect to the extended plane) is connected and <u>regular</u> in the sense that K has a Green's function $G(x,y)$ with pole at infinity: $G(x,y)$ is harmonic in K except at infinity and in a neighborhood of the point of infinity we have

$$(1.1) \qquad G(x,y) = \log\ (x^2+y^2)^{1/2} + G_o(x,y),$$

where $G_o(x,y)$ approaches a finite value at infinity; moreover, $G(x,y)$ is continuous in the closed region $\overline{K}$ except at infinity and vanishes on the boundary of K (Walsh [8]). The function

$$(1.2) \qquad t = \Phi(z) := e^{G(x,y)\ +\ i\,H(x,y)},$$

where $H(x,y)$ is conjugate to $G(x,y)$ in K, maps K onto the exterior of the unit disk. Hence, it follows by (1.1) that

$$(1.3) \qquad |\Phi(z)/z| = 1/c + O(1/z)$$

as $z \to \infty$, where the constant $c > 0$ is called the (logarithmic) <u>capacity</u> of the set E.

For each $\sigma > 1$ we consider the equipotential locus

$$(1.4) \qquad \Gamma_\sigma := \{z = x+iy \in K: G(x,y) = \log\ \sigma\}$$

with interior

$$(1.5) \qquad E_\sigma := E \cup \{z = x+iy \in K: 0 < G(x,y) < \log\ \sigma\}.$$

If K is simply connected, the function $\Phi(z)$ is single-valued in K and each locus Γ_σ is an analytic Jordan curve. If K is multiply connected, the function $\Phi(z)$ cannot be single-valued in K and has <u>critical</u> points, i.e. points where $\Phi'(z) = 0$. Each locus Γ_σ consists of a finite number of Jordan curves which are mutually exterior except for a finite number of critical points. Moreover, the normal derivative $\dfrac{\partial G}{\partial n}$ exists at every point of the locus Γ_σ except for such critical points (n being the exterior normal for $\overline{E}_\sigma$).

If a function $f(z)$ is analytic on E, there exists a largest real number σ (finite or infinite), say $\sigma = \rho$, such that $\rho > 1$ and $f(z)$ is single-valued and analytic on E. Then, denoting

73

by Π_n the collection of all complex polynomials of degree $\leq n$, there exist (cf. [8]) polynomials $p_n \in \Pi_n$, $n = 0,1,2, \ldots$, such that we have

$$(1.6) \qquad \overline{\lim_{n \to \infty}} \ \|f-p_n\|_E^{1/n} = 1/\rho,$$

where we denote by $\|\cdot\|_E$ the uniform norm on the set E. But there exist no polynomials $p_n \in \Pi_n$ such that the left-hand side of (1.6) is less than $1/\rho$. A sequence $\{p_n\}$ satisfying (1.6) is said to converge <u>maximally</u> to f(z) on E.

In [2] the following characterization of functions f, which are not analytic on E, is given in terms of the leading coefficients of the polynomials of best uniform approximation to f(z) on E: This result is analogous to the Cauchy-Hadamard formula for the radius of convergence of a power series.

<u>Theorem 1</u>: <u>Let</u> E <u>be a closed bounded point set whose complement</u> K <u>is connected and regular, and suppose that the function</u> f <u>is continuous on</u> E, <u>analytic in the interior of</u> E. <u>For each</u> $n = 0,1,2, \ldots$, <u>let</u> $p_n^*(z) = a_n z^n + \ldots \in \Pi_n$ <u>be the polynomial of best uniform approximation to</u> f <u>on</u> E. <u>Then</u> f <u>is not analytic on</u> E <u>if and only if</u>

$$(1.7) \qquad \overline{\lim_{n \to \infty}} \ |a_n|^{1/n} = 1/c,$$

<u>where</u> c <u>is the capacity of</u> E.

Just the same arguments as in the proof of the above theorem in [2] lead to an analogous result for functions f analytic on E_ρ, namely

<u>Theorem 2</u>: <u>Let</u> E <u>be a closed bounded set whose complement</u> K <u>is connected and regular, and suppose the function</u> f <u>is analytic on</u> E_ρ, $1 < \rho < \infty$, <u>but not on</u> Γ_ρ. <u>Let</u> $\{p_n\}$, $n = 0,1, \ldots$, $p_n(z) = a_n z^n + \ldots \in \Pi_n$, <u>be a sequence of polynomials converging maximally to</u> f(z) <u>on</u> E. <u>Then</u>

$$(1.8) \qquad \overline{\lim_{n \to \infty}} \; |a_n|^{1/n} = 1/c\rho,$$

<u>where</u> c <u>is</u> <u>the</u> <u>capacity</u> <u>of</u> E.

Hence, the polynomials of best uniform approximation in Theorem 1 or the maximally convergent polynomials of Theorem 2 can be considered as special cases of the following situation: There is given a closed, bounded set E of $\mathbb{C}$ such that the complement K of E is connected and regular, and a polynomial sequence $\{p_n\}$ satisfying the following properties:

(A1) $\qquad n \in \mathfrak{N} := \{n_1 < n_2 < n_3 < \ldots\}$,

(A2) $\qquad p_n(z) \in \Pi_n \smallsetminus \Pi_{n-1}$ with leading coefficient $a_n \neq 0$,

(A3) $\qquad \lim_{n \to \infty} \; |a_n|^{1/n} = \dfrac{1}{c}$,

(A4) $\qquad \lim_{n \to \infty} \; \|p_n\|_E^{1/n} = 1$.

In (A3) and (A4) the limits are considered for $n = n_1, n_2, n_3, \ldots$.

2. <u>Distribution of zeros: K simply connected</u>

In [2] the distribution of the zeros of polynomial sequences satisfying (A1) - (A4) was studied in the neighborhood of an analytic Jordan arc $J \subset \partial K$, when K is simply connected. The equation of an <u>analytic Jordan arc</u> J in $\mathbb{C}$ is given for $z \in J$ in parametric form $z = \gamma(t)$, where t runs through a real compact interval [a,b], $a < b$, $\gamma(t)$ is continuous and $\gamma(t_1) = \gamma(t_2)$ only if $t_1 = t_2$; in addition, $\gamma(t)$ is analytic in the open interval (a,b) and $\gamma'(t) \neq 0$ for all $t \in (a,b)$. Hence, there exists a region Δ, symmetric to the interval (a,b), with the property that $\gamma(t)$ is analytic for all $t \in \Delta$. If, moreover, $J \subset \partial K$ and the region Δ can be chosen in such a way that $\gamma(t) \in K$ when t lies in the upper half Δ^+ of Δ,

$$(2.1) \qquad \Delta^+ := \{t \in \Delta: \text{Im}(t) > 0\},$$

and that $\gamma(t) \notin K$ for $t \in \Delta^-$, where

(2.2) $\qquad \Delta^- := \{t \in \Delta : \text{Im}(t) < 0\}$,

then J is a __free one-sided boundary arc of__ K; if, for an appropriate Δ, $\gamma(t) \in K$ for all $t \in \Delta^+ \cup \Delta^-$, then J is a __free two-sided boundary arc of__ K.

A point $z \in \partial K$ is an __accessible boundary point of__ K if there exists a Jordan arc J with endpoint z such that all other points of J lie in K. If K is simply connected and all points of ∂K are accessible boundary points of K, then, for the inverse mapping $\psi(t)$ of $\Phi(z)$, there exists a continuous extension to $\{t : |t| \le 1\}$. Therefore, suppose J is a free one-sided boundary arc of K, then there are two arguments α and β, $\alpha < \beta < \alpha + 2\pi$, such that

(2.3) $\qquad \psi^{-1}(J) = \{t = e^{i\varphi} : \alpha \le \varphi \le \beta\}$;

if J is a free two-sided boundary arc of K, then

(2.4) $\qquad \psi^{-1}(J) = \{t = e^{i\varphi} : \alpha \le \varphi \le \beta \text{ or } \tilde{\alpha} \le \varphi \le \tilde{\beta}\}$,

where $\alpha < \beta \le \tilde{\alpha} < \tilde{\beta} \le \alpha + 2\pi$.

In stating the next theorem it is convenient to introduce the following notation: For any set C in $\mathbb{C}$ let $Z_n(C)$ be the number of zeros of the polynomial p_n in C, counted with their multiplicities, where $\{p_n\}_{n \in \mathbb{N}}$ is a given sequence of polynomials satisfying (A1) - (A4). Then, in [2], the following result was proved.

__Theorem 3: Let__ E __be a closed bounded set__ whose __complement__ K __is simply connected and__ suppose __that all boundary points of__ K __are accessible boundary points.__ Furthermore, __let__ J __be a subarc in the interior of a free one-sided boundary arc of__ K __such that the connected component__ B __of__ $\overset{\circ}{E}$, __where__ $J \subset \overline{B}$, __is a Jordan region, and assume__

(2.5) $\qquad Z_n(C) = o(n)$ as $n \to \infty$

for any compact set C in B. If D is a neighborhood of the inte-
rior of J such that $\overline{D} \cap \partial E = J$, then for the distribution of the
zeros of the polynomials p_n in D holds

$$(2.6) \qquad \lim_{n \to \infty} \frac{Z_n(D)}{n} = \frac{\beta - \alpha}{2\pi} \, ,$$

where α and β are defined by (2.3).

Theorem 4: Let E be a closed bounded point set
whose complement K is simply connected and suppose that all
boundary points of K are accessible. If J is a subarc in the interior of a free two-sided boundary arc of K and D is a neighborhood of the interior of J such that $\overline{D} \cap \partial E = J$, then

$$(2.7) \qquad \lim_{n \to \infty} \frac{Z_n(D)}{n} = \frac{\beta - \alpha + \tilde{\beta} - \tilde{\alpha}}{2\pi} \, ,$$

where $\alpha, \beta, \tilde{\alpha}, \tilde{\beta}$ are defined by (2.4).

For obtaining results in the next section when K is
connected it is useful to give a new interpretation for the
right-hand side of (2.6) and (2.7): Suppose the conditions of
Theorem 3 are satisfied and J is a free one-sided boundary arc
of K, then $\Phi(z)$ can be analytically extended to J and

$$\beta - \alpha = \int_{\alpha}^{\beta} dt = \frac{1}{i} \int_{J} (\log \Phi(z))' dz$$

$$= \frac{1}{i} \int_{J} \frac{\Phi'(z)}{\Phi(z)} \, dz = \frac{1}{i} \int_{J} \left(\frac{\partial G(x,y)}{\partial x} - i \, \frac{\partial G(x,y)}{\partial y} \right) dz$$

where J is oriented in such a way that K lies to the right. If
J has the equation $z = \gamma(t)$, $a \leq t \leq b$, the direction of the tangent
is determined by the angle $\alpha = \arg \gamma'(t)$ and we can write

$$\left(\frac{\partial G}{\partial x} - i \, \frac{\partial G}{\partial y} \right) dz = \left(\frac{\partial G}{\partial x} - i \, \frac{\partial G}{\partial y} \right) \gamma'(t) \, dt$$

$$= |\gamma'(t)| \left(\frac{\partial G}{\partial x} - i \, \frac{\partial G}{\partial y} \right) (\cos \alpha + i \sin \alpha) \, dt$$

$$= |\gamma'(t)| \left(\frac{\partial G}{\partial s} + i\,\frac{\partial G}{\partial n}\right) dt.$$

Here, $\frac{\partial G}{\partial s}$ is the tangential derivative which is identically zero in the interior of J, since $G(x,y) \equiv 0$ on the boundary of K. The expression $\frac{\partial G}{\partial n}$ is the right-hand normal derivative with respect to the curve J. Or, with other words, $\frac{\partial G}{\partial n}$ is the normal derivative where n is the normal directed into K.

Summarizing we have obtained

$$(2.8) \qquad \beta - \alpha = \int_a^b \frac{\partial G}{\partial n}\,|\gamma'(t)|\,dt = \int_J \frac{\partial G}{\partial n}\,|dz|.$$

If J is a free two-sided boundary arc of K, we consider the region Δ of section 2 such that $\gamma(t) \in K$ for all $t \in \Delta^+ \cup \Delta^-$. Then, there exists a harmonic extension $G_1(x,y)$ of the function $G(x,y)$, defined in $\gamma(\Delta^+)$, across J into some neighborhood of int(J), where

$$(2.9) \qquad int(J) = \{\gamma(t):\ a < t < b\}.$$

This follows from Schwarz's principle of reflection. Moreover, let us denote by n_1 the normal of the curve J directed into $\gamma(\Delta^+)$. Analogously, let $G_2(x,y)$ be the extension of $G(x,y)$, defined in $\gamma(\Delta^-)$, to some neighborhood of int(J) and let n_2 be the normal of the curve J directed into $\gamma(\Delta^-)$. Then we obtain

$$(2.10) \qquad \beta - \alpha + \tilde{\beta} - \tilde{\alpha} = \int_J \left(\frac{\partial G_1}{\partial n_1} + \frac{\partial G_2}{\partial n_2}\right) |dz|.$$

Hence, we may replace in Theorem 3 equation (2.6) by

$$(2.11) \qquad \lim_{n \to \infty} \frac{Z_n(D)}{n} = \frac{1}{2\pi} \int_J \frac{\partial G}{\partial n}\,|dz|$$

and equation (2.7) in Theorem 4 by

$$(2.12) \qquad \lim_{n \to \infty} \frac{Z_n(D)}{n} = \frac{1}{2\pi} \int_J \left(\frac{\partial G_1}{\partial n_1} + \frac{\partial G_2}{\partial n_2}\right) |dz|.$$

3. Distribution of zeros: K connected

The first main result for the more general case, K connected, can be stated as follows.

Theorem 5: Let E be a closed bounded set whose complement K is connected and regular, $\{p_n\}_{n\in\mathfrak{N}}$ a sequence of polynomials satisfying (A1) - (A4). Then, for any $\sigma > 1$,

$$(3.1) \qquad \lim_{n\to\infty} \frac{Z_n(K\smallsetminus \overline{E}_\sigma)}{n} = 0.$$

Moreover, the convergence relation

$$(3.2) \qquad \lim_{n\to\infty} \sum_{z_{n,k}\in\overline{E}_\sigma} \frac{1}{z-z_{n,k}} = \frac{\Phi'(z)}{\Phi(z)}$$

holds locally uniformly in $K\smallsetminus \overline{E}_\sigma$.

Now we are in position to formulate results about the distribution of the zeros of p_n where K does not have to be simply connected any more.

Theorem 6: Let E be a closed bounded point set whose complement K is connected and regular, $\{p_n\}_{n\in\mathfrak{N}}$ a sequence of polynomials satisfying (A1) - (A4), $\sigma > 1$. If J is a Jordan curve contained in Γ_σ, then

$$(3.3) \qquad \lim_{n\to\infty} \frac{Z_n(S)}{n} = \frac{1}{2\pi} \int_J \frac{\partial G}{\partial n} |dz|,$$

where S is the region interior to J.

As an application let us consider an example due to Walsh [10]: Let E be the set $|z(z-1)| \leq 1/16$ bounded by the lemniscate $|z(z-1)| = 1/16$, so $G(x,y) = \frac{1}{2} \log |z(z-1)| + \log 4$. We choose $f(z)$ identically zero in the right-hand oval of the lemniscate bounding E and $1/(1-4z)$ in the left hand oval. Then $f(z)$ is analytic in E_ρ, where $\rho = \sqrt{3}$ is maximal, and E_ρ is the lemniscate $|z(z-1)| = 3/16$ passing through $z = 1/4$. Let $p_{2n-1}(z)$ be the polynomial of degree $2n-1$ which is determined by interpolation to $f(z)$ in the points $z = 0$ and $z = 1$, each considered of multiplicity n. Define $p_{2n}(z) := p_{2n-1}(z)$, then the sequence converges maximally to $f(z)$ on E. Since $G(x,y)$ has a critical point at $z = x+iy = 1/2$ on E_2, by Theorem 5, for any $\sqrt{3} < \sigma \leq 2$ the right-hand and the left-hand oval contain $n + o(n)$ zeros of $p_{2n}(z)$, at least for a subsequence of $\{p_{2n}\}$. But this result holds for the whole sequence $\{p_{2n}\}$, since $\lim_{n \to \infty} |a_{2n}|^{1/2n} = 4$ and the capacity of E is $1/4$. For the right-hand oval this can be easily verified, since n zeros of $p_{2n}(z)$ lie at the point $z = 1$.

In the above-mentioned paper of Walsh [10] it was shown that every point z_0 of the boundary of Γ_ρ which is a limit of points of E_ρ on which $f(z) \neq 0$, is again a limit point of zeros of the polynomials p_n; therefore nothing was said about the right-hand oval in the example above.

4. Proofs

<u>Proof of Theorem 5</u>: For any $\sigma > 1$, the equation (3.1) was already proved in [2].

Let $Z_{n,\sigma} := Z_n(K \smallsetminus \overline{E}_\sigma)$ and

$$(4.1) \qquad p_n(z) = \frac{a_n}{|a_n|} \cdot \tilde{p}_n(z) \cdot q_n(z),$$

where $q_n \in \Pi_{Z_{n,\sigma}}$ is the monic polynomial whose zeros are the zeros of $p_n(z)$ in $K \smallsetminus \overline{E}_\sigma$. Let K^* be a simply connected subregion of K such that the point at infinity lies in K^*. We define a fixed branch of $\Phi(z)$ in K^* by

$$(4.2) \qquad \Phi(z) = \frac{1}{c} z + \alpha_0 + O(\tfrac{1}{z}) \quad \text{for } z \to \infty$$

and set for $z \in K^*$

$$(4.3) \qquad h_n(z) := \frac{\{\tilde{p}_n(z)\}^{1/(n-Z_{n,\sigma})}}{\Phi(z)} ,$$

where the branch of the numerator is chosen such that $h_n(\infty) > 0$.
For $z \in E$ we have

$$|\tilde{p}_n(z)| \leq \frac{|p_n(z)|}{(d_1)^{Z_{n,\sigma}}} ,$$

where d_1 is the minimal distance of Γ_σ to the set E.
Since

$$\log \frac{|\tilde{p}_n(z)|}{|\Phi(z)|^{n-Z_{n-\sigma}}}$$

is harmonic in K and continuous in $\overline{K}$, except at the zeros of
$\tilde{p}_n(z)$, we conclude from the maximum principle for harmonic functions that

$$\frac{|\tilde{p}_n(z)|}{|\Phi(z)|^{n-Z_{n,\sigma}}} \leq \frac{\|p_n\|_E}{(d_1)^{Z_{n,\sigma}}} .$$

Hence, we obtain from (3.1) and (A4) that the functions $h_n(z)$
are uniformly bounded in $K^* \smallsetminus \overline{E}_\sigma$ and satisfy

$$(4.4) \qquad \overline{\lim_{n \to \infty}} |h_n(z)| \leq \overline{\lim_{n \to \infty}} \|p_n\|_E^{1/(n-Z_{n,\sigma})} = 1.$$

Moreover, because of the normalization in (4.2) and the condition (A3), it follows that

$$(4.5) \qquad \lim_{n \to \infty} h_n(\infty) = 1.$$

Since each function $h_n(z)$ is analytic in $K^* \smallsetminus \overline{E}_\sigma$, we conclude
from (4.4), (4.5) and the maximum principle that the functions
$h_n(z)$ converge uniformly to the constant function 1 in any compact subset of $K^* \smallsetminus \overline{E}_\sigma$. Consequently the functions $\log h_n(z)$
converge uniformly to zero in any compact set of $K^* \smallsetminus \overline{E}_\sigma$, if

we take for the logarithm the branch with $\log h_n(\infty) = 0$. Then, on differentiating $\log h_n(z)$ we get

$$\lim_{n \to \infty} \frac{1}{n - Z_{n,\sigma}} \frac{\tilde{p}_n{}'(z)}{p_n(z)} = \frac{\Phi'(z)}{\Phi(z)}$$

or, using (3.1),

$$(4.6) \qquad \lim_{n \to \infty} \frac{1}{n} \sum_{z_{n,k} \in \overline{E}_\sigma} \frac{1}{z - z_{n,k}} = \frac{\Phi'(z)}{\Phi(z)}$$

locally uniformly in $K^* \smallsetminus \overline{E}_\sigma$. Now, we observe that the function

$$\frac{\Phi'(z)}{\Phi(z)} = \frac{\partial G(x,y)}{\partial x} - i \frac{\partial G(x,y)}{\partial y}$$

is independent of the branch of $\Phi(z)$. Since (4.6) holds locally uniformly in $K^* \smallsetminus \overline{E}$, where $K^* \subset K$ is any simply connected region with $\infty \in K^*$, it follows, that (3.2) is true. □

<u>Proof of Theorem 6</u>: For any $\sigma > 1$, the locus Γ_σ consists of a finite number of Jordan curves which are mutually exterior except for a finite number of critical points of $\Phi(z)$. Let us fix a function $f(z)$ analytic in E_σ and continuous in the interior and the boundary of each Jordan curve of Γ_σ, except at the critical points. Then, by Cauchy's integral formula, we obtain for any ρ, $1 < \rho < \sigma$, from (3.2)

$$\lim_{n \to \infty} \frac{1}{n} \sum_{z_{n,k} \in \overline{E}_\rho} f(z_{n,k}) = \frac{1}{2\pi i} \int_{\Gamma_\sigma} f(z) \frac{\Phi'(z)}{\Phi(z)} \, dz,$$

where Γ_σ is oriented in such a way that E_σ lies to the left. Since $Z_n(E_\sigma \smallsetminus \overline{E}_\rho) = o(n)$ as $n \to \infty$, it follows that

$$(4.7) \qquad \lim_{n \to \infty} \frac{1}{n} \sum_{z_{n,k} \in E_\sigma} f(z_{n,k}) = \frac{1}{2\pi i} \int_{\Gamma_\sigma} f(z) \frac{\Phi'(z)}{\Phi(z)} \, dz.$$

Now, let us consider the function $f(z)$ defined by $f(z) = 1$ for $z \in \bar{S}$ and $f(z) = 0$ for $z \in \bar{E}_\sigma \smallsetminus \bar{S}$. Then we obtain from (4.7) and (3.1):

$$\lim_{n\to\infty} \frac{Z_n(\bar{S})}{n} = \lim_{n\to\infty} \frac{Z_n(S)}{n}$$

$$= \frac{1}{2\pi i} \int_J \frac{\Phi'(z)}{\Phi(z)}\, dz$$

$$= \frac{1}{2\pi} \int_J \frac{\partial G(x,y)}{\partial n}\, |dz| . \quad \square$$

5. References

1. L. V. Ahlfors: Complex Analysis, McGraw-Hill Book Company, third edition, 1979.

2. H.-P. Blatt, E. B. Saff: Behaviour of zeros of polynomials of near best approximation, to appear.

3. P. Borwein: The relationship between the zeros of best approximations and differentiability, Proceedings of the Fourth Texas Conference on Approximation Theory, College Station, 1983.

4. G. M. Golusin: Geometric Theory of Functions of a Complex Variable, American Mathematical Society, Vol. 26, Providence, Rhode Island, 1969.

5. R. Jentzsch: Untersuchungen zur Theorie analytischer Funktionen, Inangural-Dissertation, Berlin 1914.

6. P. Ch. Rosenbloom: Sequences of polynomials, especially sections of power series, Dissertation, Stanford University, 1943.

7. G. Szegö: Über die Nullstellen von Polynomen, die in einem Kreis gleichmäßig konvergieren, Sitzungsberichte der Berliner Math. Gesellschaft, 21 (1922), 59-64.

8. J. L. Walsh: Interpolation and approximation by rational functions in the complex domain, Amer. Math. Soc. Colloquium Publications, vol. 20, 1935, Fifth Edition 1969.

9. J. L. Walsh: Overconvergence, degree of convergence and zeros of sequences of analytic functions, Duke Math. Journal 13 (1946), 195-234.

10. J. L. Walsh: The analogoue for maximally convergent polynomials of Jentzsch's theorem, Duke Math. Journal 26 (1959), 605-616.

International Series of
Numerical Mathematics, Vol. 74
© 1985 Birkhäuser Verlag Basel

ON A CONJECTURE ABOUT THE CRITICAL POINTS OF A POLYNOMIAL

by

B. D. Bojanov (Sofia), Q. I. Rahman (Montréal)
and J. Szynal (Lublin)

1. The conjecture

By a "circular domain" we shall mean a domain in $\hat{\mathbb{C}}$ whose boundary is a circle or a straight line. By $D(a;\eta)$ we shall denote the open disk $\left\{z \in \mathbb{C} : |z-a| < \eta\right\}$ and by $\overline{D(a;\eta)}$ its closure.

According to a well-known theorem of Gauss every convex domain containing all the zeros of a polynomial $p(z)$ also contains all the zeros of the derivative $p'(z)$. In particular, if all the zeros $z_1, \ldots,$ z_n of $p(z)$ lie in $\overline{D(0;1)}$ then so do all the zeros of $p'(z)$. It was conjectured by Bl. Sendov (for a history of the conjecture see [7]) that, for $n \geq 2$, $p'(z)$ has at least one zero in each of the disks $\overline{D(z_k;1)}$, $k = 1, 2, \ldots, n$. The example $p(z) := z^n - 1$ shows that the open disks $D(z_k;1)$, $k = 1, 2, \ldots, n$ may not contain any zero of $p'(z)$.

2. The apolarity theorem

The conjecture is obviously true for quadratics but it does not take long to realize that for polynomials of higher degree it is far from being trivial. It is natural to try to relate the problem to a fundamental result of J. H. Grace called "the apolarity theorem" which gives a sufficient condition under which a circular domain contains <u>at least one zero</u> of a given polynomial. Two polynomials

$$A(z) := \sum_{k=0}^{m} \binom{m}{k} a_k z^k, \qquad B(z) := \sum_{k=0}^{m} \binom{m}{k} b_k z^k$$

of degree m are said to be "apolar" if their coefficients satisfy the condition

$$(1) \qquad \sum_{k=0}^{m} (-1)^k \binom{m}{k} a_k b_{m-k} = 0.$$

Grace's apolarity theorem ([4]; also see [6, p. 61] and [12]) may then be stated as follows.

THEOREM A. <u>If</u> $A(z)$ <u>and</u> $B(z)$ <u>are apolar then every circular</u>

domain which contains all the zeros of one polynomial also contains at least one zero of the other.

It may be hard to believe that a result like Theorem A could be of much use since two arbitrarily given polynomials of degree m are not very likely to satisfy the apolarity condition (1). However, its significance becomes apparent when it is stated in the following form. The equivalence of the two formulations is obvious.

THEOREM A$'$. <u>Let</u> λ_0, λ_1, ..., λ_m <u>be given constants which do not all vanish and let</u>

$$P(z) := \sum_{k=0}^{m} p_k z^k$$

<u>be a polynomial of degree</u> m <u>whose coefficients satisfy the linear condition</u>

$$L(P) := \sum_{k=0}^{m} \lambda_k p_{m-k} = 0.$$

<u>Then</u> P(z) <u>has at least one zero in each circular domain which contains all the zeros of the polynomial</u> $L([z \mapsto (z-w)^m])$ <u>in</u> w.

Here we will like to mention certain consequences of Theorem A which will be needed in § 8.

If a real-valued function which is differentiable on the interval [-1,1] assumes the same value at -1 and 1, then according to Rolle's theorem its derivative must vanish at least once in (-1,1). The conclusion does not remain true if the function is allowed to assume complex values on [-1,1] even if it may be entire. This is shown by the example $f(z) := e^{\pi i z}$.

Let us consider a polynomial p(z) of degree n which assumes the same value at the points -1 and 1. This means that

$$P(z) := p'(z) := \sum_{k=0}^{n-1} p_k z^k$$

is a polynomial of degree n-1 such that

$$L(P) := \int_{-1}^{1} P(t)\, dt = 2\, p_0 + \frac{2}{3} p_2 + \frac{2}{5} p_4 + \ldots = 0.$$

Since the zeros of the polynomial

$$L([z \mapsto (z-w)^{n-1}]) = \int_{-1}^{1} (t-w)^{n-1}\, dt = \frac{(1-w)^n - (-1-w)^n}{n}$$

are

$$\sigma_k := i \cot(k\pi/n) \quad (k = 1, \ldots, n-1)$$

we may apply Theorem A' to conclude the following

LEMMA 1 [12, Satz 5]. <u>Let</u> $p(z)$ <u>be a polynomial assuming the same value at</u> -1 <u>and</u> 1. <u>Then</u> $p'(z)$ <u>has at leas at least one zero in each of the two half-planes</u> $\{z : \text{Re } z \leq 0\}$ <u>and</u> $\{z : \text{Re } z \geq 0\}$. <u>Besides,</u> $p'(z)$ <u>has at least one zero in</u> $D(0;\cot(\pi/n))$ <u>if the degree of</u> $p(z)$ <u>happens to be</u> n.

The example

$$p(z) := \left\{ \frac{1}{2}(z+1)(e^{2\pi i/n}-1)+1 \right\}^n + 1$$

shows that the open disk $D(0;\cot(\pi/n))$ may not contain any zero of $p'(z)$ if $p(z)$ is a polynomial of degree n such that $p(-1) = p(1)$.

The second part of Lemma 1 is known as the Grace-Heawood theorem. The next lemma is a fairly straightforward consequence [12, §10] of this result. It is commonly known as the theorem of Alexander [1] and Kakeya [5].

LEMMA 2. <u>Let</u> $p(z)$ <u>be a polynomial of degree</u> n <u>such that</u> $p'(z)$ $\neq 0$ <u>in</u> $\overline{D(0;R)}$. <u>Then</u> $p(z_1) \neq p(z_2)$ <u>for any two distinct points</u> z_1 <u>and</u> z_2 <u>in</u> $\overline{D(0;R \sin(\pi/n))}$, <u>i.e.</u> $p(z)$ <u>is univalent in</u> $\overline{D(0;R \sin(\pi/n))}$.

Most of the progress made towards the resolution of Sendov's conjecture has been achieved with the help of the apolarity theorem in one form or the other. The conjecture still remains unresolved. The purpose of this paper is to report on the progress made so far.

3. <u>A proof of the conjecture for polynomials of degree</u> 3

Let us denote by $S_{m,k}(Z_1, Z_2, \ldots, Z_m)$ or simply $S_{m,k}$ (k = 0, 1, 2, ..., m) the elementary symmetric functions of the m variables Z_1, Z_2, ..., Z_m, i.e. $S_{m,0} := 1$,

$$S_{m,1} := Z_1+Z_2+ \cdots +Z_m, \quad S_{m,2} := Z_1Z_2+Z_1Z_3+ \cdots +Z_{m-1}Z_m, \ldots,$$

$$S_{m,m} := Z_1Z_2\cdots Z_m .$$

With each polynomial

$$A(z) := \sum_{k=0}^{m} \binom{m}{k}a_k z^k$$

of degree m we associate the "convolution equation"

$$(2) \qquad A(Z_1, Z_2, \ldots, Z_m) := \sum_{k=0}^{m} a_k S_{m,k} = 0.$$

86

Let β_1, β_2, ..., β_m be the zeros of the polynomial

$$B(z) := \sum_{k=0}^{m} \binom{m}{k} b_k z^k.$$

Then clearly $B(z)$ is apolar to $A(z)$ if and only if the system of numbers β_1, β_2, ..., β_m satisfy (2). Hence the theorem of Grace is equivalent to

THEOREM B [12]. <u>Let</u>

$$A(z) := \sum_{k=0}^{m} \binom{m}{k} a_k z^k$$

<u>be a polynomial of degree m. If the system of numbers</u> β_1, β_2, ..., β_m <u>satisfy the corresponding convolution equation</u> $A(Z_1, Z_2, ..., Z_m) = 0$, <u>then each circular domain containing all the numbers</u> β_k (k = 1, 2, ..., m) <u>contains at least one zero of</u> $A(z)$.

Returning to the conjecture of Sendov let

$$p(z) := \prod_{k=1}^{3} (z-z_k)$$

be a cubic with all its zeros in $\overline{D(0;1)}$. The conjecture will be proved if we show that the polynomial

$$A(z) := p'(z+z_1) := 3z^2 + 2(2z_1-z_2-z_3)z + (z_1-z_2)(z_1-z_3)$$

has at least one zero in $\overline{D(0;1)}$. The convolution equation associated with $A(z)$ is:

$$A(Z_1, Z_2) := 3Z_1Z_2 + (2z_1-z_2-z_3)(Z_1+Z_2) + (z_1-z_2)(z_1-z_3) = 0.$$

It is easily checked that the system of numbers β_1, β_2 defined by:

$$\beta_1 := -(1/3)(z_1 + z_2 e^{2\pi i/3} - z_3 e^{\pi i/3}),$$

$$\beta_2 := -(1/3)(z_1 - z_2 e^{\pi i/3} + z_3 e^{2\pi i/3})$$

satisfy the equation $A(Z_1, Z_2) = 0$. Since $|\beta_1| \leq 1$, $\underline{|\beta_2| \leq 1}$ it follows from Theorem B that $A(z)$ has at least one zero in $\overline{D(0;1)}$. This **proof** of Sendov's conjecture for polynomials of degree 3 was found by A. Joyal in 1966 but has never been published before.

    4. <u>A proof of the conjecture for polynomials of degree</u> ≤ 4

The **following** result known as "the composition theorem of Szegö" is obtained from Grace's apolarity theorem on noting that $z^m A(1/z)$ and $B(-\gamma z)$ are apolar if $\gamma \neq 0$ happens to be a zero of

$$\sum_{k=0}^{m} \binom{m}{k} a_k b_k z^k.$$

THEOREM C. <u>Let</u>

$$A(z) := \sum_{k=0}^{m} \binom{m}{k} a_k z^k, \quad B(z) := \sum_{k=0}^{m} \binom{m}{k} b_k z^k, \quad (A*B)(z) := \sum_{k=0}^{m} \binom{m}{k} a_k b_k z^k$$

<u>and suppose that</u> K <u>is a circular domain containing all the zeros of</u> A(z). <u>Then for every zero</u> γ <u>of</u> (A*B)(z) <u>we can find a zero</u> β <u>of</u> B(z) <u>and a point</u> κ <u>of</u> K <u>such that</u> $\gamma = -\beta\kappa$.

Using Szegö's composition theorem Rubinstein [9] proved the conjecture of Sendov for polynomials of degree 4 as well. He argued "essentially" as follows.

Let $p(z) = (z-a)q(z)$, where $0 \leq a \leq 1$. Denote the zeros of $q(z)$ by z_1, z_2, ..., z_{n-1} and those of $p'(z)$ by w_1, w_2, ..., w_{n-1}. Let

$$|a-z_k| = r_k, \quad |a-w_k| = \eta_k \quad (1 \leq k \leq n-1)$$

and assume that $r_1 \leq r_2 \leq \cdots \leq r_{n-1}$ and $\eta_1 \leq \eta_2 \leq \cdots \leq \eta_{n-1}$. Now note that

$$q(z+a) = \sum_{k=0}^{n-1} \frac{1}{k!} q^{(k)}(a) z^k$$

can be written as (A*B)(z) where

$$A(z) := p'(z+a) = \sum_{k=0}^{n-1} \binom{n-1}{k}(k+1)\frac{q^{(k)}(a)}{\binom{n-1}{k}} \frac{1}{k!} z^k$$

and

$$B(z) := \sum_{k=0}^{n-1} \binom{n-1}{k} \frac{1}{k+1} z^k = \frac{1}{nz}\left\{(1+z)^n - 1\right\}.$$

The zeros of A(z) lie in the annular region $\left\{z : \eta_1 \leq |z| \leq \eta_{n-1}\right\}$; those of B(z) are

$$\beta_k := -1 + \exp(2\pi ki/n) \quad (1 \leq k \leq n-1)$$

so that

$$(3) \qquad 2\sin(\pi/n) \leq |\beta_k| \leq \begin{cases} 2 & \text{if } n \text{ is even} \\ 2\cos(\pi/2n) & \text{if } n \text{ is odd} . \end{cases}$$

It follows from Theorem C that q(z+a) has all its zeros in

$$\left\{z : 2\eta_1 \sin(\pi/n) \leq |z| \leq 2\eta_{n-1}\right\}$$

if n is even and in

$$\left\{z : 2\eta_1 \sin(\pi/n) \leq |z| \leq 2\eta_{n-1} \cos(\pi/2n)\right\}$$

if n is odd, i.e.

$$(4) \qquad 2\eta_1 \sin(\pi/n) \leq r_k \leq \begin{cases} 2\eta_{n-1} & \text{if } n \text{ is even} \\ 2\eta_{n-1} \cos(\pi/2n) & \text{if } n \text{ is odd} . \end{cases}$$

In particular, the zeros of $q(z)$ lie in

$$E_n := \overline{D(0;1)} \cap \left\{ z : |z-a| \geq 2\eta_1 \sin(\pi/n) \right\}.$$

Our aim is to show that $\eta_1 \leq 1$. Let $n \leq 4$ and suppose that $\eta_1 > 1$. Then, as it is easily seen, $E_n \subset D(a-1;1)$, i.e. the polynomial $p(z+a-1)$ which vanishes at $z = 1$ has all its other zeros in $D(0;1)$. Hence, according to the following lemma, $p'(z+a-1)$ has at least one zero in $\overline{D(1;1)}$ which gives the desired result for polynomials of degree ≤ 4.

LEMMA 3. **In Sendov's conjecture the disk** $\overline{D(z_k;1)}$ **corresponding to a boundary zero does contain at least one critical point of the polynomial.**

We omit the proof of this lemma since we shall present a much stronger result (Theorem 2) in § 6.

REMARK 1. From $p'(a) = q(a)$ it follows that

$$(5) \qquad n\, \eta_1\, \eta_2 \cdots \eta_{n-1} = r_1\, r_2 \cdots r_{n-1} .$$

Hence in the case $a = 0$ we have

$$\eta_1 \leq (\eta_1\, \eta_2 \cdots \eta_{n-1})^{1/n-1} \leq (1/n)^{1/n-1} ,$$

which proves

LEMMA 4. **If a polynomial** $p(z)$ **has all its zeros in** $\overline{D(0;1)}$ **and** $p(0) = 0$ **then** $p'(z)$ **has at least one zero in**

$$\overline{D(0;(1/n)^{1/n-1})}.$$

5. Quintics

Meir and Sharma [8] showed that the conjecture of Sendov holds for quintics too. They proved

THEOREM 1. **If all the zeros of a quintic** $p(z)$ **lie in** $\overline{D(0;1)}$ **and** $p(a) = 0$ **then at least one zero of** $p'(z)$ **lies in** $D(a;\eta(a))$ **where**

$$\eta(a) := (|a| + \sqrt{2 - |a|^2})/2.$$

Proof. We use the notations of the preceding section keeping in mind that $n = 5$. In view of Lemma 4 we may assume $0 < a \leq 1$. In addition, we may suppose that $q(a) \neq 0$ since otherwise $p'(a) = 0$ and there is nothing to prove. From $p'(a) = q(a)$ and $p''(a) = 2q'(a)$ it follows that

$$(6) \qquad \sum_{k=1}^{4} \operatorname{Re} \frac{1}{a - w_k} = \operatorname{Re} \frac{p''(a)}{p'(a)} = 2 \operatorname{Re} \frac{q'(a)}{q(a)} = 2 \sum_{k=1}^{4} \operatorname{Re} \frac{1}{a - z_k} .$$

Simple geometrical considerations show that if $0 < a \leq 1$ and $z \in \overline{D(0;1)}$, then

$$\text{Re } \frac{1}{a-z} \geq \frac{1}{2a} - \frac{1-a^2}{2a} \frac{1}{|a-z|^2} \, .$$

In particular

$$\text{Re } \frac{1}{a-z_k} \geq \frac{1}{2a} - \frac{1-a^2}{2a} \frac{1}{r_k^2} \qquad (1 \leq k \leq 4).$$

Using this and the trivial estimate

$$\text{Re } \frac{1}{a-w_k} \leq \frac{1}{\eta_1} \qquad (1 \leq k \leq 4)$$

in (6) we obtain

$$(7) \qquad \frac{4}{\eta_1} \geq 2 \left(\frac{2}{a} - \frac{1-a^2}{2a} \sum_{k=1}^{4} \frac{1}{r_k^2} \right).$$

$\underline{\text{Now we proceed to find an upper bound for}}$ $\sum_{k=1}^{4} \frac{1}{r_k^2}$. We assume η_1, η_2, η_3, η_4 to be given and seek to maximize this quantity under the conditions

$$(4') \qquad \eta_1 \left| \beta_1 \right| \leq r_k \leq \eta_4 \left| \beta_2 \right|$$

and

$$(5') \qquad 5 \, \eta_1 \, \eta_2 \, \eta_3 \, \eta_4 = r_1 \, r_2 \, r_3 \, r_4$$

where

$$\left| \beta_1 \right| := 2 \sin(\pi/5), \quad \left| \beta_2 \right| := 2 \cos(\pi/10).$$

For $\sum_{k=1}^{4} \frac{1}{r_k^2}$ to be maximum, at most one of the numbers r_k can lie in the open interval $(\eta_1 \left| \beta_1 \right|, \eta_4 \left| \beta_2 \right|)$. Indeed, if for some j we had $\eta_1 \left| \beta_1 \right| < r_j \leq r_{j+1} < \eta_4 \left| \beta_2 \right|$ then replacing r_j by $\frac{1}{1+\varepsilon} r_j$ and r_{j+1} by $(1+\varepsilon) r_{j+1}$ with suitable ε such that $(5')$ remains valid, the sum $\sum_{k=1}^{4} \frac{1}{r_k^2}$ would be increased by

$$\frac{(1+\varepsilon)^2 - 1}{r_j^2} + \frac{(1+\varepsilon)^{-2} - 1}{r_{j+1}^2}$$

which is strictly positive.

Now we note that if all or even three of the numbers r_k were equal to $\eta_1 \left| \beta_1 \right|$ then $(5')$ could not be satisfied. Hence the quantity

$\sum_{k=1}^{4} \dfrac{1}{r_k^2}$ can never be larger than the value of this expression for

$$r_1 = r_2 = \eta_1 |\beta_1|, \quad r_2 = \eta_4 |\beta_2|,$$

$$r_3 = (5\, \eta_1\, \eta_2\, \eta_3\, \eta_4)/(r_1\, r_2\, r_4) = (5\, \eta_2\, \eta_3)/(\eta_1 |\beta_1|^2 |\beta_2|),$$

the choice of r_3 being subject to (5$'$). Since

$$\beta_1\, \beta_2\, \beta_3\, \beta_4 = 5, \quad |\beta_1|^{-2} + |\beta_2|^{-2} = 1 \text{ and } \eta_1 \leq \eta_2 \leq \eta_3 \leq \eta_4$$

we obtain

$$(8) \qquad \sum_{k=1}^{4} \dfrac{1}{r_k^2} \leq \dfrac{2}{\eta_1^2 |\beta_1|^2} + \dfrac{1}{\eta_4^2 |\beta_2|^2} + \dfrac{\eta_1^2}{\eta_2^2 \eta_3^2 |\beta_2|^2} \leq \dfrac{2}{\eta_1^2}.$$

We use this estimate for $\sum_{k=1}^{4} \dfrac{1}{r_k^2}$ in (7) to obtain

$$\dfrac{4}{\eta_1} \geq 2\left(\dfrac{2}{a} - \dfrac{1-a^2}{2a}\,\dfrac{2}{\eta_1^2}\right)$$

from which the desired result follows by elementary calculation.

The conjecture of Sendov remains unresolved for polynomials of higher degree. It has been verified in several special cases — one of the most important being

6. <u>The case of a boundary zero</u>

A more precise statement than the conjecture of Sendov can be made in the case of a boundary zero.

THEOREM 2 ([3], [10, Lemma 1], [8, Theorem 1]). <u>If all the zeros of a polynomial</u> $p(z)$ <u>lie in</u> $\overline{D(0;1)}$ <u>and a is a zero of modulus 1, then</u> $p'(z)$ <u>has at least one zero in</u> $\overline{D(a/2;1/2)}$.

<u>Proof</u>. Let $p(z) := (z-a)q(z)$. If we denote by z_1, z_2, ..., z_{n-1} the zeros of $q(z)$ and by w_1, w_2, ..., w_{n-1} those of $p'(z)$, then in the non-trivial case "$q(a) \neq 0$" we obtain

$$\sum_{k=1}^{n-1} \mathrm{Re}\, \dfrac{a}{a-w_k} \geq \mathrm{Re}\, \dfrac{a\, p''(a)}{p'(a)} = 2\, \mathrm{Re}\, \dfrac{a\, q'(a)}{q(a)} = 2 \sum_{k=1}^{n-1} \mathrm{Re}\, \dfrac{a}{a-z_k} \geq n-1.$$

Hence $\mathrm{Re}\, \dfrac{a}{a-w_k} \geq 1$ for some k ($1 \leq k \leq n-1$) and so $\overline{D(a/2;1/2)}$ must contain at least one of the numbers w_k.

REMARK 2. Consider the particular polynomial

$$p(z) := (z-a)(z-ae^{i\alpha})(z-ae^{-i\alpha})$$

$$= z^3 - (1+2\cos\alpha)az^2 + (1+2\cos\alpha)a^2 z - a^3.$$

Then $p'(z)$ has zeros at

$$w_1, \ w_2 = (a/3)(1+2 \cos \alpha \pm i \sqrt{2(1+2 \cos \alpha)(1-\cos \alpha)}),$$

where the quantity under the radical sign is positive if $0 < \alpha < 2\pi/3$. It is an easy matter to show that as α runs from 0 to $2\pi/3$ the points w_1 and w_2 describe the boundary of $D(a/2;1/2)$.

7. <u>Some consequences of Theorem 2</u>

7.1. Let $z_{n_1}, \ z_{n_2}, \ \dots, \ z_{n_m}$ be the corners of the closed convex hull H_p of the zeros of $p(z)$. If all the corners z_{n_μ} $(1 \leq \mu \leq m)$ lie on the unit circle, then

$$H_p \subseteq \bigcup_{\mu=1}^{m} D(\tfrac{1}{2} z_{n_\mu} ; \tfrac{1}{2}),$$

i.e. <u>each</u> zero z_k of $p(z)$ lies in at least one of the disks $D(\tfrac{1}{2} z_{n_\mu} ; \tfrac{1}{2})$ — call it D. By Theorem 2, the disk D must contain a zero w of $p'(z)$, and we have $|w-z_k| \leq 1$. This is how Schmeisser [10, Theorem 1] proved

THEOREM 3. <u>The conjecture of Sendov holds for the polynomial</u> $p(z)$ <u>if the corners of the convex hull of its zeros lie on the unit circle.</u>

7.2. Let $D(a;\eta)$ be the <u>smallest</u> disk containing all the zeros of $p(z)$. Then at least <u>three</u> of the zeros must lie on the boundary $\{z : |z-a| = \eta\}$ unless two of them lie at the extremities of a diameter. Hence from Theorem 2 it follows [11, p. 409] that if all the zeros z_1, $z_2, \ \dots, \ z_n$ of a polynomial $p(z)$ lie in $D(0;1)$ then $p'(z)$ has at least one zero in each of the disks $D(z_k;(1+\sqrt{5})/2)$ $(k = 1, 2, \dots, n)$. Indeed, Schmeisser [11, Satz 2] proved more. He showed that each of the disks $D(z_k;1,568)$ $(k = 1, 2, \dots, n)$ contains at least one zero of $p'(z)$. Now the question arises:

If we cannot settle Sendov's conjecture, how well can we improve upon the number 1,568?

8. <u>An answer to the preceding question</u>

The consideration of this problem led Bojanov, Rahman and Szynal [2] to prove the following

THEOREM 4. <u>If</u>

$$p(z) := \prod_{k=1}^{n} (z-z_k)$$

<u>has all its zeros in</u> $D(0;1)$, <u>then each of the disks</u>

$$D(z_k; (1 + |z_1\, z_2\, \cdots\, z_n|)^{1/n}) \quad (k = 1, 2, \ldots, n)$$

<u>contains at least one zero of $p'(z)$.</u>

The proof of Theorem 4 depends on Lemma 2 and the first part of Lemma 1. If there exists a z_k, call it a, such that $p'(z) \neq 0$ in $\overline{D(a; \delta_n)}$ where $\delta_n := (1 + |z_1\, z_2\, \cdots\, z_n|)^{1/n}$, then by Lemma 2, $p'(z+a)$ is univalent in $\overline{D(a; \delta_n \sin(\pi/n))}$. This, in conjunction with the fact that $|p'(z+a)| > n(\delta_n - |z|)^{n-1}$ for $|z| \leq \delta_n$ leads to the conclusion that $p(z)$ must assume the value $(-1)^n (z_1\, z_2\, \cdots\, z_n) = p(0)$ at some point z^* in $D(a; \delta_n - 1)$. Hence, in the non-trivial case "$|a| > \delta_n - 1$", the first part of Lemma 1 implies that $p'(z)$ has at least one zero w in

$$\left\{ z : |z - z^*| \leq |z| \right\} \cap \overline{D(0; 1)}.$$

It is geometrically evident that $|w - z^*| \leq 1$ and so $|w - a| \leq \delta_n$.

The following fact discovered by Schmeisser [10, Satz 3] is an immediate consequence of Theorem 4.

COROLLARY 1. <u>Sendov's conjecture holds if</u> $p(0) = 0$.

Theorem 4 also leads to the following improvement upon the number 1,568 appearing at the end of § 7.

COROLLARY 2. <u>If</u>

$$p(z) := \prod_{k=1}^{n} (z - z_k)$$

<u>has all its zeros in</u> $\overline{D(0; 1)}$, <u>then</u> $p'(z)$ <u>has at least one zero in each of the disks</u> $\overline{D(z_k; \eta)}$ <u>where</u> $\eta := 1{,}08331641\ldots\, .$

REFERENCES

1. J. W. Alexander, <u>Functions which map the interior of the unit circle upon simple regions</u>, Ann. of Math. 17 (1915), 12-22.

2. B. D. Bojanov, Q. I. Rahman and J. Szynal, <u>On a conjecture of Sendov about the critical points of a polynomial</u>, Math. Z. (to appear).

3. A. W. Goodman, Q. I. Rahman and J. S. Ratti, <u>On the zeros of a polynomial and its derivative</u>, Proc. Amer. Math. Soc. 21 (1969), 273-274.

4. J. H. Grace, <u>The zeros of a polynomial</u>, Proc. Cambridge Philos. Soc. 11 (1902), 352-357.

5. S. Kakeya, <u>On zeros of a polynomial and its derivative</u>, Tôhoku Math. J. 11 (1917), 5-16.

6. M. Marden, _Geometry of Polynomials_, Math. Surveys, No. 3, Amer. Math.
   Soc., Providence, R. I., 1966.
7. ──────, _Conjectures on the critical points of a polynomial_, Amer.
   Math. Monthly 90 (1983), 267-276.
8. A. Meir and A. Sharma, _On Ilyeff's conjecture_, Pacific J. Math. 31
   (1969), 459-467.
9. Z. Rubinstein, _On a problem of Ilyeff_, Pacific J. Math. 26 (1968),
   159-161.
10. G. Schmeisser, _Bemerkungen zu einer Vermutung von Ilieff_, Math. Z.
    111 (1969), 121-125.
11. ──────, _Zur Lage der kritischen Punkte eines Polynoms_, Rend.
    Sem. Mat. Univ. Padova 46 (1971), 405-415.
12. G. Szegö, _Bemerkungen zu einem Satz von J. H. Grace über die Wurzeln_
    algebraischer Gleichungen, Math. Z. 13 (1922), 28-55.

Dr. B. D. Bojanov
Department of Mathematics
University of Sofia
Boul. Ivanov 5
1126 Sofia
Bulgaria,

Prof. Q. I. Rahman
Département de Mathématiques et
de Statistique
Université de Montréal
Montréal H3C 3J7
Canada,

Dr. J. Szynal
Instytut Matematyki
Univ. Marie Curie-Skłodowskiej
ul. Nowotki 10
Lublin
Poland.

International Series of
Numerical Mathematics, Vol. 74
© 1985 Birkhäuser Verlag Basel

Inclusion of solutions of certain types of linear and nonlinear delay-equations

L. Collatz

Summary Approximation and optimization give often numerical procedures for calculation of solutions of differential- and other functional equations. If there monotonicity principles hold, these methods are often (in not too complicated cases) the only ones for obtaining lower and upper (pointwise) bounds for the approximate solution , which one can guarantee. The method is illustrated on a simple nonlinear integral equation and then applied to (linear and nonlinear) delay problems, to an initial value problem with an integro-differential equation and a boundary value problem for a nonlinear ordinary differential equation.

Zusammenfassung: In einfachen Modellproblemen kann man, wenn Monotoniesätze gelten, oft mit Hilfe von Approximation und Optimierung die Lösung in garantierbare Schranken einschließen, und punktweise gültige Schranken für den absoluten Fehler einer Näherungslösung berechnen. Die Methode wird zunächst an einer einfachen nichtlinearen Integralgleichung vorgeführt und dann auf (lineare und nichtlineare) Delay-Probleme angewendet, auf eine Anfangswertaufgabe mit einer Integrodifferentialgleichung und auf eine Randwertaufgabe mit einer gewöhnlichen Differentialgleichung.

1. Monotonicity-Principles

We describe the iteration procedure for a (linear or nonlinear) functional equation

$$(1.1) \qquad u = Tu.$$

u may be a (wanted) function $u(x_1,\ldots,x_n)$ in a given bounded domain B of the real point-space $\mathbb{R}^n$. Let $R=C[B]$ be the Banachspace of continuous functions $f(x)$ with the maximum norm. Then holds the

Theorem of Schauder [30]: Suppose: The completely continuous ope-

rator T maps a closed convex bounded set M of the Banachspace R
into a relatively compact set TM which belongs to M. Then exist at
least one fixed point u of T with u=Tu and u∈TM.

The applications base on the idea of monotonicity. We suppose the
space R as partially ordered. We write for two function $f(x)$,
$g(x) \in C[B]$:
(1.2) $f \leq g$,   iff. $f(x) \leq g(x)$ for all $x \in B$;
($f(x) \leq g(x)$ means the classical ordering for real number). If
$f \leq g$, the intervall $J=[f,g]$ is defined as the set of all elements
h with $f \leq h \leq g$. Every interval is closed, convex and bounded.

(1.3) $\begin{cases} T \text{ is syntone,} & \text{if } f \leq g \text{ implies } Tf \leq Tg \\ & \text{for all } f,g \in D. \\ T \text{ is antitone,} & \text{if } f \leq g \text{ implies } Tf \geq Tg \end{cases}$

        Here D is the domain D of definition of T.
T is monotonically decomposible (J. Schröder [56]) if T can be
written as $T=T_1+T_2$, where T_1 is syntone and T_2 is antitone. We
iterate with
(1.4) $v_1 = T_1 v_0 + T_2 w_0$, $w_1 = T_1 w_0 + T_2 v_0$,
starting with two elements v_0, w_0 of D with $v_0 \leq w_0$, we suppose D
as convex.
We suppose for the starting elements:
(1.5) $v_0 \leq v_1 \leq w_1 \leq w_0$ (J. Schröder [80], Bohl [74]);
the condition $v_1 \leq w_1$ can be omitted (J. Albrecht).

If T is completely continuous, the Schauder theorem can be applied
to the interval $M=[v_0,w_0]$ and assures the existence of a fixed
point $u = Tu$ and $u \in [v_1,w_1]$.

Another application of ordering is the idea of an operator T of
monotonic type defined by the property:
(1.6) $Tf \leq Tg$ implies $f \leq g$ for all $f,g \in D$.

Then one has a simple possibility for inclusion of a solution
Tu=r with given r. If one can determine two function $\underline{v}$ and $\overline{v}$
with $T\underline{v} \leq r \leq T\overline{v}$, then the inclusion holds $v \leq u \leq \overline{v}$. We illu-

strate the power of the monotonicity principles on a simple

Example: (many other examples are given in Collatz [81])
of a nonlinear integral equation of Urysohn-type with an antitone
completely continuous operator T:

$$(1.7) \qquad u(x) = Tu(x) = \lambda \int_0^1 \frac{dt}{\sqrt{1+x+u(t)}} \;.$$

Here is $T = T_2$, T_1 = "zero operator".
λ may be a given positive constant. We ask for a positive solu-
tion $u(x)$. We start with $v_0 \equiv 0$ and get from (1.4) $w_1 = Tv_0 = \frac{\lambda}{\sqrt{1+x}}$; here
is $w_1 \leq \lambda$; therefore we try to choose $w_0 = \lambda$ and get from (1.4)
$v_1 = Tw_0 = \frac{\lambda}{\sqrt{1+x+\lambda}}$; the inequalities

$$v_0 \leq w_0, \; v_0 \leq v_1, \; w_1 \leq w_0 \text{ are satisfied,}$$

fig. 1

and therefore exists a solution $u(x)$ of (1.7)
in the strip

$$(1.8) \qquad \frac{\lambda}{\sqrt{1+x+\lambda}} \leq u(x) \leq \frac{\lambda}{\sqrt{1+x}} \;.$$

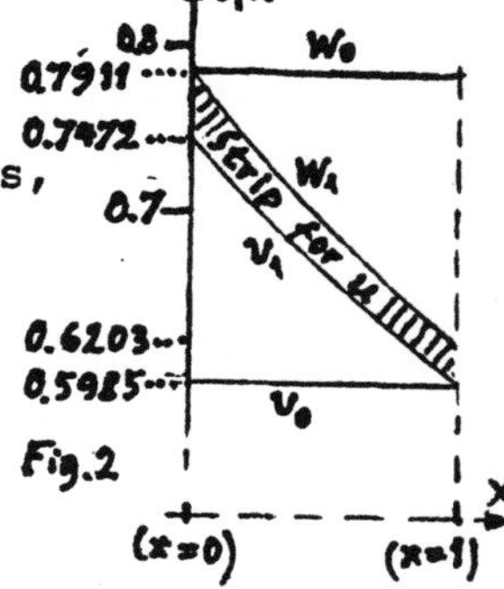

The existence is assured for every $\lambda > 0$; the classical contracting
mapping theorem (see f.i. Collatz[68]p.166) is in the usual way only
for a finite λ-interval applicable. In this example the monotoni-
city-principle is more powerful as the mentioned Banach-theorem.

Working with constant v_0, w_0 one gets for $\lambda = 1$ with $v_0 = 0.5985$ and
$w_0 = 0.7911$ fig. 2

$$v_1 = \frac{1}{\sqrt{1.7911+x}} \leq u(x) \leq \frac{1}{\sqrt{1.5985+x}} = w_1$$

Better bounds on can get with linear functions,
using

$$z(t) = a - bt, \quad Tz = \frac{2}{b} (\sqrt{1+a+x} - \sqrt{1+a-b+x})$$

2. Parabolic delay equation

It is well known that monotonicity properties
for nonlinear parabolic equations hold under
weak conditions.

We consider the operator for a function $u = u(x,t) = u(x_1, \ldots, x_n, t)$:

(2.1) $Tu = u_t - f(x,t,u,u_j,u_{jk})$

As usual the symbols u_j, u_{jk}, u_t represent partial derivatives.
In the hyperplane $t=0$ let B be an open, bounded simply connected
domain with piecewise smooth boundary Γ, Fig. 3. The inner normal
may exist at every point of Γ (not necessarily unique). Let B_t
be the cylinder $x \in B$, $0 < t < t_a$ and Γ_t the cylinder surface $x \in \Gamma$,
$0 < t < t_a$ and

$\hat{B}_t = B_t + \Gamma_t + \hat{B}$ with $\hat{B} = B + \Gamma$.

We suppose, that f is "monotonically
nondecreasing in (u_{jk})", that means

$(2.2) \begin{cases} (u_{jk}) \leq (v_{jk}) & \text{implies} \\ f(x,t,u,u_j,u_{jk}) \leq f(x,t,u,u_j,v_{jk}); \end{cases}$

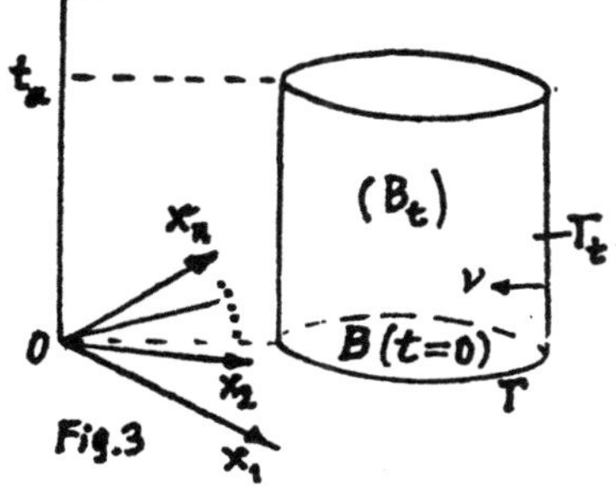

here the arguments of f are considered
as independent variables and $(u_{jk}) \leq (v_{jk})$
means that the matrix $(v_{jk}) - (u_{jk})$ is positive semidefinite. Then
holds the

<u>Theorem</u> (Westphal [49], Walter [61], Redheffer [67] u.a., compare
the books Collatz [66] p.394, Schröder [80] a.o.)
The operator (Tu,Ru) with (2.1) (2.2) and

(2.3) $Ru = \left\{ u \text{ on } \hat{B}, u-k(x,t,u,u_\nu) \text{ on } \Gamma_t, k \text{ in } u_\nu \text{ monotone nondecreasing} \right\}$

is of strictly monotone kind, that means, for two functions
$u,v \in C(\hat{B}_t) \cap C^2(B_t)$

(2.4) $Tu < Tv$ on B_t, $Ru < Rv$ on $\hat{B} + \Gamma_t$ implies $u < v$ on $\hat{B}_t$.

For numerical purposes, if is important, to admit the equalities
in (2.4)

<u>Theorem</u> (compare Redheffer [62], Collatz [66] p.395) If in Tu and
in k the unknown function u does not occur explicitly, (that means
f and k depend only on the derivatives of u, but not on u itself),
one can substitute in (2.4) all signs < by $\leq$:
(2.5) $(Tu,Ru) \leq (Tv,Rv)$ implies $u \leq v$ on $\hat{B}_t$.

Application to the heat conduction equation with delay.

Let us consider the operator L for a function $u(x,t)$

$$(2.6) \quad Lu = \frac{\partial u}{\partial t}(x,t) - \frac{\partial^2 u}{\partial x^2}(x,t) - \int_0^t \varphi(x,s,u_x(x,s),u_{xx}(x,s))\,ds \neq 0,$$

that means, that the time-derivative of $u(x,t)$ does depend not only on u_{xx} at the time t, but also on the values of u_{xx} at ealier times s; u can be interpreted f.i. as temperature or as concentration of a **gas**. A common assumption for φ is

$$(2.7) \qquad \varphi = \frac{s}{t}\,\Phi(u_{xx}(x,s))$$

with a monotone nondecreasing function Φ; φ is small for $s \ll t$ (the values of u_{xx} long time ago have small influence) and greater, if s is near to t (short time ago).

Linear example

$$(2.8) \quad Lu - u_t - u_{xx} - k \int_0^t \frac{s}{t} u_{xx}(x,s)\,ds = 0 \quad \text{in B:} \left\{|x|<\frac{\pi}{2},\ t>0\right\}$$

with k as given constant. We take as boundary conditions, Fig. 4,

$$(2.9) \quad u(x,0) = \cos x \text{ for } |x|\leq\frac{\pi}{2};\quad u(\pm\frac{\pi}{2},t) = 0 \text{ for } t \geq 0.$$

We try to choose as approximate solution (with a as constant)

$$u \approx v = e^{-at}\cos x, \text{ and we get } L = \psi\cos x$$

with

$$\psi = e^{-at}\left[1 - a - \frac{k}{a}\right] + \frac{k}{a^2}\frac{1-e^{-at}}{t}$$

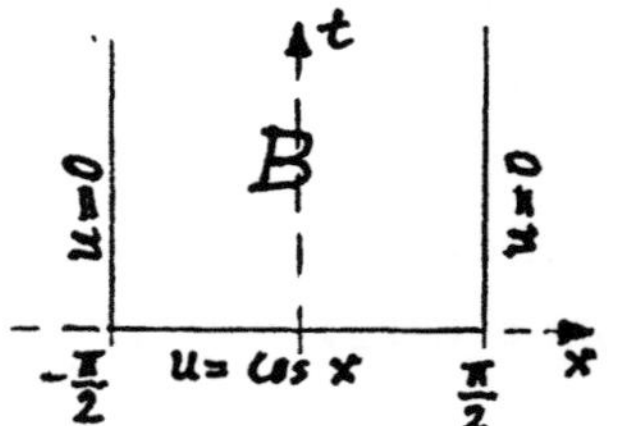

For $a=1$, we have $\psi = \frac{k}{t}[t-e^{-t}-te^{-t}] \geq 0$ for every k and every t; therefore

$$e^{-t}\cos x \geq u(x,t) \quad \text{in B.}$$

We can get a lower bound for $a>1$ for values of t, which are not too big, because for every $a>1$ and every $k>0$ and sufficiently big t we have $\psi>0$, fig. 5, because

$$\lim_{t\to+\infty} t\psi(x,t) = \frac{k}{a^2} > 0$$

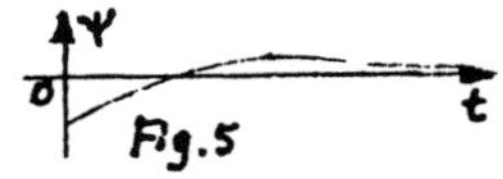

3. Periodic solutions of delay equations.

The results of Bellen and Zennaro [83] for linear delay equations can be generalized to certain nonlinear delay equations of the form (3.1) (see also Zennaro [82],Collatz [83],an der Heiden [83], Workshop [83], Hoffmann-Sprekels [83] a.o.

Theorem Let $u(t)$, $c(t)$ be realvalued function of the period T defined for all real t. We consider the operator

(3.1) $Lu = u''(t) + f(u(t)) + b\, u(t-\tau(t)) + c\, u'(t)$ where $b \leq 0$ and $c \leq 0$

are given real constants and $f(z)$ is a nondecreasing function, fig. 6, with

(3.2) $\text{sgn}\,(f(z)+bz) = \text{sgn}\, z$ for real z,

(3.3) $f(z_1)-f(z_2) \leq \sigma\,(z_1-z_2)$ with $\sigma>0$ (as Lipschitz-constant).

If

(3.4) $(1+|c|T)\,\dfrac{T^2}{8} \cdot \sigma < 1,$

then $Lu \geq 0$ implies $u \geq 0$.

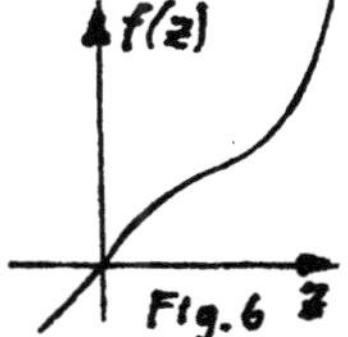

Proof. We follow the proof given by Bellen-Zennaro [83] for linear problems. Let m and M be the minimum and the maximum of $u(t)$, attained at the points t_m, T_M resp., which can be chosen such that

$|t_M - t_m| \leq \frac{1}{2}T$, fig. 7.

Taylor's formula with $u'(t_M)=0$ yields

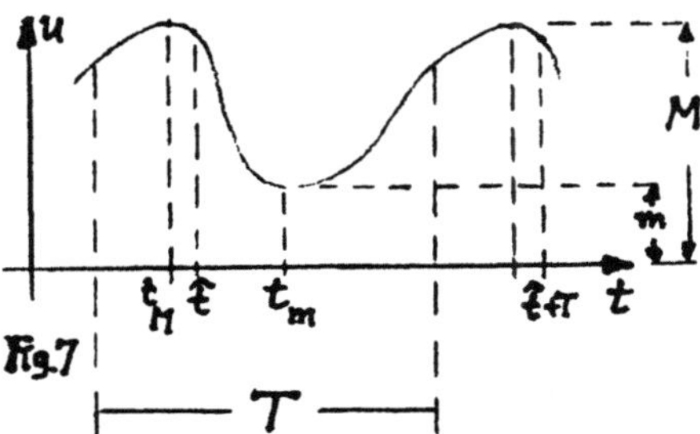

$$m = u(t_m) = u(t_M) + (t_m-t_M)u'(t_M) + \tfrac{1}{2}(t_m-t_M)^2 u''(\hat{t}),$$

$$M - m = -\tfrac{1}{2}(t_m-t_M)^2 u''(\hat{t}),$$

with an unknown intermediate value $t = \hat{t}$. Therefore $-u''(\hat{t}) \geq 0$ and we obtain

(3.5) $M-m \leq \dfrac{1}{8}\,T^2\,(-u''(\hat{t})).$

Now we consider (3.1) with $Lu \geq 0$. Observing $f(u(t)) \leq f(M)$ and $u(t-\tau(t)) \geq m$ we obtain

$$(3.6) \qquad -u''(t) \leq f(M) + bm + cu'(t) = \psi + cu'(t)$$

with $\psi = f(M) + bm$. Thus

$$cu''(t) \leq |c| \psi - c^2 u'(t).$$

We integrate this inequality from t_m to $\hat{t}$. Since $u(t)$ is a periodic function, we can assume $0 \leq \hat{t} - t_m \leq T$ (if $\hat{t} < t_m$, choose $\hat{t} + T$ instead of $\hat{t}$) and we get with $\hat{t} - t_m = \tilde{T} \leq T$

$$(3.7) \quad c\, u'\,(\hat{t}) = c \int_{t_m}^{t} u''(t)\,dt \leq |c|\tilde{T}\psi - c^2 \zeta \leq |c|\tilde{T}\psi;$$

where we have used $\zeta = \int_{t_m}^{\hat{t}} u'(\hat{t})\,dt = u(\hat{t}) - u(t_m) = u(\hat{t}) - m \geq 0.$

(3.6) together with (3.7) yields

$$-u''(\hat{t}) \leq \psi + |c|\tilde{T}\,\psi = (1+|c|\tilde{T})\psi.$$

Therefore (3.5) implies $M - m \leq Q \cdot \psi$ with

$$Q = \frac{1}{8}\, T^2\, (1+|c|\tilde{T}) > 0.$$

Since

$$\psi = f(M) - f(m) + f(m) + bm \leq \sigma(M-m) + f(m) + bm$$

by (3.3), we deduce

$$(M-m)\, [1-\sigma Q] \leq Q \cdot (f(m)+bm).$$

It is $(1-\sigma Q) > 0$ by (3.4) and we obtain $f(m) + bm \geq 0$, since $(M-m) \geq 0$ and $Q > 0$. Hence (3.2) implies $u(t) \geq 0$ for all t.

For error bounds one can use comparison theorems. We need for this purpose a slight modification of the theorem of Bellen-Zennaro [83]:

<u>Lemma</u>: Let $u(t)$, $a(t)$ be realvalued functions with the period T. We consider the linear operator

$$(3.8) \qquad Lu = u''(t) + a(t)\, u(t) + bu(t-\tau(t)),$$

and the nonnegative coefficient $a(t)$ may be bounded by a positive constant A:

$0 \leq a(t) \leq A$ for all t; b is a real constant with $b \leq 0$, $A+b>0$; $T^2A<8$. Then $Lu \geq 0$ has the consequence $u \geq 0$.

Proof: We introduce m, M, t_m, T_M as in the foregoing proof and we get as above (3.5):
$$M - m \leq \frac{1}{8} T^2 (-u''(\hat{t}))$$
(3.8) gives $-u''(\hat{t}) \leq a(\hat{t})u(\hat{t}) + b(u(\hat{t}-\tau(\hat{t})) \leq AM + bm$
$$M - m \leq \frac{1}{8} T^2 [A(M-m) + (A+b)m]$$
$$(M-m) (1-\frac{1}{8} T^2A) \leq \frac{1}{8} T^2 (A+b)m.$$

The left side is ≥ 0 and $A + b>0$; therefore $m \geq 0$ and $u(t) \geq 0$.

Comparison-theorem: Let $v(t)$, $w(t)$, $g(t)$ realvalued functions with the period T and with the ranges $|v| \leq k$, $|w| < k$, $0<g(t) \leq G$ for all t.

We consider the operator
$$(3.9) \qquad Lv = v''(t) + g(t)f(v(t)) + bv(t-\tau(t))$$
with $b \leq 0$ as given constant; $f(z)$ may be a nondecreasing continuously differentiable function for $|z| \leq k$ with

(3.10) $\quad 0 \leq \dfrac{df}{dz} \leq F$ and we suppose $A+b>0$, $T^2A<8$ with $A = F \cdot G$.

Then $Lw \geq Lv$ (for all t) has the consequence $w(t) \geq (t)$ for all t.

Proof. We introduce the difference $\varepsilon = w - v$; we get by Taylor's theorem
$$g(t) \left[f(w(t))-f(v(t)) \right]= g(t)\varepsilon \frac{df}{dz}(\zeta) = \varepsilon \cdot a(t)$$
where $\zeta = \zeta(t)$ is a value between $v(t)$ and $w(t)$ and therefore $|\zeta| \leq k$; the function $a(t)$ is bounded by $0 \leq a(t) \leq GF = A$. It follows
$$Lw - Lv = \varepsilon''(t) + a(t)\varepsilon(t) + b \varepsilon(t-\tau(t)).$$
We have by assumption $L\varepsilon \geq 0$; the Lemma gives then $\varepsilon(t) \geq 0$ or $w(t) \geq v(t)$; q.e.d.

Inclusion of a solution. We consider the equation

$$(3.11) \quad Lu(t) = r(t),$$

where Lu is given by (3.9) and r(t) is a given continuous function with the period T. We suppose that the assumptions of the comparison theorem are satisfied and that a solution u(t) of (3.11) exists. We try to find functions $v(t)$, w(t) with $Lv \leq r \leq Lw$ satisfy, then we have the inclusion

$$v \leq u \leq w.$$

Numerical example

We look for a periodic solution of period T = 2 of the delay equation

$$(3.12) \qquad Lu = u''(t) + f(u(t)) - u(t-1) - h(t) = 0$$

with

$$(3.13) \qquad \begin{cases} f(z) = \dfrac{3}{2}z + \dfrac{1}{4}z^3, \\[2mm] h(t) = \dfrac{1}{4} - \dfrac{1}{2}\cos(\pi t) \end{cases}$$

and ask for approximate solutions of the form

$$(3.14) \qquad u(t) \approx v(t) = \sum_{\nu=0}^{m} b_\nu \cos(\nu\pi t);$$

and for a lower bound $\underline{v}$ with coefficients $b_\nu = \underline{b}_\nu$ resp. an upper bound $\overline{v}$ with coefficients $b_\nu = \overline{b}_\nu$. We try to use the domain of functions $C[0,2] = \left\{ g(x); \; |g(x)| \leq \gamma, \; g(x) \in C^2[0,2] \text{ of period } 2 \right\}$
The optimization problem

$$(3.15) \qquad 0 \leq \overline{v} - \underline{v} \leq \delta, \; L\underline{v} \leq 0 \leq L\overline{v}, \; \delta = \text{Min}$$

is then calculated with 101 equidistant points $t_j = \dfrac{2j}{100}$ (j=0,1,...,100). The following table gives the bound δ for the difference $\overline{v}-\underline{v}$ in dependence of the number m+1 of terms in (3.14):

m	number (m+1) of terms	$\overline{v} - \underline{v} \leq \delta, \; \delta =$
0	1	1.453
1	2	0.002 485
2	3	0.000 065 065
3	4	0.000 000 075 608
4	5	0.000 000 000 5802

We give as example the values of the coefficients for m=2

ν	$\underline{b}_\nu$	$\overline{b_\nu}$
0	0.452 184 716	0.452 119 651
1	0.069 296 545	0.069 296 545
2	0.000 020 934	0.000 020 934

Fig. 8 shows the graph of
$$v^*(t) = \frac{1}{2} (\overline{v} + \underline{v})$$
which we take as approxima-
tion for u(t). It makes no
difference for the dawing of
v*(t) whether we use the
values for m=1,2,3 or 4.

Now we have to check
whether the conditions
(3.10) are satisfied:

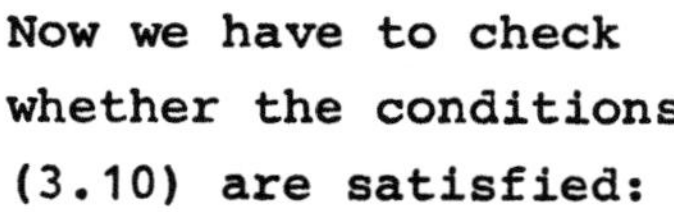

From (3.13) follows: $0 < \frac{df}{dz} \le \frac{3}{2} + \frac{3}{4}\gamma^2 = F.$

We have $T = 2$ and $g(t) = G = 1$ and $A = F \cdot G = \frac{3}{2} + \frac{3}{4}\gamma^2$,
$T^2 A = 6 + 3\gamma^2.$

For $\gamma < \sqrt{\frac{2}{3}} \approx 0.8165$ or for $\gamma < 0.8$ we see that $T^2 A < 8$ is
satisfied; $u, \underline{v}, \overline{v}, v^*$ are contained in the strip of functions
$|g(x)| < 0.8.$

I thank Mr. Qinghua Zheng from Shanghai very much for the careful
calculations of the example (3.12) (3.13) on a computer.

References

Bellen A.-Zennaro [83] Maximum principles for periodic solutions
    of linear delay differential equations, "Internat.Ser.Numer.
    Math. Vol. 62, Birkhäuser" 1983, 19-24

Bohl, E. [74] Monotonie, Lösbarkeit und Numerik bei Operator-
    gleichungen, Springer, 1974, 255 S.

Collatz, L. [68] Funktional Analysis und Numerische Mathematik,
    Springer 1968, 371 S.

Collatz, L. [81] Anwendung von Monotoniesätzen zur Einschließung
    der Lösungen von Gleichungen, Jahrbuch Überblicke Mathem.
    1981, 189-225.

Collatz, L. [83] Einschließung periodischer Lösungen bei einer
    Klasse von Differenzen-Differentialgleichungen, Internat.Ser.
    Numer.Math. Vol. 62, Birkhäuser 1983, 49-54

an der Heiden, U. [83], Periodic, aperiodic and stochastic beha-
    vior...modeling biological and economical processes, "Internat.
    Ser.Numer.Math. Vol. 62, Birkhäuser 1983, 91-108

Hoffmann, K.-H. and J. Sprekels [83] Automatic Delay-Control in
    a Two Phase Stefan-Problem, Internat.Ser.Numer.Math. Vol.62
    Birkhäuser 1983, 119-135

Redheffer, R.M.[67]Differentialungleichungen unter schwachen Vor-
    aussetzungen, Abhandl. Math.Sem.Univ.Hamburg, 31 (1967) 33-50.

Schauder, J. [30] Der Fixpunktsatz in Funktionenräumen, Studia
    Math. 2 (1930) 171-182

Schröder, J. [56] Das Iterationsverfahren bei allgemeinerem Ab-
    standsbegriff, Math.Z. 66 (1956) 111-116

Schröder, J. [80] Operator-Inequalities, Academic Press 1980,
    367 p.

Walter, W. [70] Differential and Integral Inequalities
    Springer 1970, 352 S.

Westphal, M. [49] Zur Abschätzung der Lösungen nichtlinearer
    parabolischer Differentialgleichungen, Math.Z.51 (1949),
    690-695

Workshop Oberwolfach [83] Differential-Difference Equations,
    ed. Collatz, Meinardus, Wetterling, Birkhäuser 1983, 196 p.

Zennaro, M. [82] Maximum Principles for linear Difference-
    Differential Operators in Periodic Function Spaces.
    Universita di Trieste, Quaderno n. 47, 1982, 30 p.

Lothar Collatz
Institut für Angewandte Mathematik
der Universität Hamburg
Bundesstraße 55
D 2000 Hamburg 13
Germany

International Series of
Numerical Mathematics, Vol. 74
© 1985 Birkhäuser Verlag Basel

INTERPOLATION OF ODD PERIODIC FUNCTIONS

ON UNIFORM MESHES

Franz-Jürgen Delvos

Dedicated to

Prof. Dr. K. ZELLER

on the occasion of his 60 th birthday

Introduction

Let $g \in C_{2\pi}$ have an absolutely convergent Fourier series.
For $n \in \mathbb{N}$ we define the uniform mesh $t_k = 2\pi k/n$, $k \in \mathbb{Z}$,
and the translates $g_k = g(\cdot - t_k)$, $0 \leq k < n$, of g . Locher
[4] presented a method of interpolation of periodic functions
f at the uniform mesh t_k, $k \in \mathbb{Z}$, by functions h from
the linear space $V_n(g) = \lim \{ g_0, g_1, \ldots, g_{n-1} \}$ of translates of
g . Locher's method is only applicable if $B_k(0) \neq 0$ for
$k = 0, 1, \ldots, n-1$ where the functions B_k , $k = 0, 1, \ldots, n-1$,
are defined by $\quad B_k(t) = \sum_{j=0}^{n-1} g(t - t_j) \exp(ikt_j)$. In [1] we
derived a modified method of interpolation by translation which
is applicable under the hypothesis $\hat{B}_k(0) \neq 0$ for $k = 1, \ldots, n-1$.
It is the objective this paper to develop a related method of
interpolation of odd periodic functions which works under the
assumption $B_k(0) \neq 0$, $\quad 0 < k < m = n/2$.

1. Interpolation by translation

In this section we briefly recall the method of interpolation by translation [4] . For the sequel we assume

$$\dim V_n(g) = n \ . \tag{1.1}$$

Proposition 1

The functions $B_0, B_1, \ldots, B_{n-1}$ form a basis of $V_n(g)$:

$$V_n(g) = \lin\{ B_0, \ldots, B_{n-1} \} \ . \tag{1.2}$$

Proof: The translates $g_0, \ldots, g_{n-1}$ are linearly independent by assumption . The defining equations

$$B_k = \sum_{j=0}^{n-1} \exp(ikt_j) g_j \ , \quad k = 0, \ldots, n-1 \ , \tag{1.3}$$

may be considered as a basis transformation whence relation (1.2) follows.

Proposition 2

The following relations hold :

$$B_j(t_k) = \exp(ijt_k) B_j(0) \ , \quad 0 \leq j < n \ , \quad k \in \mathbb{Z} \ ; \tag{1.4}$$

$$B_{n-j}(t) = \overline{B_j(t)} \ , \quad 0 \leq j < n \ , \quad t \in \mathbb{R} \ . \tag{1.5}$$

Here we assume that g is real valued.

Proof: For (1.4) we refer to [1] . Since

$$\overline{B_j(t)} = \sum_{k=0}^{n-1} g(t-t_k) \exp(-ijt_k) = \sum_{k=0}^{n-1} g(t-t_k)\exp(i(n-j)t_k)$$

relation (1.5) follows .

Locher's result may be formulated as follows.

<u>Theorem 1</u> [2]

For any $f \in C_{2\pi}$ there is a unique function $P_n(f) \in V_n(g)$ satisfying $P_n(f)(t_k) = f(t_k)$, $k \in \mathbb{Z}$, if and only if

$$B_j(0) \neq 0 \quad , \quad k = 0,1,\ldots,n-1 \quad . \tag{1.6}$$

In view of of (1.6) and (1.4) the function

$$L(t) \;=\; \frac{1}{n} \sum_{j=0}^{n-1} B_j(t)/B_j(0) \quad , \; t \in \mathbb{R} \; , \tag{1.7}$$

is an element of $V_n(g)$ which satisfies

$$L(t_k) = \delta_{0,k} \quad , \quad 0 \leq k < n \quad . \tag{1.8}$$

Then we obtain

$$P_n(f) \;=\; \sum_{k=0}^{n-1} f(t_k) L(\cdot - t_k) \tag{1.9}$$

and

$$V_n(g) \;=\; V_n(L) \quad . \tag{1.10}$$

Note that any function $L \in C_{2\pi}$ satisfying (1.8) defines a method of interpolation by translation . The classical methods of trigonometric interpolation are generated by the trigonometric polynomials

$$g(t) = \sin(nx/2) / (n \sin(x/2)) \quad , \quad n = 2m+1 \quad ,$$
$$g(t) = \sin(nx/2) / (n \tan(x/2)) \quad , \quad n = 2m \quad . \tag{1.11}$$

In view of subsequent applications we consider the Bernoulli functions which are given by

$$P_q(t) = \sum_{m \neq 0} (im)^{-q} \exp(imt) \quad , \quad t \in \mathbb{R} \quad , \quad q \in \mathbb{N} \quad . \tag{1.12}$$

Since

$$\lim_{\substack{x \to 0 \\ x > 0}} (D^{q-1}P_q(x) - D^{q-1}P_q(-x)) = \pi \quad , \quad q \in \mathbb{N} \quad , \tag{1.13}$$

we obtain

$$\dim V_n(P_q(\cdot -a)) = n \quad , \quad q \geq 2 \quad , \quad 0 \leq a < 2\pi \quad . \tag{1.14}$$

It follows from the well known properties of the Bernoulli functions that $V_n(P_q(\cdot -a))$ is a space of periodic polynomial splines of degree q with spline knots $z_k = t_k + a$, $k \in \mathbb{Z}$, of multiplicity 2 . Moreover, we have

$$\int_0^{2\pi} h(t)\,dt = 0 \quad , \quad h \in V_n(P_q(\cdot -a))$$

in view of (1.12) . It follows from (1.12) that theorem 1 is applicable to P_{2r} but not to P_{2r+1} since for any odd function g the equality $B_0(0) = 0$ holds.

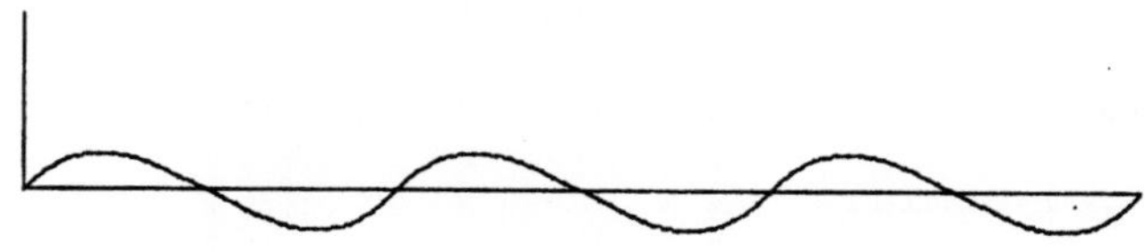

Fig. 1   The function B_0 from $V_3(P_3)$

In $[1]$ we developed a modification of Locher's method of interpolation by translation which is applicable to odd functions g . Instead of $V_n(g)$ we considered the space

$$V_n^1(g) \; = \; \text{lin } \{1,g_1-g,\ldots,g_{n-1}-g\} \quad . \tag{1.15}$$

<u>Proposition 3</u>

The functions $1,B_1,\ldots,B_{n-1}$ form a basis of $V_n^1(g)$:

$$V_n^1(g) \; = \; \text{lin } \{1,B_1,\ldots,B_{n-1}\} \quad . \tag{1.16}$$

<u>Proof</u>: Since

$$\sum_{k=0}^{n-1} \exp(ijt_k) \; = \; 0 \quad , \quad j = 1,\ldots,n-1 \quad ,$$

we have

$$B_j \; = \; \sum_{k=1}^{n-1} \exp(ijt_k)\,(g_k - g) \; e \; V_n^1(g) \quad , \quad j = 1,\ldots,n-1 \quad .$$

Assume now

$$d_0\, 1 \; + \; d_1\, B_1(t) \; + \; \ldots \; + \; d_{n-1}\, B_{n-1}(t) \; = \; 0 \quad , \quad t \; e \; \mathbb{R} \quad ,$$

and choose $t = t_k$, $k = 0,\ldots,n-1$. Using proposition 2 we get

$$d_0 \exp(i0t_k) \; + \; \sum_{j=1}^{n-1} \exp(ijt_k)\, B_j(0)\, d_j \; = \; 0 \quad , \quad k = 0,\ldots,n-1 \quad ,$$

which implies $d_0 = 0$. Since $B_1,\ldots,B_{n-1}$ are linearly independent by proposition 1 we also obtain $d_j = 0$, $j = 1,\ldots,n-1$. This completes the proof of proposition 3 .

<u>Theorem 2</u> $[1]$

For any $f \; e \; C_{2\pi}$ there is a unique $Q_n(f) \; e \; V_n^1(g)$ satisfying $Q_n(f)(t_k) = f(t_k)$, $k \; e \; \mathbb{Z}$, if $B_j(0) \neq 0$ for $j = 1,\ldots,n-1$.

The function M defined by

$$M(t) = \frac{1}{n} \left(1 + \sum_{k=1}^{n-1} B_k(t)/B_k(0) \right) \quad , \quad t \in \mathbb{R} \quad , \tag{1.17}$$

is an element of $V_n^1(g)$ which satisfies

$$M(t_k) = \delta_{k,0} \quad , \quad k = 0,\ldots,n-1 \quad . \tag{1.18}$$

Then we obtain

$$Q_n(f) = \sum_{j=0}^{n-1} f(t_j) \, M(\cdot - t_j) \tag{1.19}$$

and

$$V_n^1(g) = V_n(M) \quad . \tag{1.20}$$

The function M was constructed by Golomb [2] for the space $V_n^1(P_{2r})$. It was shown in [1] that (1.19) remains true for the space $V_n^1(P_{2r+1})$ if $n = 2m+1$.

In general $V_n^1(P_q(\cdot - a))$, $q \geq 2$, $0 \leq a < 2\pi$, is the space of periodic polynomial splines of degree $q-1$ with spline knots $z_j = t_j + a$, $j \in \mathbb{Z}$. The basis $1, P_q(\cdot - z_j) - P_q$, $0 < j < n$, was constructed by Meinardus [5] .

Proposition 4

Let g be odd and $n = 2m$. Then $B_m(0) = 0$.

Proof: This follows from

$$B_m(t) = \sum_{k=0}^{n-1} g(t-t_k) \cos(\pi k) \tag{1.21}$$

since

$$B_m(0) = \sum_{k=1}^{m-1} \left(g(-t_k) + g(-t_{n-k}) \right) \cos(\pi k) = 0 \quad .$$

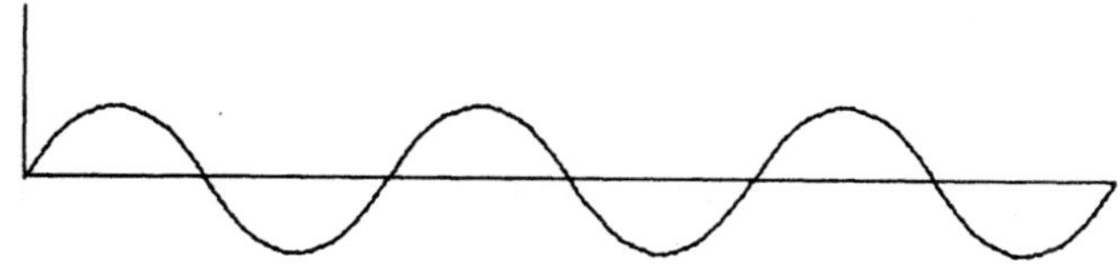

Fig. 2 The function B_3 from $V_6^1(P_3)$

2. Interpolation of odd periodic functions

In this section we will develop a method of interpolation of odd periodic functions at the points t_k, $k \in \mathbb{Z}$, by functions from $V_n^1(g)$, $n = 2m$, $m \in \mathbb{N}$.

Proposition 5

Suppose that

$$B_j(0) \neq 0 \quad , \quad j = 1,\ldots,m-1 \quad . \tag{2.1}$$

The functions S_j , $j = 1,\ldots,m-1$, defined by

$$S_j(t) = \left(B_j(t)/B_j(0) - B_{n-j}(t)/B_{n-j}(0) \right)/(2i) \quad , \quad n = 2m , \tag{2.2}$$

are elements of $V_n^1(g)$ which satisfy the interpolation conditions

$$S_j(t_k) = \sin(jt_k) \quad , \quad k \in \mathbb{Z} \quad . \tag{2.3}$$

Proof: This is an immediate consequence of proposition 2 .

We will denote the linear space spanned by the functions

S_j , $j = 1,\ldots,m-1$ by $V_n^S(g)$, $n = 2m$:

$$V_n^S(g) \;=\; \text{lin}\{\, S_1,\ldots,S_{m-1} \,\} \qquad . \tag{2.4}$$

Since g is real valued and $n = 2m$ we obtain the following

real representation formula for S_j , $j = 1,\ldots,m-1$:

$$B_j'(t)$$

$$= g(t) + g(t-\pi)\cos(\pi j) \;+\; \sum_{k=1}^{m-1} \cos(jt_k)\,\big(\, g(t-t_k) + g(t+t_k) \,\big) \;,$$

$$B_j''(t) \;=\; \sum_{k=1}^{m-1} \sin(jt_k)\,\big(g(t-t_k) - g(t+t_k)\big) \;,$$

$$S_j(t) \;=\; \frac{-B_j''(0)\,B_j'(t) + B_j'(0)\,B_j''(t)}{B_j'(0)^2 + B_j''(0)^2} \;,\qquad 0 < j < m \quad . \tag{2.5}$$

In view of relation (2.3) we apply the classical method
of Lagrange of interpolation by sine functions (see
Lanczos $[3]$, Tonelli $[6]$) to interpolate odd periodic functions by functions from $V_n^S(g)$.

<u>Theorem 3</u>

Assume $B_j(0) \neq 0$, $0 < j < m$. Then for any odd $f \in C_{2\pi}$
there is a unique function $R_m(f) \in V_n^S(g)$, $n = 2m$, which interpolates f at the points t_k , $k \in \mathbb{Z}$.

<u>Proof:</u>   Define

$$A_j(f) \;=\; \frac{2}{m} \sum_{k=1}^{m-1} f(t_k)\,\sin(jt_k) \;,\qquad 0 < j < m \quad . \tag{2.6}$$

Then the sine interpolant of f is given by

$$\tilde{R}_m(f)(t) \;=\; \sum_{j=1}^{m-1} A_j(f)\,\sin(jt) \qquad , \; t \in \mathbb{R} \tag{2.7}$$

(see $\begin{bmatrix}3\end{bmatrix}$). Next we define

$$R_m(f)(t) \;=\; \sum_{j=1}^{m-1} A_j(f)\,S_j(t) \qquad , \; t \in \mathbb{R} \; . \tag{2.8}$$

Taking into account proposition 5 we can conclude

$$R_m(f)(t_k) \;=\; \tilde{R}_m(f)(t_k) \;=\; f(t_k) \quad , \; k \in \mathbb{Z} \; .$$

This completes the proof of theorem 3 .

<u>Proposition 6</u>

The functions $\;L_r\;$, $\;0 < r < m\;$, defined by

$$L_r(t) \;=\; \frac{2}{m}\sum_{j=1}^{m-1} \sin(jt_r)\,S_j(t) \quad , \; t \in \mathbb{R} \; , \tag{2.9}$$

form a Lagrange basis of $\;V_n^s(g)\;$, i. e.,

$$L_r(t_k) \;=\; \delta_{k,r} \quad , \; 0 < k,r < m \quad . \tag{2.10}$$

<u>Proof</u>: This follows from (2.6) and (2.8) .

In the sequel we consider special classes of functions g
for which theorem 3 is applicable .

<u>Proposition 7</u>

Assume that

$$g(t) \;=\; \sum_{k=1}^{\infty} d_k\,\sin(kt) \quad , \; t \in \mathbb{R} \; , \tag{2.11}$$

satisfies

$$d_k > d_{k+1} > 0 \quad , \quad k \in \mathbb{N} \quad . \tag{2.12}$$

Then we have $B_j(0) \neq 0$, $0 < j < m$, i. e. , theorem 3 is applicable to g .

<u>Proof</u>: Note first that

$$g(t) = \sum_{r=-\infty}^{\infty} c_r \exp(irt) \quad ,$$

$$c_0 = 0 \quad , \quad c_r = -i d_r/2 \quad , \quad c_{-r} = -c_r \quad , \quad r \in \mathbb{N} \quad . \tag{2.13}$$

Moreover, we have

$$B_j(0) = n \left(\sum_{k=0}^{\infty} c_{j+kn} + \sum_{k=1}^{\infty} c_{j-kn} \right) \, , \, 0 < j < n \, , \, n = 2m \, . \tag{2.14}$$

Using (2.13) we get

$$B_j(0) = (-i\,n/2) \sum_{k=0}^{\infty} \left(d_{j+kn} - d_{n-j+kn} \right) \tag{2.15}$$

which implies $B_j(0) \neq 0$ for $0 < j < m$.

<u>Corollary 7.1</u>

Assume that $f \in C_{2\pi}$ is odd . Then there is a unique spline function $R_m(f)$ in $V_{2m}^s(P_{2r+1})$ which interpolates f at the points t_k, $k \in \mathbb{Z}$.

<u>Proof</u>: Note that

$$g(t) = (-1)^r P_{2r+1}(t) = \sum_{k \neq 0} (-i)\,k^{-2r-1} \exp(ikt)$$

$$= \sum_{k=1}^{\infty} 2\,k^{-2r-1} \sin(kt) \quad .$$

115

Thus, proposition 7 is applicable and corollary 7.1 follows from theorem 3 .

A second example is the periodic analytic function g considered by de la Vallée Poussin $[7]$,p.113 :

$$g(t) = \frac{\sin(t)}{2(\,\mathrm{ch}(b) - \cos(t))} = \sum_{k=1}^{\infty} e^{-kb} \sin(kt) \ , \ t \ e \ \mathbb{R} \ ,$$

$$(\ b > 0\ ) \ . \tag{2.16}$$

It follows from (2.16) that proposition 7 is applicable.

<u>Proposition 8</u>

Assume that g satisfies the conditions of proposition 7. Let $G = g(\cdot - \pi/n)$, $n = 2m$, denote the shifted function of g. Then

$$B_j^G(0) = \sum_{k=0}^{n-1} G(-t_k) \exp(ijt_k) \neq 0 \ , \ j = 1,\ldots,m-1 \ . \tag{2.17}$$

Thus, theorem 3 is applicable to the shifted function G of g.

<u>Proof</u>: Using (2.13) we get

$$G(t) = \sum_{r=-\infty}^{\infty} c_r \exp(-ir\pi/n) \exp(irt) \quad .$$

Applying (2.14) to G we obtain

$$B_j^G(0) = n \exp(-ij\pi/n)/(2i) \sum_{k=0}^{\infty} (-1)^k (d_{j+kn} + d_{n-j+kn}) \ . \tag{2.18}$$

It follows from (2.12) that

$$d_{j+kn} + d_{n-j+kn} > d_{j+(k+1)n} + d_{n-j+(k+1)n} > 0 \ , \ k \ e \ \mathbb{N} \ ,$$

which implies (2.17) .

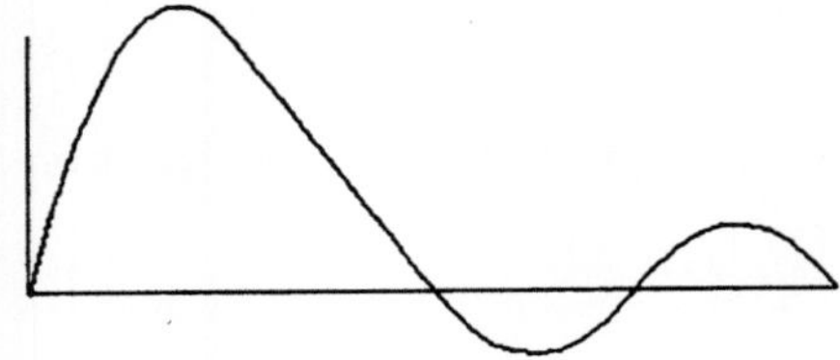 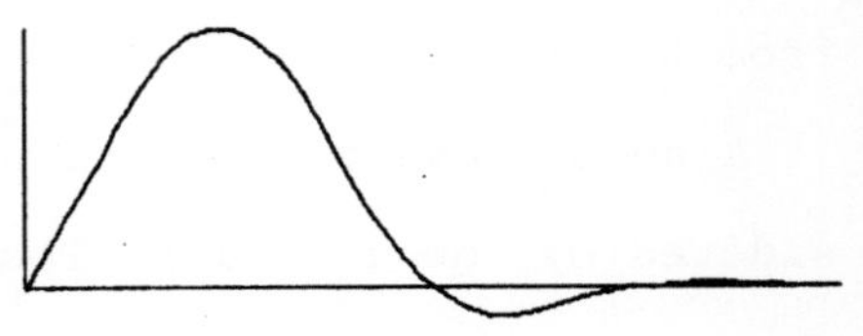

Fig. 3 $L_1 \ e \ V_8^S(P_3)$ Fig. 4 $L_1 \ e \ V_8^S(P_3(\cdot-\pi/8))$

<u>Remark</u>: It follows from (2.18) that $B_j^G(0) \neq 0$, $0 < j < n$, holds for arbitrary n . Thus, theorem 2 is applicable to the shifted function G of g .

<u>Proposition 9</u> $\begin{bmatrix} 4 \end{bmatrix}$

Assume that the even function

$$g(t) \ = \ \sum_{k=0}^{\infty} d_k \cos(kt) \quad , \quad t \ e \ \mathbb{R} \ , \tag{2.19}$$

satisfies

$$d_k > d_{k+1} > 0 \quad , \quad k \geq 1 \ . \tag{2.20}$$

Then we have $B_j(0) \neq 0$, $0 \leq j < n$, i. e., the theorems 1, 2, 3 are applicable.

<u>Proof</u>: Note that

$$g(t) \ = \ \sum_{r=-\infty}^{\infty} c_r \exp(irt) \quad ,$$

$$c_0 = d_0 \quad , \quad c_r = d_r/2 \quad , \quad c_{-r} = c_r \quad , \quad r \ e \ \mathbb{N} \ . \tag{2.21}$$

Then we compute

$$B_j(0) \ = \ n \sum_{k=-\infty}^{\infty} c_{j+kn} \ = \ (n/2) \sum_{k=0}^{\infty} (d_{j+kn} + d_{n-j+kn}) \tag{2.22}$$

117

which implies $B_j(0) \neq 0$, $0 \leq j < n$.

Proposition 9 holds for the following functions :

$$g(t) \;=\; \sum_{k=1}^{\infty} 2\,k^{-2r}\cos(kt) \;=\; (-1)^r\,P_{2r}(t) \quad , \qquad (2.23)$$

$$g(t) \;=\; \frac{1}{2} + \sum_{k=1}^{\infty} e^{-kb}\cos(kt) \;=\; \frac{\operatorname{sh}(b)}{2(\operatorname{ch}(b)-\cos(t))} \;,\; b > 0 \quad (2.24)$$

(see de la Vallée Poussin $\begin{bmatrix}7\end{bmatrix}$, p.113).

Again we consider the shifted function $G = g(\cdot - \pi/n)$.

<u>Proposition 10</u>

Assume that the even function g given by (2.19) satisfies (2.20) and the relation

$$d_{j+kn} - d_{n-j+kn} \;>\; d_{j+(k+1)n} - d_{n-j+(k+1)n} \quad ,$$

$$0 < j < m \quad , \quad n = 2m \quad , \quad k \geq 0 \quad . \qquad (2.25)$$

Then $B_j^G(0) \neq 0$, $0 < j < m$, i. e., theorem 3 holds for the shifted function G of g .

<u>Proof</u>: Taking into account (2.21) and (2.22) we obtain

$$B_j^G(0) \;=\; \frac{n}{2}\exp(-ij\pi/n)\sum_{k=0}^{\infty} (-1)^k(d_{j+kn} - d_{n-j+kn}) \;,$$

$$0 < j < m \quad . \qquad (2.26)$$

In view of (2.25) it follows from the basic facts on alternating series that $B_j^G(0) \neq 0$ for $j = 1,\ldots,m-1$.

Next we will show that the functions (2.23) and (2.24) satisfy the additional conditions (2.25).

We consider the function

$$F(x) = 2 ((x+a)^{-2r} - (x+b)^{-2r}) \ , \quad x \geq 1 \ ,$$

$$a = kn \ , \quad b = (k+1)n$$

which is strictly decreasing . Now we can conclude

$$d_{j+kn} - d_{n-j+kn} > d_{j+(k+1)n} - d_{n-j+(k+1)n} \qquad <=>$$

$$d_{j+kn} - d_{j+(k+1)n} > d_{n-j+kn} - d_{n-j+(k+1)n} \qquad <=>$$

$$F(j) > F(n-j) \ , \quad 0 < j < m \ .$$

Thus, theorem 3 is applicable to $P_{2r}(\cdot - \pi/n)$ and we obtain the

Corollary 10.1

Assume that $f \in C_{2\pi}$ is odd. Then there is a unique spline function $R_m(f) \in V^s_{2m}(P_{2r}(\cdot - \pi/n))$ which interpolates f at the points t_k , $k \in \mathbb{Z}$.

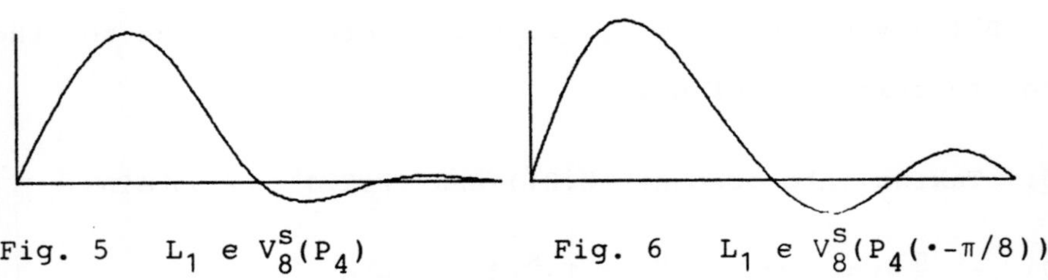

Fig. 5 $L_1 \in V^s_8(P_4)$ Fig. 6 $L_1 \in V^s_8(P_4(\cdot - \pi/8))$

To establish proposition 10 for g given by (2.24) we consider the function   ($A = kn$, $B = (k+1)n$)

$$F(x) = \exp(-b(x+A)) - \exp(-b(x+B)) \ , \quad x \geq 1 \ ,$$

and proceed as above .

Finally we consider an alternative method of interpolation of odd periodic functions. Note that the function N defined by

$$N(t) = \frac{1}{n} \sum_{j=1}^{m-1} (B_j(t)/B_j(0) + B_{n-j}(t)/B_{n-j}(0)) , \quad t \in \mathbb{R}, \quad (2.27)$$

is an element of $V_{2m}^1(g)$.

Proposition 11

The odd functions $N_1,\ldots,N_{m-1}$ defined by translation

$$N_k(t) = N(t_k-t) - N(t_k+t) , \quad 0 < k < m , \quad (2.28)$$

satisfy the interpolation conditions

$$N_k(t_r) = \delta_{k,r} , \quad 0 < k,r < m . \quad (2.29)$$

Proof: Using proposition 2 we can conclude

$$N_k(t_r)$$

$$= \frac{1}{n} \sum_{j=1}^{m-1} (\exp(ijt_{k-r}) + \exp(i(n-j)t_{k-r}))$$

$$- \frac{1}{n} \sum_{j=1}^{m-1} (\exp(ijt_{k+r}) + \exp(i(n-j)t_{k+r}))$$

$$= \frac{1}{n} (\sum_{j=0}^{n-1} \exp(ijt_{k-r}) - 1 - \exp(imt_{k-r}))$$

$$- \frac{1}{n} (\sum_{j=0}^{n-1} \exp(ijt_{k+r}) - 1 - \exp(imt_{k+r}))$$

$$= \delta_{k,r} - \frac{1}{n} \exp(imt_k) (\exp(-imt_r) - \exp(imt_r)) = \delta_{k,r}$$

which completes the proof of proposition 11 .

Thus, for any odd function $f \in C_{2\pi}$ the odd function

$$T_m(f) = \sum_{j=1}^{m-1} f(t_j) N_j \ e \ V_{2m}^1(g) \quad , \qquad (2.30)$$

satisfies the interpolation conditions

$$T_m(f)(t_k) = f(t_k) \quad , \quad k \in \mathbb{Z} \quad . \qquad (2.31)$$

It can be shown that for even g one has $T_m = R_m$.

References

1. F.-J. DELVOS, Periodic interpolation on uniform meshes, (accepted for publication in Journal of Approximation Theory).

2. M. GOLOMB, Approximation by periodic splines on uniform meshes, Journal of Approximation Theory $\underline{1}$(1968), 26-65 .

3. C. LANCZOS, Trigonometric interpolation of empirical and analytic functions, Journal of Mathematics and Physics $\underline{17}$(1938), 123-199 .

4. F. LOCHER, Interpolation on uniform meshes by translates of one function and related attenuation factors, Mathematics of Computation $\underline{37}$(1981), 403-416 .

5. G. MEINARDUS, Periodische Splinefunktionen, in "Spline functions, Karlsruhe 1975" (K. Böhmer, G. Meinardus, W. Schempp), pp. 177-199, LNM 501, Springer Verlag, Berlin, 1976 .

6.   L. TONELLI, Serie trigonometriche, N. Zanichelli,
     1928, Bologna 1928 .

7.   C. J. de la VALLÉE POUSSIN, Leçons sur l'approximation
     des fonctions d'une variable réelle, Gauthier-Villars,
     Paris 1952 .

apl. Prof. Dr. F.-J. Delvos

Lehrstuhl für Mathematik I
Universität Siegen

Hölderlin-Str. 3
D-5900 Siegen (W.-Germany)

International Series of
Numerical Mathematics, Vol. 74
© 1985 Birkhäuser Verlag Basel

SIMULTANEOUS INTERPOLATION
AND NORM-PRESERVATION

Frank Deutsch

Department of Mathematics, The Pennsylvania State
University, University Park, PA, U.S.A.

ABSTRACT

A geometric characterization of property SIN is
given. This property is more general than property SAIN but
is equivalent to it in a strictly convex space.

1 INTRODUCTION

Let Y be a dense subspace of the normed linear space
X and let Γ be a finite dimensional subset of X^*, the dual
space of X. The triple (X, Y, Γ) is said to have <u>property
SAIN</u> if and only if for each $x \in X$ and $\varepsilon > 0$, there exists a
$y \in Y$ such that

$$(1) \quad \|x-y\| < \varepsilon,$$
$$(2) \quad x^*(y) = x^*(x) \quad \text{for every } x^* \in \Gamma, \text{ and}$$
$$(3) \quad \|y\| = \|x\|.$$

(SAIN is an acronym for "simultaneous approximation and interpolation with norm-preservation".) This property was introduced
and studied by the writer and Morris in [3]. This definition
was motivated by a result of Wolibner [17] who essentially
proved that the triple $(C[a,b], P, \{e_{t_1}, e_{t_2}, \ldots, e_{t_n}\})$ has

property SAIN   for any choice of the $t_i \in [a,b]$. (Here
C[a,b] denotes the Banach space of all real continuous functions
on the interval [a,b] and endowed with the supremum norm, P
denotes the set of all polynomials, and e_t denotes the functional "evaluation at t": $e_t(x) = x(t)$ for every $x \in C[a,b]$.)
That is, he showed that every continuous function can be approximated arbitrarily well by polynomials which interpolate the
function at a prescribed finite set of points and which have the
same norm as the function. The writer and Morris [3] showed,
more generally, that $(C(T), A, \{e_{t_1}, e_{t_2}, \ldots, e_{t_n}\})$ has property
SAIN for any choice of $t_i \in T$, where C(T) denotes the Banach
space of all real continuous functions on the compact Hausdorff
space T and endowed with the supremum norm, A is any dense
subalgebra (or any dense linear sublattice containing constants),
and e_t again denotes point evaluation at t. (It was also
observed that this result is false in general if either A is
replaced by any dense linear subspace or if the point evaluation
functionals are replaced by other bounded linear functionals.)
In [3] it was further proved that a <u>necessary</u> condition that the
triple (X,Y,Γ) possess property SAIN is that every $x^* \in \Gamma$
either attains its norm at a point in Y or not at all. In the
case when X is a Hilbert space, it was shown [3] that (X,Y,Γ)
has property SAIN if and only if each functional $x^* \in \Gamma$
attains its norm at a point in Y.

    Šmatkov [16] and, independently, Holmes and Lambert [6]
gave geometrical characterizations of those triples which have
property SAIN. In both of these papers it was also shown that,
in a strictly convex reflexive space X, the necessary condition
of [3] is always sufficient for each choice of Y and Γ if
and only if X is a Hilbert space.

    Many other writers have contributed to the study of
property SAIN. An incomplete list might mention the papers of
Šmatkov [16], McLanghlin and Zaretski [13], Holmes and Lambert
[6], Lambert [9], [10], [11], Luna [12], Chui et al [2], Johnson
[7], Keener [8], and the survey papers of Rao [15] and Chalmers

124

and Taylor [2].   (In addition, a more up-to-date and comprehensive survey is being prepared by Hiremath [4].)

It should be mentioned that if the norm-preserving condition (3) of property SAIN is dropped, then _every_ triple (X,Y,Γ) satisfies conditions (1) and (2) or, briefly, has property SAI. This result is due to Yamabe [18]. Thus it is the presence of the norm-preserving condition (3) which complicates matters. Moreover, it is easy to verify that if the interpolating condition (2) were dropped, then _every_ triple satisfies conditions (1) and (3) or, briefly, has property SAN. This naturally leads us to ask: what happens when condition (1) is dropped? That is, when does a triple satisfy conditions (2) and (3) or, briefly, have property SIN?

In this paper we will characterize those triples which have property SIN (Theorem 2.2). While SIN and SAIN are different properties in general (Example 2.5), they turn out to be equivalent properties in strictly convex spaces (Theorem 2.4).

Historically, the first result of SIN-type appears to be that of Wolibner [17] who proved that the triple $(C[a,b],\ P,\ \{e_{t_1},e_{t_2},\ldots,e_{t_n}\})$ has property SAIN and, a fortiori, property SIN. Paszkowski [14] refined Wolibner's result by verifying the following surprising fact: there exists a positive integer $N = N(t_1,t_2,\ldots,t_n)$, _depending_ _only_ _on_ _the_ _nodes_ $\{t_1,t_2,\ldots,t_n\}$ in $[a,b]$, such that for each continuous function x on $[a,b]$ a polynomial p of degree $\leq N$ can be found for which $p(t_i) = x(t_i)$ $(i=1,2,\ldots,n)$ and $\|p\| = \|x\|$. Briefly, the triple $(C[a,b],\ P_N,\ \{e_{t_1},e_{t_2},\ldots,e_{t_n}\})$ has property SIN, where P_N denotes the set of all polynomials of degree $\leq N$.

2 PROPERTY SIN. Unless otherwise specified, throughout this paper X will denote a (real) normed linear space, Y a dense linear subspace, and Γ a finite dimensional subset of

the dual space X^*.

  2.1 DEFINITION. The triple (X,Y,Γ) is said to have <u>property SIN</u> if and only if, for each $x \in X$, there exists $y \in Y$ with

(1) $x^*(y) = x^*(x)$ for every $x^* \in \Gamma$, and

(2) $\|y\| = \|x\|$.

  SIN is an acronym for "simultaneous interpolation and norm-preservation". Note that (X,Y,Γ) has property SIN if and only if $(X,Y,\text{Span}(\Gamma))$ has property SIN if and only if (X,Y,Γ_b) has property SIN, where $\text{Span}(\Gamma)$ denotes the linear subspace spanned by Γ and Γ_b denotes any basis for $\text{Span}(\Gamma)$.

  For any subset Γ of X^*, we define the annihilator of Γ in X by

$$\Gamma_\perp := \{x \in X \mid x^*(x) = 0 \text{ for every } x^* \in \Gamma\}.$$

Note that if Γ has dimension n, then $\Gamma_\perp$ is a closed subspace of codimension n. Also, (X,Y,Γ) has property SIN if and only for for each $x \in X$, there exists $y \in Y$ with $x-y \in \Gamma_\perp$ and $\|y\| = \|x\|$.

  For any subspace M of X, the <u>metric projection</u> onto M is the set-valued mapping $P_M: X \to 2^M$ defined by

$$P_M(x): = \{y \in M \mid \|x-y\| = d(x,M)\},$$

where $d(x,M) := \inf\{\|x-m\| \mid m \in M\}$. That is, $P_M(x)$ is the set of all best approximations to x from M. The kernel of P_M is the set

$$P_M^{-1}(0) := \{x \in X \mid 0 \in P_M(x)\}$$

$$= \{x \in X\} \mid \|x\| = d(x,M)\}.$$

Note that $P_M(\alpha x) = \alpha P_M(x)$ holds for each $\alpha \in \mathbb{R}$. Also,

126

$P_M^{-1}(0)$ is a closed nonempty (since $0 \in P_M^{-1}(0)$) subset of X such that $\alpha x \in P_M^{-1}(0)$ whenever $x \in P_M^{-1}(0)$ and $\alpha \in \mathbb{R}$.

A functional $x^* \in X^*$ is said to <u>attain</u> <u>its</u> <u>norm</u> if there exists an $x \in X$ with $\|x\| = 1$ such that $x^*(x) = \|x^*\|$. In this case we say that x^* attains its norm at x. Finally, the <u>unit</u> <u>sphere</u> in X is defined by

$$S(X) := \{x \in X \mid \|x\| = 1\}.$$

The first result is a geometric characterization of those triples which have property SIN.

2.2 THEOREM. The following statements are equivalent.

(1) (X,Y,Γ) has property SIN;

(2)   For each $x \in S(X)$, $Y \cap S(X) \cap (x+\Gamma_\perp) \neq \phi$;

(3)   For each $x \in S(X) \cap P_{\Gamma_\perp}^{-1}(0)$, $Y \cap S(X) \cap (x+\Gamma_\perp) \neq \phi$;

(4)   For each $x \in S(X) \cap P_{\Gamma_\perp}^{-1}(0)$, $Y \cap (x-P_{\Gamma_\perp}(x)) \neq \phi$;

(5)   For each $x \in P_{\Gamma_\perp}^{-1}(0)$, $Y \cap (x-P_{\Gamma_\perp}(x)) \neq \phi$.

PROOF. $(1) \Rightarrow (2)$. If (1) holds and $x \in S(X)$, choose $y \in Y$ with $\|y\| = \|x\| = 1$ and $y-x \in \Gamma_\perp$. Then $y \in Y \cap S(X) \cap (x+\Gamma_\perp)$ so (2) holds.

$(2) \Rightarrow (3)$. This is obvious.

$(3) \Rightarrow (4)$. Assume (3) holds and let $x \in S(X) \cap P_{\Gamma_\perp}^{-1}(0)$. Select any $y \in Y \cap S(X) \cap (x+\Gamma_\perp)$. Then $y = x - z$ for some $z \in \Gamma_\perp$ and

$$\|x-z\| = \|y\| = 1 = \|x\| = d(x,\Gamma_\perp).$$

Thus $z \in P_{\Gamma_\perp}(x)$ implies $y \in x - P_{\Gamma_\perp}(x)$ and (4) holds.

$(4) \Rightarrow (5)$. Assume (4) holds and $x \in P_{\Gamma_\perp}^{-1}(0)$. If $x = 0$, then $0 \in Y \cap (x - P_{\Gamma_\perp}(x))$ and (5) holds. Thus we may assume $x \neq 0$. By scaling, we may further assume $\|x\| = 1$. Then $x \in S(X) \cap P_{\Gamma_\perp}^{-1}(0)$ so by (4), $Y \cap (x - P_{\Gamma_\perp}(x)) \neq \phi$ and (5) holds.

$(5) \Rightarrow (1)$. Assume (5) holds and let $x \in X$. We must show that there exists $y \in Y$ with $\|y\| = \|x\|$ and $x - y \in \Gamma_\perp$. We may assume $\|x\| = 1$. If $x \in P_{\Gamma_\perp}^{-1}(0)$, select any $y \in Y \cap (x - P_{\Gamma_\perp}(x))$. Then $y = x - z$ for some $z \in P_{\Gamma_\perp}(x)$ so $x - y \in \Gamma_\perp$ and

$$\|y\| = \|x - z\| = d(x, \Gamma_\perp) = \|x\|.$$

If $x \notin P_{\Gamma_\perp}^{-1}(0)$, then $d(x, \Gamma_\perp) < \|x\|$. Choose $z \in \Gamma_\perp$ such that $\|x - z\| < \|x\|$. Fix any $\rho > 2\|x\| \cdot \|z\|^{-1}$ and set $x_1 = \frac{1}{2}(x - z) + \frac{1}{2} x$ and $x_2 = \rho(x - z) + (1 - \rho)x$. Then $x_i \in x + \Gamma_\perp$ $(i = 1, 2)$,

$$\|x_1\| \leq \tfrac{1}{2}\|x - z\| + \tfrac{1}{2}\|x\| < \|x\|, \quad \text{and}$$

$$\|x_2\| \geq \rho\|z\| - \|x\| > \|x\|.$$

Since $\Gamma_\perp$ has finite codimension, $Y \cap (x + \Gamma_\perp)$ is dense in $x + \Gamma_\perp$ (cf. e.g. [5; p. 49]) so we can choose $y_i \in Y \cap (x + \Gamma_\perp)$ $(i = 1, 2)$ such that $\|y_1\| < \|x\| < \|y_2\|$. Thus for some $0 < \lambda < 1$, the element $y = \lambda y_1 + (1 - \lambda)y_2$ satisfies $\|y\| = \|x\|$. Since $Y \cap (x + \Gamma_\perp)$ is convex, $y \in Y \cap (x + \Gamma_\perp)$ also. ∎

Next a _necessary_ condition for property SIN is given.

2.3 THEOREM. If (X,Y,Γ) has property SIN, then each $x^* \in \Gamma$ either attains its norm at a point in Y or not at all.

PROOF. If some $x^* \in \Gamma$ attained its norm at a point $x \in S(X)$ but at no point in Y, then for every $y \in Y$ with $\|y\| = 1$ $(=\|x\|)$, $x^*(x) = \|x^*\| > x^*(y)$. Hence (X,Y,Γ) fails property SIN. ∎

In strictly convex spaces, the above necessary condition is also sufficient. In fact, properties SIN and SAIN are equivalent in this case.

2.4 THEOREM. Let X be strictly convex. Then the following statements are equivalent.

(1) (X,Y,Γ) has property SAIN;
(2) (X,Y,Γ) has property SIN;
(3) Each $x^* \in \Gamma$ either attains its norm at a point in Y or not at all.

PROOF. The implication $(1) \Rightarrow (2)$ is obvious while the implication $(2) \Rightarrow (3)$ is just Theorem 2.3.

$(3) \Rightarrow (1)$. Suppose (3) holds and let $x \in P_{\Gamma_\perp}^{-1}(0) \setminus \{0\}$. By a well-known corollary of the Hahn-Banach Theorem, there exists $x^* \in (\Gamma_\perp)^\perp = \overline{\text{span}(\Gamma)} = \text{span}(\Gamma)$ with $\|x^*\| = 1$ and $x^*(x) = \|x\|$. Thus x^* attains its norm at $x/\|x\|$. By strict convexity, x^* attains its norm at a unique point in $S(X)$. By assumption, it follows that $x/\|x\| \in Y$ and so $x \in Y$. This proves that $P_{\Gamma_\perp}^{-1}(0) \subset Y$. But $x - P_{\Gamma_\perp}(x) \subset P_{\Gamma_\perp}^{-1}(0)$ for every x so

$$Y \cap (x - P_{\Gamma_\perp}(x)) = x - P_{\Gamma_\perp}(x)$$

for every $x \in P_{\Gamma_\perp}^{-1}(0)$. By the characterization of SAIN due
to Holmes and Lambert [6; Lemma 1], it follows that (X,Y,Γ)
has property SAIN. ∎

The equivalence of (1) and (3) in Theorem 2.4 had been
proved earlier by Šmatkov [16] and Holmes and Lambert [6]
independently.

In view of Theorem 2.4, it is natural to ask whether
properties SAIN and SIN are always equivalent. The
following example from [3] of a triple which fails to have
property SAIN will be now shown to have property SIN. We
are indebted to G.D. Taylor for making this astute observation.

2.5 EXAMPLE. Let $x = C[0,1]$,
$Y = \{y \in C[0,1] \mid y'(\tfrac{1}{2})$ exists, $y'(\tfrac{1}{2}) = y(0) - y(1)\}$, and
$\Gamma = \{e_{1/2}\}$. Then as noted in [3], (X,Y,Γ) fails to have
property SAIN.

However, for any $x \in X$, define y on $[0,1]$ by

$$y(t) = 4[\|x\| - x(\tfrac{1}{2})](t - \tfrac{1}{2})^2 + x(\tfrac{1}{2}).$$

Then $y \in Y$, $e_{1/2}(y) = y(\tfrac{1}{2}) = x(\tfrac{1}{2}) = e_{1/2}(x)$, and $\|y\| = \|x\|$.
Thus (X,Y,Γ) has property SIN. (In fact, we have even shown
that (X,P_2,Γ) has property SIN!)

Another (indirect) proof that (X,Y,Γ) has property
SIN can be based on Theorem 2.2 by observing that for any
$x \in P_{\Gamma_\perp}^{-1}(0)$, $Y \cap (x - P_{\Gamma_\perp}(x))$ contains the constant function
$y(t) = x(\tfrac{1}{2})$.

In a reflexive Banach space X, the unit ball is weakly
compact so every $x^* \in X^*$ attains its norm. From Theorem 2.4,
we immediately obtain the following corollary.

2.6 COROLLARY. Let X be strictly convex and reflexive. Then the following statements are equivalent.

(1) (X,Y,Γ) has property SAIN;
(2) (X,Y,Γ) has property SIN;
(3) Each $x^* \in \Gamma$ obtains its norm at a point in Y.

The equivalence of (1) and (3) had been established earlier by Šmatkov [16] and Holmes and Lambert [6] independently.

If X is a Hilbert space, then the Fréchet-Riesz representation theorem implies that each $x^* \in X^*$ has a unique "representer" in X, i.e. an element $x \in X$ such that

$$x^*(y) = \langle y,x\rangle \quad \text{for every} \quad y \in X$$

and $\|x^*\| = \|x\|$. It is easy to verify that x^* attains its norm at the normalized representer: $x/\|x\|$. Since Hilbert space is strictly convex and reflexive, we immediately obtain the following consequence of Corollary 2.6.

2.7 COROLLARY. Let X be a Hilbert space. Then the following statements are equivalent.

(1) (X,Y,Γ) has property SAIN;
(2) (X,Y,Γ) has property SIN;
(3) Each $x^* \in \Gamma$ has its representer in Y.

The equivalence of (1) and (3) was first proved by the writer and Morris [3].

REFERENCES

1. B.L. Chalmers and G.D. Taylor, Uniform approximation with constraints, Jber. d. Dt. Math.-Verein., 81(1979), 49-86.

2. C.K. Chui, E.R. Rozema, P.W. Smith, and J.D. Ward, Simultaneous spline approximation and interpolation preserving norms, Proc. Amer. Math. Soc., 54(1976), 98-100.

3. F. Deutsch and P.D. Morris, On simultaneous approximation and interpolation which preserves the norm, J. Approx. Theory, 2(1969), 355-373.

4. K. Hiremath, A survey of simultaneous approximation and interpolation with norm-preservation, in preparation.

5. R.B. Holmes, Geometric Functional Analysis and its Applications, Graduate Texts in Mathematics 24, Springer-Verlag, New York, 1975.

6. R. Holmes and J.M. Lambert, A geometrical approach to property SAIN, J. Approx. Theory, 7(1973), 132-142.

7. D.J. Johnson, SAIN approximation in C[a,b], J. Approx. Theory, 17(1976), 14-34.

8. L.L. Keener, Characterizing local best SAIN approximations, J. Approx. Theory, 36(1982), 55-63.

9. J.M. Lambert, Simultaneous approximation and interpolation in ℓ_1, Proc. Amer. Math. Soc., 32(1972), 150-152.

10. J.M. Lambert, Simultaneous approximation and interpolation in L_1 and C(T), Pacific J. Math., 45(1973), 293-296.

11. J.M. Lambert, Conditions for simultaneous approximation and interpolation with norm preservation in C[a,b], Pacific J. Math., 66(1976), 173-180.

12. G. Luna, The dual of a theorem of Bishop and Phelps, Proc. Amer. Math. Soc., 47(1975), 711-714.

13. H.W. McLaughlin and P.M. Zaretski, Simultaneous approximation and interpolation with norm preservation, J. Approx. Theory, 4(1971), 54-58.

14. S. Paszkowski, On approximation with nodes, Rozprawy Mathematyczne 14, Warsaw (1957).

15.   G.S. Rao, Simultaneous approximation, interpolation and norm preservation (SAIN), Jber. d. Dt. Math.-Verein, 81(1979), 189-198.

16.   V.A. Šmatkov, On simultaneous approximation and interpolation in Banach spaces, Dokl. Akad. Nauk. Armyanskoi SSR, 53(1971), 65-70 (Russian).

17.   W. Wolibner, Sur un polynome d'interpolation, Colloq. Math., 2(1951), 136-137.

18.   H. Yamabe, On an extension of the Helly's theorem, Osaka Math. J., 2(1950), 15-17.

Department of Mathematics
The Pennsylvania State University
University Park, PA 16802

International Series of
Numerical Mathematics, Vol. 74
© 1985 Birkhäuser Verlag Basel

LEBESGUE CONSTANTS AND BEST CONDITIONS FOR

THE NORM CONVERGENCE OF FOURIER SERIES

B. Dreseler

Fachbereich 6-Mathematik

Universität-GH-Siegen

Siegen

1. INTRODUCTION

Let $T = [-\pi,\pi)$ be the one dimensional torus and suppose X to
be one of the spaces $C(T^N)$ or $L^p(T^N)$, $1 \leq p < \infty$. The k-th modulus
of smoothness of a function f in X is defined by

$$\Omega_k(t;f;X) = \sup_{0 < |h| \leq t} \| \Delta_h^k f \|_X ,$$

$$\Delta_h^k f(x) := \sum_{j=0}^{k} (-1)^j \binom{k}{j} f(x + jh),$$

$x, h \in T^N$. The Fourier coefficients $f^\wedge(m)$, $m \in \mathbf{Z}^N$, of f in X are
defined by

$$f^\wedge(m) = 1/(2\pi)^N \int_{T^N} f(x) e^{-imx} dx.$$

Thus, the formal Fourier series of f in X is

$$(1) \qquad \sum_{m \in \mathbf{Z}^N} f^\wedge(m) e^{imx} .$$

Define the projections $P_k : X \to X$, $k \in \mathbb{P} = \{0,1,2,\ldots\}$, by

$$P_k f = \sum_{|m|^2 = k} f^\wedge(m) e^{imx}.$$

Hence, the spherical partial sums of (1) can be written as

$$(2) \qquad S_n f = \sum_{k=0}^{n} P_k f \qquad\qquad (n \in \mathbb{P}).$$

Consider the case $n \geq 2$. Then the norm convergence of (2) in the Hilbert space $L^2(T^N)$ is trivial but it follows from a deep result of C.Fefferman [14] that $\sup_n \| S_n \|_X = \infty$, in the case $X = L^p(T^N)$, $p \neq 2$. Take $p_n \in \Pi_n^X = \bigcup_{k \leq n} P_k(X)$. One gets for f in X

$$\| S_n f - f \|_X \leq \| S_n(f - p_n) - (f - p_n) \|_X$$

$$\lesssim (1 + \| S_n \|_X) \| f - p_n \|_X ,$$

which shows the well-known fact that Jackson's theorem combined with an estimate from above of the Lebesgue constants $\{ \| S_n \|_X \}$ leads to a Dini-Lipschitz type theorem of the norm convergence of $\{ S_n \}$ on X. In the two cases $X = L^1(T^N)$ and $X = C(T^N)$ one has $\| S_n \|_X \sim n^{(N-1)/2}$ as n goes to infinity (cf. [25], [17], [4]).

This leads to the sufficient condition

$$\Omega_k(t;f;X) = o(t^{(n-1)/2}) \qquad (t \to 0+)$$

(for some $k > (N-1)/2$). In a recent paper of Cartwright and Soardi [6] it is shown that the condition

$$\Omega_k(t;f;X) = O(t^{(N-1)/2}) \qquad (t \to 0+)$$

is not sufficient for the norm convergence on X. The method of proving this in [6] is quite general, and may be used to obtain analogous results for the divergence of Fourier series with respect to some other summation methods. Also analogous results for expansions in zonal spherical functions on symmetric spaces of rank one are obtained in [6], [5] by the same method. The two cornerstones of the results in [6] are good estimates from below of $\{\| S_n \|_X\}$ and an inequality of Bernstein type. In this paper we prove a general theorem (Section 2) for systems $\{P_k\}$ of orthogonal projections on Banach spaces X which implies all the negative results above and many other applications (Section 3).

2. A GENERAL DIVERGENCE THEOREM

Let $\omega : [o,\infty) \to \mathbb{R}_+$ be continuous and monotonely increasing with the properties $\omega(o) = o$, $\omega(t) > o$ for $t > o$, $\omega(t_1 + t_2) \leq \omega(t_1) + \omega(t_2)$, and $\lim_{t \to 0+} \omega(t)/t = \infty$ (see [7]). The most important example in the following is $\omega(t) = t^\alpha$, $o < \alpha < 1$.

Let X be a Banach space with norm $\| \ \|_X$ and let $U \subset X$ be a linear subspace of X with seminorm $| \ |_U$. For f in X and $t \geq o$ the K-functional is defined by

$$K(t,f) := K(t,f;X,U)$$

$$:= \inf_{g \in U} \{\| f - g \|_X + t |g|_U\}.$$

The space $X_\omega := \{f \in X : K(t,f) = O(\omega(t)), t \to 0+\}$ is intermediate between U and X.

Let [X] denote the continuous endomorphisms on X and let $\{P_k\} = \{P_k : k \in \mathbb{P}\}$ be a total sequence of mutually orthogonal projections on X, i.e.

(i) $P_k f = 0$ for all k in $\mathbb{P}$ implies $f = 0$;

(ii) $P_j P_k = \delta_{jk} P_k$, δ_{jk} being the Kronecker symbol;

(see [30]). Then to each f in X one may associate its unique Fourier series expansion

$$f \sim \sum_{k=o}^{\infty} P_k f \qquad\qquad (f \in X).$$

The sequence $\{P_k\}$ is said to be fundamental if the set $\Pi^X = \bigcup_{k \in \mathbb{P}} P_k(X)$ is dense in X. A complex sequence $\psi = \{\psi_k\}_{k \in \mathbb{P}}$ is called a multiplier on X (corresponding to $\{P_k\}$), if for each f in X there exists an element f^ψ in X such that $\psi_k P_k f = P_k f^\psi$ for all k in $\mathbb{P}$. An operator T in [X] is called a multiplier operator if there exists a sequence ψ such that $P_k T f = \psi_k P_k f$ for all f in X and k in $\mathbb{P}$. Finally, let us consider the Cesàro means of order $\kappa \geq o$

$$(C,\kappa)_n f := (A_n^\kappa)^{-1} \sum_{k=o}^{n} A_{n-k}^\kappa P_k f,$$

$$A_n^\kappa := \frac{\Gamma(n+\kappa+1)}{\Gamma(n+1)\Gamma(\kappa+1)} \qquad\qquad (n \in \mathbb{P}).$$

$(C,\kappa)_n f$ coincides for $\kappa = o$ with (2). Then, following Stein [26], the system $\{P_k\}$ is said to be regular on X if for some $\kappa \geq o$ there exists a constant $C_\kappa > o$ such that

$$(3) \qquad \| (C,\kappa)_n f \|_X \leq C_\kappa \| f \|_X \qquad (f \in X, \ n \in \mathbb{P}).$$

THEOREM. Let the system $\{P_k\}$ be total, mutually orthogonal, fundamental and regular on X. Let $U \subset X$ be a linear subspace with seminorm $| \ |_U$ and let $\{\phi_n\}$ be a positive monotone sequence with $\lim_{n \to \infty} \phi_n = o$ such that the inequality

$$(4) \qquad |p|_U \leq (C/\phi_n) \| p \|_X \qquad (p \in \Pi_{2n}^X \subset U, \ n \in \mathbb{P})$$

holds. Let $S_n : X \to X$ be a multiplier operator with $S_n(X) \subset \Pi_n^X$ and $\lim \sup_{n \to \infty} \| S_n \|_X = \infty$. Then there exists an f_ω in X_ω such that

$$\| S_n f_\omega - f_\omega \|_X \neq o(\omega(\phi_n) \| S_n \|_X) \qquad (n \to \infty).$$

REMARK 1. The regularity (3) and the Bernstein type inequality (4) hold in many interesting special cases. Some typical examples are studied in Section 3. The value of Theorem 1 in special situations depends on the knowledge of the asymptotic behaviour of the Lebesgue constants $\{\|S_n\|_X\}$ and satisfactory characterizations of the space X_ω or at least of an upper class of X_ω. The proof of Theorem 1 follows by a simple application of Theorem 2 in [7] and is up to technicalities as simple as the analogous proof of Corollary 2 in [7] in the classical Fourier case.

PROOF OF THEOREM 1. By definition of the usual norm in $[X]$ there are elements f_n in X with $\|f_n\|_X = 1$ and $\|S_n f_n\|_X \geq (1/2)\|S_n\|_X$. Now we need a substitute for the classical delayed means of de La Vallée Poussin. Let λ be a smooth function on $[0,\infty)$ such that $\lambda(t) = 1$ for $0 \leq t \leq 1$ and $= 0$ for $t \geq 2$ and $0 \leq \lambda(t) \leq 1$ for all t. Then the means $L_n f := \sum_{k \in \mathbb{P}} \lambda(k/n) P_k f$ possess the following properties

(i) $L_n f \in \Pi_{2n}^X$ for all f in X;

(ii) $L_n p_n = p_n$ for all p_n in Π_n^X;

(iii) The sequence $\{L_n\}$ is uniformly bounded in $[X]$.

(i) and (ii) are obvious and (iii) follows from the regularity of $\{P_k\}$ by a multiplier criterion as given in [30]. Now choose $h_n = L_n h_n$. Then h_n in Π_{2n}^X by (i), $\|h_n\|_X \leq C_1$ by (iii), $|h_n|_U \leq C_2(\phi_n^{-1})$ by (4) and (iii) and from (ii) and (iii) we get for sufficiently large n (at least for a subsequence of $\{S_n\}$)

$$\| (S_n h_n - h_n)/\|S_n\|_X \|_X = \| S_n L_n f_n - L_n f_n \|_X / \|S_n\|_X$$

$$\geq [\| S_n f_n \| - C]/\|S_n\|_X \geq C_3 > 0.$$

Here the positive constants C, C_1, C_2, C_3 are independent of n. Now Theorem 1 follows from Theorem 2 in [7].-

3. APPLICATIONS

a) We first go back to our motivating example in the introduction. Let X be one of the Banach spaces $C(T^N)$ or $L^p(T^N)$, $1 \leq p < \infty$. Let U be the subspace of all elements f in X with derivatives $D^m f$ (in the distributional sense) up to the order $\| m \| = m_1 + m_2 + \ldots + m_N \leq r \in \mathbb{P}$ belonging to X. Here D^m denotes the usual differential operator $(\partial/\partial x_1)^{m_1} \ldots (\partial/\partial x_N)^{m_N}$ with the multiindex m in $\mathbb{P}^N$. We endow U with the seminorm

$$|f|_U = \sup_{\| m \| = r} \| D^m f \|_X .$$

From the classical inequality of Bernstein one gets

$$|p|_U \leq (2n)^r \| p \|_X \qquad (p \in \Pi^X_{2n}, \ n \in \mathbb{P}).$$

Thus, we can choose $\phi_n = n^{-r}$. Condition (3) holds if $\alpha > (N-1)/2$ (see [30]).

As a first example we consider the spherical partial sum operators (2). As we mentioned before in the two cases $X = L^1(T^N)$ and $X = C(T^N)$, $N \geq 2$, one has $\| S_n \|_X \sim n^{(N-1)/2}$. Now, Theorem 1 tells us the existence of a function f_ω in X_ω such that $\| S_n f_\omega - f_\omega \|_X \neq o(\omega(n^{-r}) n^{(N-1)/2})$. We take $r > (N-1)/2$, $\omega(t) = t^\alpha$ with $o < \alpha < 1$ and $\alpha r = (N-1)/2$. This gives the existence of an f_ω in X_ω with $\| S_n f_\omega - f_\omega \|_X \neq o(1)$. The easy part of Theorem 1 in [19] gives

$$M_1 \Omega_r(t;f;X) \leq K(t^r,f;X,U)$$

which implies $\Omega_r(t;f_\omega;X) = O(t^{(N-1)/2})$.

Let's look at the Bochner-Riesz means on T^N now. Put $\lambda_\delta(t) = (1 - |t|^2)^\delta$ when $|t| < 1$, $t \in \mathbb{R}^N$, and $\lambda_\delta(t) = o$ when $|t| > 1$. In the cases $X = C(T^N)$, $L^1(T^N)$ the multiplier operator S_n^δ belonging to the multiplier $\psi_\delta(k) = \lambda_\delta(k/n)$, $k \in \mathbb{P}$, allows the estimate $\|S_n^\delta\|_X \sim n^{(N-1)/2 - \delta}$ as n tends to infinity ($N \geq 2$, $o \leq \delta < (N-1)/2$) (see [1]), while Stein [27] proved the estimate $\|S_n^\delta\|_X \sim \log n$ if $\delta = (N-1)/2$.

COROLLARY 1. Let $\{S_n^\delta\}$ be the Bochner-Riesz means on T^N and let X be either $L^1(T^N)$ or $C(T^N)$. Then there is a function f in X such that (for all $r > (N-1)/2 - \delta$) $\Omega_r(t;f;X) = O(t^{(N-1)/2-\delta})$ in the case $o \leq \delta < (N-1)/2$ and $\Omega_1(t;f;X) = O(1/\log(1/t))$ in the case $\delta = (n-1)/2$ as $t \to O+$, but $S_n^\delta f$ does not converge to f in X.

In the case $X = C(T^N)$ the negative results of Il'in (see [1]) are weaker.

b) As a typical example of classical orthogonal expansions we look at Jacobi expansions. For indices $\alpha \geq \beta \geq -1/2$, let $R_n^{(\alpha,\beta)}$, $n=0,1,2,\ldots$, be the Jacobi polynomials(see [31]) defined by the requirements that they are pairwise orthogonal relative to the weight $w(x) = w^{\alpha,\beta}(x) = (1-x)^\alpha(1+x)^\beta$ on $[-1,1]$ and satisfy $R_n^{(\alpha,\beta)}(1) = 1$ and $\deg R_n^{(\alpha,\beta)} = n$. Let L_w^p, $1 \leq p < \infty$, denote the L^p space with respect to the measure $d\mu = w(x)dx$ on $[-1,1]$. Let X be either the Banach space $C[-1,1]$ ($p=\infty$) or L_w^p ($1 \leq p < \infty$). It follows that the projections

$$P_k^{(\alpha,\beta)} f(x) = f^\wedge(k) h_k^{\alpha,\beta} R_k^{(\alpha,\beta)}(x) \qquad (k \in \mathbb{P}),$$

where $f^\wedge(k) = \int_{-1}^{1} f(y) R_k^{(\alpha,\beta)}(y) w(y) dy$ and $h_k^{\alpha,\beta} = [\int_{-1}^{1} |R_k^{(\alpha,\beta)}(y)|^2 d\mu]^{-1}$, are mutually orthogonal, total, and fundamental on X. The regularity condition (3) is satisfied if $\kappa > (\alpha+1/2)|1-2/p|$ (cf. [2]).

One can define a generalized translation operator τ_t by

$$(\tau_t f)^\wedge(k) \quad := \quad R_k^{(\alpha,\beta)}(t) f^\wedge(k) \quad (k \in \mathbb{P}, \ t \in (-1,1)).$$

It has been shown in [2] and [15] that this translation is
bounded and linear on X and satisfies

$$\| \tau_t f \|_X \ \leq \ M \| f \|_X \qquad (f \in X, \ t \in (-1,1)),$$

$$\lim_{t \to 1-} \| \tau_t f - f \|_X \ = \ 0 \qquad (f \in X).$$

Finally the (strong) Jacobi derivative $D^1 f$ of f in X is
defined by

$$D^1 f \quad := \quad \lim_{h \to 1-} \frac{\Delta_h f}{1-h} \ , \quad \Delta_h f := f - \tau_h f,$$

the limit being understood in X-norm. Higher derivatives are
defined inductively, $D^r f = D^1 D^{r-1} f$, r=2,3,... . Let U be the
space

$$W_X^r \quad := \quad \{ f \in X : D^r f \text{ exists as an element of } X \}$$

endowed with the seminorm $|f|_U := \| D^r f \|_X$.
From Stein [26] we get

$$|p|_U \ \leq \ M(r)(2n)^{2r} \| p \|_X \qquad (p \in \Pi_{2n}^X, \ n \in \mathbb{P}).$$

Pollard [23], [24] has shown that $\{\| S_n \|_X \}$ is bounded provided
$p_o' < p < p_o$ where $p_o = 4(\alpha+1)/(2\alpha+1)$ and p_o' is the index dual to
p_o. Furthermore, the norm convergence does not hold on X if
$p \notin (p_o', p_o)$ (see [22] and also [13]). Since $\| S_n \|_{p'} = \| S_n \|_p$
for p outside the Pollard interval (p_o', p_o) we assume $p_o \leq p \leq \infty$.
Let $\delta := (2\alpha+1)/2 - 2(\alpha+1)/p$. Then there are numbers A,B > 0,

$$141$$

independent of n, such that

$$(5) \qquad An^{\delta} \leq \| S_n \|_X \leq Bn^{\delta} \qquad \text{for } p_O < p \leq \infty,$$

and, when $n \geq 2$

$$(6) \qquad A\log n \leq \| S_n \|_X \leq B\log n \qquad \text{for } p = p_O.$$

For $p = \infty$ see [20]. The left-hand inequality in (5) is a special case of a result in [13]. The other cases can be found in [5].

Assume $p > p_O$. We take $2r > \delta$, $\omega(t) = t^{\alpha}$ with $o < \alpha < 1$, and $2\alpha r = \delta$. Theorem 1 gives the existence of a function f_{ω} in X_{ω} with $\| S_n f_{\omega} - f_{\omega} \|_X \neq o(1)$. In the case $p = p_O$ we take $r = 1$ and $\omega(t) = 1/\log(1/t)$ and get the existence of an f_{ω} in X_{ω} with $\| S_n f_{\omega} - f_{\omega} \|_X \neq o(1)$.

Now, we use the high order modulus of continuity from [3], which is defined by

$$\omega_r(\eta; f; X) := \sup \{ \| \Delta_{h_1} \cdots \Delta_{h_r} f \|_X : \eta \leq h_j < 1, j = 1, \dots, r \}$$

for f in X and r in $\mathbb{N}$. From Theorem 1 in [3] we know the existence of a number $m(r) > O$ such that for all f in X, $o \leq \eta < 1$

$$m(r) \omega_r(\eta; f; X) \leq K((1-\eta)^r, f; X, U).$$

Thus we have

$$X_{\omega} \subset \{ f \in X : \omega_r(\eta; f; X) = O((1-\eta)^{\delta/2}), \eta \to 1- \}$$

for $p > p_O$ and

$$X_{\omega} \subset \{ f \in X : \omega_r(\eta; f; X) = O(1/\log(1/(1-\eta))), \eta \to 1- \}$$

for $p = p_O$.

Defining the best approximation of f in X by algebraic polynomials p_n of degree $\leq n$ by

$$E_n(f;X) \quad := \quad \inf_{P_n} \| f - P_n \|_X$$

one has the Jackson type theorem [3]: there exists a constant $M(r) > 0$ such that for all f in X and n in $\mathbb{N}$

$$E_n(f;X) \quad \leq \quad M(r)\omega_r(1-n^{-2};f;X).$$

By the right hand inequalities in (5) and (6) we get

COROLLARY 1. a) Suppose $f \in X$ and that (for some $r > \delta/2$)

$$\omega_r(\eta;f;X) \quad = \quad o((1-\eta)^{\delta/2}) \quad \text{as } \eta \to 1-$$

in the case $p > p_o$ and

$$\omega_r(\eta;f;X) \quad = \quad o(1/\log(1/(1-\eta))) \quad \text{as } \eta \to 1-$$

in the case $p = p_o$. Then $S_n f \to f$ in X as $n \to \infty$.
b) There is a function f in X such that (for all $r > \delta/2$)
$\omega_r(\eta;f;X) = O((1-\eta)^{\delta/2})$ in the case $p > p_o$ and $\omega_1(\eta;f;X) = O(1/\log(1/(1-\eta)))$ in the case $p = p_o$ as $\eta \to 1-$, but $S_n f$ does not converge in X.

REMARK 2. The condition $f \in X$ and $\omega_r(\eta;f;X) = O((1-\eta)^{\delta/2})$ as $\eta \to 1-$ is equivalent to the condition $f \in W_X^s$ and $\omega_2(\eta;D^s f;X) = O((1-\eta)^{\delta/2 - s})$ where s is the largest integer $< \delta/2$. This follows from Theorem 2 in [3].
If $X = C[-1,1]$ another equivalent condition is

$$\Omega_{2r}(t;x;f;C[-1,1]) \quad = \quad O\left(\left(\frac{t^2}{1-x^2}\right)^{\alpha/2 + 1/4}\right) \quad (t \to 0+, x \in (-1,1)),$$

the large-O constant being independent of δ and x. The (pointwise) modulus of continuity is defined by

$$\Omega_{2r}(t;x;f;C[-1,1]) \;=\; \sup|\Delta_h^{2r}f(x)|,$$

where the supremum is taken for fixed x in $[-1,1]$ in such a way that $x + rh \in [-1,1]$ (see [3] and [28]).

Let $p > p_o$ and $S_n^\kappa f := (C,\kappa)_n f$, $f \in X$. From the theorem in [13] (see also [21]) one gets

$$\| S_n^\kappa \|_X \;\geq\; An^{\delta-\kappa} \qquad\qquad (n \in \mathbf{N}).$$

Hence for $\kappa < \delta$ our Theorem leads to divergence results for the Cesàro means $\{S_n^\kappa\}$.

c) Similar applications to Fourier series on compact Lie homogeneous spaces follow by the same method. For instance, the regularity condition (3) follows for some spherical summation method from section 5 in [9], the Bernstein type inequality (4) from Lemma 2 in [6] and the estimate of the K-functional by a natural modulus of continuity from [18]. Thus, for symmetric spaces of rank 1 Theorem 1 implies Theorem B in [6] and the Corollary in [5]. Sharp estimates for the relevant Lebesgue constants are still an open problem in the general case (see [8], [10], [11], [12], and [16]).

REFERENCES

1. Sh. A. Alimov, V.A. Il'in, and E.M. Nikishin, Convergence problems of multiple trigonometric series and spectral decompositions, I and II, Uspehi Mat. Nauk 31 (1976), 28-83, ibidem 32 (1977), 107-130. (English translation: Russian Math. Surveys 31 (1976), 29-86, ibidem 32 (1977), 115-139.)

2. R. Askey and S. Wainger, A convolution structure for Jacobi series, Amer. J. Math. 91 (1969), 463-485.

3. P.L. Butzer, R.L. Stens, and M. Wehrens, Higher order moduli of continuity based on the Jacobi translation operator and best approximation. C. R. Math. Rep. Acad. Sci. Canada 2 (1980), 83-88.

4. M. Carenini and P.M. Soardi, Sharp estimates for Lebesgue constants (preprint).

5. D.I. Cartwright, Lebesgue constants for Jacobi expansions, Proc. Amer. Math. Soc. 87 (1983), 427-433.

6. D.I. Cartwright and P.M. Soardi, Best Conditions for the norm convergence of Fourier series, J. Approx. Th. 38 (1983), 344-353.

7. W. Dickmeis and R.J. Nessel, A unified approach to certain counterexamples in approximation theory in connection with a uniform boundedness principle with rates, J. Approx. Theory 31 (1981), 161-174.

8. B. Dreseler, Lebesgue constants for spherical partial sums of Fourier series on compact Lie groups, in: Proceedings of the Colloquium on Fourier Analysis and Approximation Theory, Budapest 1976.

9. B. Dreseler, On summation processes of Fourier expansions for spherical functions, Lecture notes in mathematics 571, Springer, Berlin 1977, 65-84.

10. B. Dreseler, Lebesgue constants for certain partial sums of Fourier series on compact Lie groups, in: Linear spaces and approximation, P.L. Butzer and B.Sz.-Nagy (eds.), 203-211, ISNM 40, Birkhäuser, Basel 1978.

11. B. Dreseler, Estimates from below for Lebesgue constants for Fourier series on compact Lie groups, Manuscripta math. 31 (1980), 17-23.

12. B. Dreseler, Norms of zonal spherical functions and Fourier series on compact symmetric spaces, J. Funct. Anal. 44 (1981), 74-86.

13. B. Dreseler and P.M. Soardi, A Cohen type inequality for Jacobi expansions and divergence of Fourier series on compact symmetric spaces, J. Approx. Theory 35 (1982), 214-221.

14. C. Fefferman, The multiplier problem for the ball, Amer. J. Math. 94 (1971), 330-336.

15. G. Gasper, Banach algebras for Jacobi series and positivity of a kernel. Ann. of Math. 95 (1972), 261-280.

16. S. Giulini, P.M. Soardi, and G. Travaglini, Norms of characters and Fourier series on compact Lie groups, J. Funct. Anal. 46 (1982), 88-101.

17. V.A. Il'in, Problems of localization and convergence for Fourier series with respect to fundamental systems of functions of the Laplace operator, Uspehi Mat. Nauk. 23 (1968), 61-120.

18. H. Johnen, Darstellungen von Liegruppen und Approximationsprozesse auf Banachräumen, J. Reine Angew. Math. 254 (1972), 160-187.

19. H. Johnen and K. Scherer, On the equivalence of the K-functional and moduli of continuity and some applications, in: Constructive theory of functions of several variables, Lecture notes in Mathematics 571, Springer, Berlin 1971, 119-140.

20. W.A. Light, Jacobi projections, Approximation Theory and Applications (Zvi. Ziegler, ed.), Academic Press, New York, 1981.

21. C. Markett, Cohen type inequalities for Jacobi, Laguerre and Hermite expansions, Siam. J. Math. Anal. 14 (1983), 819-833.

22. J. Newman and W. Rudin, Mean convergence of orthogonal series, Proc. Amer. Math. Soc. 3 (1952), 219-222.

23. H. Pollard, The mean convergence of orthogonal series II, Trans. Amer. Math. Soc. 63 (1948), 355-367.

24. H. Pollard, The mean convergence of orthogonal series III, Duke Math. J. 16 (1949), 189-191.

25. H.S. Shapiro, Lebesgue constants for spherical partial sums, J. Approx. Theory 13 (1975), 40-44.

26. E.M. Stein, Interpolation in polynomial classes and Markoff's inequality, Duke Math. J. 24 (1957), 467-476.

27. E.M. Stein, On certain exponential sums in multiple Fourier series, Ann. of Math. 73 (1961), 87-109.

28. R.L. Stens, Charakterisierung der besten algebraischen Approximation durch lokale Lipschitzbedingungen, Lecture notes in Mathematics 556, Springer, Berlin 1976, 403-415.

29. R.L. Stens and M. Wehrens, Legendre transform methods and best algebraic approximation, Comment. Math. Prace Mat. 21 (1979), 351-380.

30. W. Trebels, Multipliers for (C,α)-bounded Fourier expansions and approximation theory, Lecture notes in mathematics 329, Springer, Berlin, 1973.

31. G. Szegö, Orthogonal polynomials, 4 th ed., Amer. Math. Soc. Colloq. Publ., vol. 23, Amer. Math. Soc., Providence, R.H., 1975.

International Series of
Numerical Mathematics, Vol. 74
© 1985 Birkhäuser Verlag Basel

STOCHASTIC PROPERTIES
OF SIMPLE DIFFERENTIAL - DELAY EQUATIONS

U. an der Heiden

University of Bremen, Federal Republic of Germany

1. Introduction

The logistic difference equation

$$x_{n+1} = \alpha\, x_n\, (1 - x_n), \quad n = 0,\ 1,\ 2,\ldots \tag{1}$$

where α is a constant parameter satisfying $0 \leq \alpha \leq 4$, has been vastly and intensively studied in the last decade because of the immense variety of its different solution types and the infinity of bifurcations occuring when α varies from 0 to 4. The solutions, corresponding to initial conditions $x_0 \in [0,1]$, may be periodic, but nevertheless very complicated, or aperiodic, in some sense "chaotic". The solutions are most erratic in the case of $\alpha = 4$. This can be seen most directly by observing that the function

$$G(x) = \alpha\, x\, (1 - x)$$

is topologically conjugate to the "roof map" given by

$$\tilde{G}(x) = 2x \qquad (0\ \leq\ x\ \leq 1/2)$$
$$\tilde{G}(x) = 2 - 2x \qquad (1/2 \leq\ x\ \leq 1),$$

namely

148

$$G = H^{-1} \circ \tilde{G} \circ H \qquad \text{with } H(x) = 2/(\pi \sin \sqrt{x}\,), \qquad (2)$$

(as noticed by Ulam and von Neumann, see (Stein, Ulam 1964)). The difference equation

$$x_{n+1} = \tilde{G}(x_n), \qquad x_0 \in [0,1]\ , \ n = 0, 1, 2,\ldots \qquad (3)$$

is a model for the Markov chain of an infinite sequence of fair coin tosses: In fact, given any random sequence of 0`s (heads) and 1`s (tails), a binary number $b = 0.\ b_1 b_2 \ldots$ is defined by

$$b_k = 0 \qquad \text{if k-th toss is a head}$$
$$b_k = 1 \qquad \text{if k-th toss is a tail.}$$

From this another binary number $x_0 = 0.\ a_1 a_2 a_3 \ldots$ is constructed by

$$a_1 = b_0, \ a_{k+1} = a_k + b_k \ (\text{mod } 2).$$

Then the solution $(x_n)_{n=0,1,2,\ldots}$ corresponding to the initial condition x_0 is given by

$$x_n = 0.\ c_1 c_2 c_3 \ldots$$

with

$$c_1 = b_n, \ c_{k+1} = c_k + b_{n+k} \ (\text{mod } 2), \ k = 1,2,\ldots$$

In particular for all $n \in \mathbb{N}$ the first binary digit of x_n is just b_n. The solution (x_n) thus fairly well reproduces the random sequence (b_n), and, consequently, if an initial condition for the system (3) is chosen at random a random sequence of events will occur. Of course, the probability distribution induced by this process on the interval $[0,1]$ is uniform.

Since G is topologically conjugate to $\tilde{G}$ there is a one-to-one correspondence between the solutions of Eq.(1), with $\alpha = 4$, and the solutions of Eq.(3), and the random behavior of the second system carries over to the first one. However, because of the homeomorphism (2), for the logistic equation the probability density for solution points in the interval $[0,1]$ is not uniform, but is given by

$$f^*(x) = d(2\pi^{-1} \sin (\sqrt{x}))/dx = 1/(\pi \sqrt{x(x-1)}\,).$$

This means, roughly speaking, that the asymptotic likelihood of finding x_n (given some arbitrary initial condition x_0) between a and b, $0 \leq a < b \leq 1$, equals

$$\int_a^b f^*(x)\ dx.$$

f^* is called an invariant density for G, a term that will be precisely defined below.

The logistic and the roof map are two examples of deterministic systems the behavior of which allows a true stochastic description. A description in terms of probabilities appears much more adequate than to follow and predict the time course of single solutions. In these systems each single trajectory is unstable. This follows immediately from the fact, that the behavior of a solution depends significantly on <u>all</u> the infinitely many digits of its initial condition. Two initial conditions very close to each other, but not identical, generally differ in infinitely many digits. Hence the time course of the corresponding solutions is completely different. This "critical dependence on initial conditions" excludes any precise prediction of the solutions over a longer time range, if the initial value is not known absolutely precisely. The asymptotic behavior can therefore only reasonably be described in terms of a statistic theory.

Of course, generally systems cannot be analysed in such a simple and complete way as in the special cases given by (1) and (3) (there are still many unsolved questions with respect to (1) when $\alpha < 4$). Generally, abstract concepts of a theory have to be used that describe stochastic properties of deterministic systems. One such theory has been strongly developed in recent years and some of its features will be presented below. For a detailed exposition of this theory the reader is referred to the book of Lasota and Mackey (1985).

In the following, after summarizing some essential concepts and results of this theory I shall apply it to a more difficult problem, namely the behavior of a differential delay equation that has its origin in a variety of feedback processes in biology and economy. In this way it will be demonstrated that nonlinear feedback systems have an inherent tendency to behave very irregular and even stochastic. Moreover the present approach is completely analytical and does not require any use of computers or numerical simulations.

2. Invariant Measures, Ergodicity, Mixing

We first restrict our attention to discrete dynamical systems given by a map

$$F : X \to X$$

on some state space X (which e.g. might be an interval, as above, a subset of $\mathbb{R}^n$, or a differentiable manifold). An initial condition is an element $x_0 \in X$, and the corresponding solution is defined by the iteration

$$x_n = F^n(x_0), \quad n \in \mathbb{N}.$$

The main idea is not to investigate individual solutions corresponding to single initial conditions (as done ordinarily with deterministic systems) but to study the fate of distributions of infinitely many initial conditions. Distributions or densities can only be introduced if X is endowed with some measure such that functions (or densities) $f:X \to \mathbb{R}$ may be integrated. Hence we assume throughout that a measure space $(X, \mathcal{O}l, \mu)$ is given, where $\mathcal{O}l$ denotes the σ - Algebra of all measurable subsets of X, and $\mu : \mathcal{O}l \to \mathbb{R}_+$ gives the measure $\mu(A)$ for all $A \in \mathcal{O}l$. In applications $\mathcal{O}l$ most often consists of the Borel sets and μ is the Borel measure (which may be extended to the Lebesgue measure). A distribution of initial conditions is described by some density function $f \in L^1(X)$ (= set of all functions integrable with respect to μ):

$$f(x) \geq 0 \text{ for all } x \in X, \qquad \int_X f \, d\mu = 1.$$

Via the map $x \mapsto F(x)$, which from now on is assumed to be measurable (a weaker condition than integrability) with respect to μ, the density f is transformed into a density $g \in L^1(X)$ obeying

$$\int_A g \, d\mu = \int_{S^{-1}(A)} f \, d\mu \qquad \text{for all } A \in \mathcal{O}l. \tag{4}$$

This means that if x_0 is distributed according to f then x_1 is distributed according to g. It may be easily seen that the association $f \mapsto g$ may be linearly extended to a transformation

$$P: L^1(X) \to L^1(X), \quad Pf = g,$$

under the (not very restrictive) condition that F is nonsingular, i.e. that

$$\mu(A) = 0 \text{ implies } \mu(F^{-1}(A)) = 0 \text{ for all } A \in \mathcal{O}\mathfrak{l},$$

(see (Lasota and Mackey, 1985)).

P is called the <u>Frobenius - Perron operator</u> associated with F, and it turns out that P is a Markov operator. Most important is the asymptotic behavior of the Frobenius - Perron operator, i.e. the properties of the iterates $P^n f$ as $n \to \infty$ for densities $f \in L^1(X)$. If $P^n f \to f^* \in L^1$ then the equation

$$\int_A f^* \, d\mu = \int_{S^{-1}(A)} f^* \, d\mu \qquad \text{for all } A \in \mathcal{O}\mathfrak{l} \tag{5}$$

is satisfied (an immediate consequence of (4)). A density f^* obeying relation (5) is called an <u>invariant density under F</u>. A probability measure $\nu: \mathcal{O}\mathfrak{l} \to [0, 1]$ is defined by

$$\nu(A) = \int_A f^* \, d\mu. \tag{6}$$

Evidently

$$\nu(A) = \nu(F^{-1}(A)) \text{ for all } A \in \mathcal{O}\mathfrak{l}. \tag{7}$$

A measure obeying (7) is called <u>invariant under F</u>. Moreover, a measure $\nu: \mathcal{O}\mathfrak{l} \to [0, 1]$ obeying (6) with some density $f^* \in L^1(X)$ is called <u>absolutely continuous with respect to the measure</u> μ.

There are rather simple mappings $F: X \to X$ having an absolutely continuous invariant measure. For example if X is some interval, μ the Borel measure in X and F the identity, $F(x) = x$, then there are infinitely many such measures since in this example all sets are invariant, i.e. $F^{-1}(A) = A$ for all $A \in \mathcal{O}\mathfrak{l}$. The existence of an absolutely continuous invariant measure does not imply any interesting stochastic, erratic or turbulent - like behavior of the discrete dynamical system defined by the mapping F. These types of behavior have to be investigated by other, more restrictive concepts.

The behavior is already much more interesting if there is exactly one absolutely continuous invariant measure, i.e. if the operator P has exactly one fixed point f^*, which is a density. Namely in this situation it can be shown that F is ergodic (see (Lasota and Mackey, 1985). Ergodicity means that there are no nontrivial invariant sets under F; more precisely F is called <u>ergodic</u> if for all A

$$F^{-1}(A) = A \text{ implies } \mu(A) = 0 \text{ or } \mu(X - A) = 0.$$

As an example let us consider the rotation on the unit circle $X = S^1$, $F(x) = x + \phi$ with some constant angle $\phi \in [0, 2\pi]$. Obviously the measure induced by the arc length is invariant under F. But, depending on whether $\phi/(2\pi)$ is a rational or irrational number, F is not ergodic or F is ergodic, respectively. (In the first case the reader may find other invariant measures.)

        This example shows that ergodic transformations are not necessarily very irregular, "turbulent", or "chaotic". In particular, if two initial conditions are rather close to each other the correspoding solutions may stay together very closely for all time (in the example above the distance between related solution points even remains constant). This kind of "tameness" is entirely upset by the concept of "mixing":

<u>Definition.</u> Let $(X, \mathcal{A}, \nu)$ be a measure space with $\nu(X) = 1$ and let ν be invariant under the transformation F: $X \to X$. F is called <u>mixing</u> if

$$\lim_{n \to \infty} \nu(A \cap F^{-n}(B)) = \nu(A)\,\nu(B) \quad \text{for all A, B} \in \mathcal{A} \ . \tag{8}$$

Condition (8) roughly means that if one starts with a set A of initial conditions then after many iterations the fraction of solution points lying in some (arbitrarily given measurable) set B equals just the product of the measures of A and B. Mixing is one very strong notion for what is often loosely called irregular or chaotic behavior. In particular, it implies the phenomenon of "critical dependence of solutions on their initial conditions". Of course, mixing transformations are also ergodic. For some of the well known chaotic transformations, like the baker-, Anosov-, the roof-transformation (and hence for the logistic with $\alpha = 4$) it may be shown quite directly that they are mixing. However, in general this is a very difficult problem. Fortunately, in recent years some rather general sufficient criteria have been developed by Rochlin, Renyi, Lin, Lasota, Yorke, and others. For a coherent representation see (Lasota and Mackey, 1985). The most general criteria are derived from properties of the Frobenius - Perron operator P. They are based on the fact, that F is mixing if and only if the sequence $(P^n f)$ is weakly convergent to 1 for all densities $f \in L^1(X)$; (presupposing μ to be invariant under F). Both, the existence of a unique invariant measure for F and the property that F is mixing are guaranteed if P is aymptotically stable, i.e. if there is a unique density $f^* \in L^1(X)$ such that $Pf^* = f^*$ and

$$\lim_{n \to \infty} \| P^n f - f^* \| = 0 \text{ for every density } f \in L^1(X).$$

For one - dimensional mappings the follwowing theorem can be proved (see (Lasota and Mackey, 1985)).

Theorem 1. Let $F: I \to I$ be a mapping on some compact interval $I \subset \mathbb{R}$ satisfying all of the following conditions.

(i) There is a partition $a_0 < a_1 < a_2 < \ldots < a_r$ of $I = [a_0, a_r]$ such that for each integer 1, 2, ..., r the restriction of F to the interval $[a_{i-1}, a_i)$ is twice continuously differentiable;

(ii) $F([a_{i-1}, a_i)) = [a_0, a_r)$, i.e. F is surjective on each subinterval;

(iii) There is a constant $\lambda > 1$ such that $|F'(x)| \geq \lambda$ whenever $x \neq a_i$ for $i = 1, 2, \ldots, r$.

(iv) There is a constant $c < \infty$ such that

$$|F''(x)| < c\,(F'(x))^2 \quad \text{whenever } x \neq a_i \text{ for } i = 1, 2, \ldots, r.$$

Then the Frobenius - Perron operator associated with F is asymptotically stable.

In particular the theorem implies that there is a probability measure on the interval I, which is absolutely continuous with respect to the Borel measure on I and which is invariant under F, and that F is mixing. No wonder, the roof mapping is mixing.

The theorem will be applied below to the distribution of zeros of solutions of some differential - delay equations by showing that the differences of successive zeros may in fact obey a difference equation that is mixing.

However, before doing this we want to point out that there may be even more difficult situations also in the case of one - dimensional maps if e.g. the very restrictive surjectivity condition (ii) in Theorem 1 is not satisfied (the reader should look at steep roofs the edges of which have different heights). Namely in such situations it may happen that P is not asymptotically stable, but asymptotically periodic:

Definition. The Frobenius - Perron operator $P: L^1 \to L^1$ is called asymptotically periodic if there is an integer s, densities $g_i \in L^1$, and functions $k_i \in L^\infty$, $i = 1, \ldots, s$, and an operator $Q: L^1 \to L^1$

such that Pf may be written in the form

$$Pf(x) = \sum_{i=1}^{s} \lambda_i(f)\, g_i(x) + Qf(x) \qquad (9)$$

where

$$\lambda_i(f) = \int_X f(x)\, k_i(x)\, d\mu \ .$$

The function g_i and the operator Q have the following properties:

1) $g_i(x)\, g_j(x) = 0$ for all $i \neq j$ (i.e. the functions g_i have disjoint supports);

2) for each integer i there exists a unique integer $\alpha(i)$ such that $Pg_i = g_{\alpha(i)}$. Further $\alpha(i) \neq \alpha(j)$ for $i \neq j$ and thus the operator P just serves to permute the functions g_i;

3) $\| P^n Qf \| \to 0$ as $n \to \infty$ for every $f \in L^1$.

It turns out that P is asymptotically stable if and only if P is asymptotically periodic and s = 1. The following theorem is proved in (Lasota and Mackey, 1985).

Theorem 2. Let F: $I \to I$ be a mapping on some compact interval I satisfying the conditions (i), (iii), and (iv) of Theorem 1. Then the Frobenius - Perron operator associated with F is asymptotically periodic.

Remark 1. Also in the situation of Theorem 2 generally an invariant measure does exist. It has been shown by Li and Yorke, 1978, that if r = 2 then the measure is unique and thus the transformation F is at least ergodic. However, it follows from the steepness condition on F that all solutions of the associated discrete dynamical system are unstable, hence its behavior is much more complicated than may be described by the notion of ergodicity.

Below we shall have the occasion to discuss in the context of differential - delay equations a situation where P is possibly not asymptotically stable, but asymptotically periodic.

3. Invariant Measures and Mixing for Simple Differential - Delay Equations

We want to apply the theory of Sect. 2 to a simple version of the differential - difference equation

$$dy(t)/dt = a \ (f(y(t-1))- y(t)) \tag{10}$$

Here f: $\mathbb{R} \to \mathbb{R}$ is some real - valued function and a > 0 is a constant, $0 \leq$ t in most applications measures time. The applications range from models in biology on metabolic processes, blood formation (hematopoiesis), respiration, neural interactions, and population growth (for summaries on this see (an der Heiden, 1979), (an der Heiden and Mackey, 1982), special emphasis on neural interactions is given in (Mackey and an der Heiden, 1984)) to models in economy on price development in commodity markets (for more details see (Anderson and Mackey, preprint)).

Most often y denotes some quantity which is produced according to the term f(y(t-1)) and decays or is consumed according to the term - a y(t). (In fact a is the product of a decay rate and a delay time resulting from normalizing the delay time to 1.)

Mackey and Glass (1977) and, independently, Wazewska and Lasota (1976) discovered by numerical simulations that Eq. (10) may have very complicated solutions: periodic solutions with more than 1 maximum per period and apparently aperiodic solutions appearing very irregular and "chaotic". In their simulations they used functions of type

$$f(\xi) = \lambda\xi \ /(\ \Theta + \xi^{n})$$

with constant parameters λ, Θ, n > 0. For n > 1 these functions are not monotone, but are "humped". Roughly speaking the steeper the hump the more complicated is the behavior of the solutions.

Attempts have been made to verify these findings analytically, but until now without much success (there are some results on the existence of simple periodic solutions (1 maximum / period), for a review see (an der Heiden, 1979)). The analysis is essentially simplified if the humped function is approximated rather crudely by a piecewise constant nonlinearity defined in the following way:

$$f(\xi) = \begin{cases} 0 & \text{whenever } \xi < 1 \\ c & \text{whenever } 1 \leq \xi \leq b \\ d & \text{whenever } \xi > b, \end{cases} \tag{11}$$

where b, c, d are constants satisfying c > b > 1, and d < 1.

For such simple functions f, and even for continuous functions f not too far away from such piecewise constant nonlinearities, an der Heiden and Walther (1983) and an der Heiden and Mackey (1982) could prove the existence of infinitely many different periodic orbits and of infinitely many different aperiodic orbits for Eq. (10) with fixed parameters a, b, c, d.

Despite of the success of these proofs the problem is that all of these solutions may turn out to be unstable and that almost all solutions are attracted by some asysmptotically stable periodic solution (limit cycle). In this situation only the limit cycle would have any importance for the long time behavior of the system. The present approach avoids this difficulty in showing that the behavior of solutions corresponding to a very broad class of initial conditions may be characterized by the stochastic concepts introduced in the previous section.

The basic idea is to show that there is an interval I and a transformation $F: I \to I$ such that any solution y corresponing to some initial condition from that class has the following property: there are countably many times t_1, t_2, ... such that

$$0 < t_i < t_{i+1}, \lim t_i = \infty \qquad \text{as } i \to \infty \quad , \tag{12}$$

$$y(t_i) = 1 \text{ for all } i \in \mathbb{N}, \ y(t) \neq 1 \text{ if } t \neq t_i \text{ for all } i \in \mathbb{N}, \tag{13}$$

and

$$t_{2i+2} - t_{2i+1} = F(t_{2i} - t_{2i-1}). \tag{14}$$

For some range of the parameters it will be proved that the Frobenius - Perron operator associated with F is asymptotically stable (and hence F has an absolutely continuous invariant measure and is mixing), for other parameters it will be shown that this operator is asymptotically periodic. In this way it becomes clear that the times t of a solution y where $y(t) = 1$, which we call the "zeros" of y, are distributed in a random fashion.

Before stating the main theorem let us observe that given any continuous initial condition $\phi: [-1, 0] \to \mathbb{R}$ there is a unique continuous solution $y = y_\phi$, $y: [-1, \infty) \to \mathbb{R}$ satisfying Eq. (10), with f given by (11), for all $t > 0$, and $y(t) = \phi(t)$ for all $t \in [-1, 0]$. The solution may be obtained by piecewise integration, and it is piecewise composed of segments from exponentially decreasing or increasing functions, more precisely for any two times t, t` obeying $0 < t` < t$ the solution satisfies

$$y(t) = \begin{cases} y(t`) \exp(-a(t-t`)) & \text{if } y(s) < 1 \\ c-(c-y(t`)) \exp(-a(t-t`)) & \text{if } 1 \leq y(s) \leq b \\ d-(d-y(t`)) \exp(-a(t-t`)) & \text{if } y(s) > b \end{cases} \text{for all } s \in (t`-1, t-1) \tag{15}$$

We restrict our attention to the set D of initial conditions defined in the following way:
$\phi \in D$ if and only if there is a number $w = w(\phi) \in [0, 1]$ such that $1 < \phi(t) < b$ for all $t \in [-1, -1+w)$, $\phi(t) < 1$ for all $t \in (-1+w, 0)$, and $\phi(-1+w) = 1 = \phi(0)$.

A continuous map $V: D \to [0, 1]$ is defined by $V(\phi) = w(\phi)$. It follows from (14) with $t` = 0$ and $t` = w = V(\phi)$ that

$$y_\phi(t) = c-(c-1)\exp(-at) \quad \text{for all } t \in [0, w], \qquad (16a)$$

$$y_\phi(t) = y_\phi(w)\exp(-a(t-w)) \quad \text{for all } t \in [w, 1]. \qquad (16b)$$

Hence for any two initial conditions $\phi, \psi \in D$ satisfying $V(\phi) = V(\psi)$ the two corresponding solutions coincide on the interval $[0,1]$ and therefore $y_\psi(t) = y_\phi(t)$ for all $t \geq 0$. This means that solutions associated with initial conditions from the set D are uniquely determined by the single parameter $w = V(\phi)$. Therefore we write y_w instead of y_ϕ.

The following inequalities play an important role for our main result:

$$c/(c-1) < b < 2 - 1/c, \qquad (17)$$

$$d(c-1)(b-c(b-1)) < c(b-1)(bc(c-2)+b-c(c-1)) \qquad (18)$$

$$c^3(b-1)^2(c-b) < (c(b-1) - d(c-1))b^2(c-1) \qquad (19)$$

We can now state the main theorems of this paper:

<u>Theorem 3.</u> Let b, c, d be three paramater values satisfying the inequalities (17), (18), and (19). Then there is a value a^* (depending on b, c, d) such that for all $a > a^*$ the folllowing is true:
There is a compact interval $I \subset [0, 1]$ and a mapping $F:I \to I$ such that the associated Frobenius - Perron operator is asymptotically periodic. For each initial condition $\phi \in D$ obeying $V(\phi) \in I$ the corresponding solution y_ϕ of Eq.(10) has infinitely many times t_1, t_2, ... such that the conditions (12) - (14) hold.

<u>Remark 2.</u> The proof of this theorem relies on Theorem 2 with $r = 2$, hence according to Remark 1 the transformation F of Theorem 3 has a unique invariant measure and F is at least ergodic.

Theorem 4. Let the parameters b, c obey the inequalities (17). Then there is a number a^* such that for all $a > a^*$ and for some value of d (= $d(a,b,c)$) the following is true:
There is a compact interval $I \subset [0, 1]$ and a mapping $F:I \to I$ which is mixing and such that each solution y_ϕ of Eq.(10), corresponding to initial conditions $\phi \in D$ with $V(\phi) \in I$, has infinitely many times t_1, t_2, ... obeying the relationships (12) - (14).

Proof of Theorems 3 and 4:   Let $\phi \in D$ with $w = V(\phi)$. It follows from (16) that

$$y_w(w) = c - (c-1) \exp(-aw), \tag{20}$$

$$y_w(1) = (c \exp(aw) - (c-1)) \exp(-a). \tag{21}$$

$y_w(1)$ is an increasing function of w and $y_0(1) = \exp(-a) < 1$, $y_1(1) = c-(c-1) \exp(-a) > 1$. Hence there is a unique $w=w_2$ obeying $x_w(1)=1$. The number w_2 is determined by $\exp(aw_2)=(\exp(a)+c-1)/c$. Moreover there is a unique $w\grave{} = w\grave{}(a,b,c)$ such that $x_{w\grave{}}(w\grave{})=b$. Obviously $w\grave{}$ is determined by

$$\exp(aw\grave{}) = (c-1)/(c-b). \tag{22}$$

We have $w\grave{} < w_2$ if and only if

$$b(c-1) < (c-b) \exp(a).$$

Note that this relation is satisfied if a is sufficiently large with respect to b and c.
For all $w < w\grave{}$ the solution y_w is increasing on $[0,w]$, decreasing on $[w,1]$, and $y_w(1) < 1$. Hence there is a unique $t_1=t_1(w)$ where $y_w(t_1) = 1$. The number t_1 is given by

$$\exp(a\,t_1) = y_w(w) \exp(aw) = c \exp(aw) - (c-1). \tag{23}$$

Since $1 \le y_w(t) \le b$ for $t \in [0,t_1(w)]$, the solution y_w increases on the interval $[1,1+t_1(w)]$, and

$$\begin{aligned} y_w(1+t_1(w)) &= c - (c-y_w(1)) \exp(-a\,t_1(w)) \\ &= c + \exp(-a) - c/(c \exp(aw)-(c-1)). \end{aligned} \tag{24}$$

$y_w(1+t_1(w))$ is an increasing function of w, and there exists a unique value $w_1 = w_1(a,c) \in (0,1)$ such that

$$y_{w_1}(1 + t_1(w_1)) = 1,$$

which is given by

$$\exp(a\ w_1) = (c-1)/c + 1/(c-1+\exp(-a)). \tag{25}$$

Since we assumed $w < w`$ it must be guaranteed that $w_1 < w`$. Comparing Eqs.(22) and (25) it is obvious that this inequality is satisfied iff

$$b(c-1) < c^2(b-1) + b(c-1)\ \exp(-a).$$

This inequality follows from the first part of (17).

Whenever $w < w_1$ then $y_w(t) < 1$ for all $t \in (t_1(w), t_1(w)+1)$, and hence $y_w(t) \to 0$ as $t \to \infty$. These solutions are not interesting to us, and in the following we assume $w \in [w_1, w`]$.

There is a unique time $t_2(w) \in [1, t_1(w)+1]$, where $y_w(t_2(w))=1$. The function $\psi_w : [-1,0] \to I\!R$ defined by

$$\psi_w(t) = y_w(t_2(w) + t) \tag{26}$$

is an element of D. Of course, for the future development of y_w it only matters what the value of

$$F(w): = V(\psi_w) = 1 - (t_2(w)-t_1(w)) \tag{27}$$

is, more precisely

$$y_w(t) = y_{F(w)}(t-t_2(w))\ \text{holds for all}\ t \geq t_2(w). \tag{28}$$

In particular the function $F: [w_1, w`] \to [0,1]$ defined in this way, and its iterates F^i, as long as they exist, determine the differences $t_{2i}(w) - t_{2i-1}(w)$ between the 2i-th and the (2i-1)-th time t where $y_w(t) = 1$:

$$t_{2i}(w) - t_{2i-1}(w) = 1 - F^i(w)$$

for all $i \in I\!N$ such that $F^{i-1}(w) \in [w_1, w`]$. We shall explicitely calculate the function F. Because of $1=c-(c-y_w(1))\ \exp(-a(t_2(w)-1))$ and (23) we have

$$\exp(a\ F(w)) = \exp(a(1-t_2+t_1)) = y_w(w)\ \exp(aw)\ (c-1)/(c-y_w(1))$$

$$= (c-1)(c \exp(aw)-(c-1))/(c-(c \exp(aw)-(c-1))\exp(-a)). \quad (29)$$

By definition of w_1 we have

$$F(w_1) = 0. \quad (30)$$

Using (22) and (29) we obtain

$$F(w\grave{}) > w\grave{} \quad (31)$$

if and only if

$$c^2(b-1) > b^2(c-1) - b(c-1) \exp(-a), \quad (32)$$

an inequality that follows from the first part of (17).

Logarithmizing the second equality of (29) it is easy to see that

$$dF(w)/dw > 1 \text{ for all } w \in [w_1,w\grave{}] . \quad (33)$$

The relations (30) - (33) imply that there is a unique fixed point $u \in (w_1,w\grave{})$ of F, and that this fixed point is unstable. Obviously all solutions y_w with $w \in [w_1,u)$ converge to 0 as $t \to \infty$, since there is some $i \in \mathbb{N}$ such that $F^i(w) < w_1$. Since u is an unstable fixed point of F, the relation (28) implies that y_u is an unstable periodic solution of Eq.(10).

If $w \in (u,w\grave{}]$ then it is not yet clear what the ultimate fate of the corresponding solutions will be. Associated with such a value of w there is an iteration number i obeying $F^i(w) \in (w\grave{},F(w\grave{})]$. Let us call $F(w\grave{}) = w_3$. Then we have to study solutions y_w with $w \in (w\grave{},w_3]$. We shall do this with the goal to extend the function F beyond $w\grave{}$.

By the definition of $w\grave{}$ for such solutions we have $y_w(w) > b$. Hence there is a unique time $\tau_1(w)$ obeying

$$0 < \tau_1(w) < w \text{ and } y_w(\tau_1(w)) = b.$$

From (15) it follows that

$$\exp(a \tau_1(w)) = (c-1)/(c-b). \quad (34)$$

The solution increases on $[1,1+\tau_1(w)]$ and from (15) and (21)

$$y_w(1+\tau_1(w)) = c-(c-y_w(1)) \exp(-a \tau_1(w))$$

161

$$= c((b-1)+(c-b)\exp(-a(1-w)))/(c-1) - (c-b)\exp(-a). \qquad (35)$$

The relation

$$y_w(1+\tau_1(w)) < 1 \quad \text{for all } w \in (w\grave{\,},w_3] \qquad (36)$$

holds iff it is valid for $w=w_3=F(w\grave{\,})$. Eqs. (29) and (22) imply that

$$\exp(aw_3) = c(c-b)/(b(c-1)^2) - \exp(-a)/(c-1). \qquad (37)$$

Therefore (36) holds if $c(2-b) > 1$ (which follows from the second inequality of (17)) and a is sufficiently large.

Since $y_w(w) > b$ and $y_w(1) < 1$ there is a unique time $\tau_2(w)$ obeying $y_w(\tau_2(w))=b$. Since $y_w(t) > b$ for all $t \in (\tau_1(w),\tau_2(w))$ the solution obeys

$$y_w(t) = d - (d-y_w(\tau_1(w)+1)) \exp(-a(t-\tau_1(w)-1))$$

for all $t \in (\tau_1(w)+1,\tau_2(w)+1]$. In particular

$$y_w(\tau_2(w)+1) = d + (y_w(\tau_1(w)+1) -d) \exp(-a(\tau_2(w)-\tau_1(w))).$$

The term $\exp(-a(\tau_2-\tau_1))$ can be calculated from

$$y(w) = c - (c-b) \exp(-a(w-\tau_1)) \quad \text{and}$$

$$b = y(w) \exp(-a(\tau_2-w)), \quad \text{resulting in}$$

$$\exp(-a(\tau_2(w)-\tau_1(w)))\grave{\,} = b(c-1)\exp(-aw)/((c-b)(c-(c-1)\exp(-aw)))$$

Together with (35) we obtain

$$y_w(\tau_2(w)+1) = d + (c-(c-1)\exp(-aw))^{-1} b(\exp(-aw)(c(b-1)/(c-b)$$
$$- d(c-1)/(c-b) - (c-1)\exp(-a)) + c \exp(-a)). \qquad (38)$$

In the interval $(\tau_2(w)+1,t_1(w)+1)$ the solution y_w is again increasing exponentially towards c. Assume $y_w(t_1(w)+1) > 1$ (this is the case for $w=w\grave{\,}$ and hence also at least for w near to $w\grave{\,}$, because the solutions depend continuously on w). Then there is a unique number $t_2(w) \in (\tau_2(w)+1,t_1(w)+1)$ obeying $y_w(t_2(w))=1$. In this case the segment $\psi_w: [-1,0] \to \mathbb{R}$ of the solution y_w defined by Eq.(26) is an element of the initial set D. Defining F(w) for these w by the expression (27) leads to a continuous extension of F beyond $w\grave{\,}$, and the relation (28) is again valid.

F(w) can be calculated exactly as following.

$$\exp(-aF(w)) = \exp(-a((t_1-\tau_2)-(t_2-(\tau_2+1)))) = \exp(a(t_2-(\tau_2+1)))/b$$
$$= (c-y_w(\tau_2(w)+1))/(b(c-1)). \tag{39}$$

Obviously the extension covers the interval $[w',w_3]$ if and only if the expression (39), now used as a definition for F(w), evaluated for $w=w_3$ results in $F(w_3) > 0$. As we shall see below this inequality follows from the relation (18).

In order to simplify the calculations we introduce the following transformation. Let

$$v = h(w) = \exp(-aw), \quad v_1 = h(w_1), \quad v` = h(w`), \quad v_3 = h(w_3).$$

In stead of F we study the topologically conjugate mapping

$$G: [v_3,v_1] \to [v_3,1] \ , \quad G(v) = h(F(w)) = h \circ F \circ h^{-1}(v). \tag{40}$$

There is a one-to-one correspondence between the iterates of F and G since $F^i = (h^{-1} \circ G \circ h)^i = h^{-1} \circ G^i \circ h$.

It follows from (39), (38), and (29) that G is given by

$$G(v) = \begin{cases} (c-d)/(b(c-1))+((d/(c-b)-c(b-1)/((c-1)(c-b))+e^{-a})v \\ \qquad -c\,\exp(-a)/(c-1))/(c-(c-1)v) \text{ if } v \in [v_3,v`] \quad (41a) \\ \\ cv/((c-1)(c-(c-1)v)) - \exp(-a)/(c-1) \text{ if } v \in [v`,v_1]. \quad (41b) \end{cases}$$

In the interval $[v`,v_1]$ the function G represents an arc of a monotone increasing hyperbola which has its vertical asymptote at $c/(c-1) > 1$. The slope of this arc is minimal at $v=v`$, and with (22) we conclude that

$$dG(v)/dv > 1 \text{ for all } v \in [v`,v_1] \text{ iff } c^2 > (c-1)\,b.$$

The last inequality is satisfied because of the assumption (17).
It follows from (37) and $v_3 = G(v`) = \exp(-a\,F(w`))$ that

$$v_3 = c(c-b)/(b(c-1)^2) - \exp(-a)/(c-1). \tag{42}$$

Because of (31) we have $G(v`) < v`$. Hence there is a unique fixed point $\bar{v}$ of G in the interval $(v`,v_1)$, $G(\bar{v}) = \bar{v}$. (41b) implies

$$\bar{v} = \kappa + \sqrt{\kappa^2 + c\,\exp(-a)/(c-1)^2}, \text{ where}$$

163

$$\kappa = c(c-2)/(2(c-1)^2) - \exp(-a)/(2(c-1)).$$

For all sufficiently large values of a one obtains the estimate

$$1 - 1/(c-1)^2 = c(c-2)/(c-1)^2 < \bar{v} < 1. \tag{43}$$

Let us now discuss the other arc of G defined on $[v_3,v\grave{}]$. We are interested in a situation where $dG(v)/dv < -1$ and $G(v) \leq \bar{v}$ for all $v \in [v_3,v\grave{})$. In such a situation there is an interval I $\subset [v_3,\bar{v}]$ which is mapped by G into itself (namely I = $[v_3, G(v_3)]$) and such that the restriction $G|_I$ is a mapping satisfying all conditions of Theorem 2 (using the symbol G in stead of F). Hence the Frobenius - Perron operator associated with G (and also of F, since F and G are topologically conjugate) is asymptotically periodic. If moreover $G(v_3) = \bar{v}$ then the conditions of Theorem 1 are satisfied and G (and also F) is mixing. (Note that the relevant invariant domain of F is $h^{-1}(I)$).

The r. h. s. of (41a) defines a decreasing hyperbola with vertical asymptote at $c/(c-1) > 1$. Hence the maximal value of G(v) and the maximal value of $dG(v)/dv$ in the interval $[v_3,v\grave{})$ is located at $v = v_3$. In calculating $G(v_3)$ from (41a) and (42) we combine all terms containing the factor $\exp(-a)$ into a term $\lambda(a,b,c)$ arriving at

$$G(v_3) = (c-d)/(b(c-1))+d/(c(c-1)(b-1))-(c-1)^{-2} + \lambda(a,b,c) \tag{44}$$

where for any fixed values of b and c we have $\lambda(a,b,c) \to 0$ as $a \to \infty$.

Comparing the expressions (43) and (44) it becomes obvious that $G(v_3) < \bar{v}$ if the inequality (18) is satisfied and a is sufficiently large (with respect to b, c, and d).

Finally we observe that for sufficiently large a the inequality $dG(v_3)/dv < -1$ is equivalent to

$$dc/(c-b) - c^2(b-1)/((c-1)(c-b)) < -(c-(c-1)v_3)^2.$$

Using (42) it turns out that this inequality follows from (19) if a is again sufficiently large. This completes the proof of Theorem 3.

In order to obtain Theorem 4 we choose a special set of parameters a, b, c, d, say $a=a_1$, $b=b_1$, $c=c_1$, $d=d_1$ such that the conditions of Theorem 3 are satisfied. If a_1, b_1, c_1 are held fixed, and if d decreases from d_1 to $-\infty$ then the inequality (19) is always satisfied (since decreasing d results in increasing the r.h.s.) and hence $dG(v_3)/dv < -1$ for all $d < d_1$. On the other hand there will be a $d=d^* < d_1$ such that $G(v_3) = \bar{v}$, as may be seen from the representation (44) of $G(v_3)$. For the parameters $a=a_1$, $b=b_1$, $c=c_1$, and $d=d^*$ all the conditions of Theorem 1 are satisfied (using the symbol G in stead of F), completing the proof of Theorem 4.

4. References

ANDERSON, R.F.V., M.C. MACKEY: Commodities, cycles, and multiple time delays. (preprint)

AN DER HEIDEN, U. (1979) Delays in physiological systems. J. Math. Biol. $\underline{8}$, 345-364

AN DER HEIDEN, U., M.C. MACKEY (1982) The dynamics of production and destruction: Analytic insight into complex behavior. J. Math. Biol. $\underline{16}$, 75-101

AN DER HEIDEN, U., H.-O. WALTHER (1983) Existence of chaos in control systems with delayed feedback. J. Diff. Equs. $\underline{47}$, 273-295

LASÓTA, A., M.C. MACKEY (1985) Probabilistic properties of deterministic systems. Cambridge University Press (in press)

LI, T.Y., J.A. YORKE (1978) Ergodic transformations from an interval into itself. Trans. Amer. Math. Soc. $\underline{235}$, 183

MACKEY, M.C., L. GLASS (1977) Oscillation and chaos in physiological control systems. Science $\underline{197}$, 287-289

MACKEY, M.C., U. AN DER HEIDEN (1984) The dynamics of recurrent inhibition. J. Math. Biol. $\underline{19}$, 211-225

STEIN, P., S. ULAM (1964) Non-linear transformation studies on electronic computers. Rozprawy Metamatyczne $\underline{39}$, 401-484

WAZEWSKA, M., A. LASOTA (1976) Matematyczne problemy dynamiki ukladu krwinek czerwonych. Matematyka Stosowana $\underline{VI}$, 23-40

U. an der Heiden, Universität Bremen NW2, D-28 Bremen 33, W.Germany

International Series of
Numerical Mathematics, Vol. 74
© 1985 Birkhäuser Verlag Basel

Bivariate Natural Spline Smoothing

by

Chui Li Hu

Larry L. Schumaker

§1. Introduction.

The purpose of this paper is to develop a highly efficient stable algorithm for smoothing data given on a grid in the plane. Our approach is to use the natural B-splines introduced for the univariate case in [6], coupled with the tensor technique of [3].

The paper is organized as follows. In §2 we introduce some notation and review bivariate interpolation using natural splines. In the following section we discuss a smoothing problem closely related to the interpolation problem. In §4 a method for selecting the smoothing parameter is presented. We conclude the paper with remarks and references.

§2. Bivariate Natural Interpolation.

Let $n \geq 2m$ and $\tilde{n} \geq 2\tilde{m}$ be positive integers, and let $x_1 < \ldots < x_n$ and $\tilde{x}_1 < \ldots < \tilde{x}_{\tilde{n}}$. Suppose $\{z_{ij}\}_{i=1,j=1}^{n \quad \tilde{n}}$ are given real numbers. The interpolation problem of interest is the following: find s such that

$$(2.1) \qquad s(x_i, \tilde{x}_j) = z_{ij}, \; i = 1, \ldots, n \; \text{ and } \; j = 1, \ldots, \tilde{n}.$$

One approach to solving this problem is to use tensor-product natural splines. Associated with $\{x_i\}_1^n$, let $\{B_i(x)\}_1^n$ be the natural B-spline of order $2m$ introduced in [6]. Similarly, let $\{\tilde{B}_j(y)\}_1^{\tilde{n}}$ be the natural B-splines of order $2\tilde{m}$ associated with $\{\tilde{x}_j\}_1^{\tilde{n}}$. We look for a solution of (2.1) in the form

$$(2.2) \qquad s(x,y) = \sum_{\upsilon=1}^{n} \sum_{\mu=1}^{\tilde{n}} c_{\upsilon\mu} B_\upsilon(x) \tilde{B}_\mu(y).$$

The $n \times \tilde{n}$ coefficients of s can be determined by enforcing the $n \times \tilde{n}$ interpolation conditions (2.1). This leads to the linear system

$$(2.3) \qquad \sum_{\upsilon=1}^{n} \sum_{\mu=1}^{\tilde{n}} c_{\upsilon\mu} B_\upsilon(x_i) \tilde{B}_\mu(\tilde{x}_j) = z_{ij}, \qquad \begin{array}{l} i = 1,\ldots,n \\[4pt] j = 1,\ldots,\tilde{n}. \end{array}$$

This system can be rewritten in the form

$$(2.4) \qquad B\,C\,\tilde{B}^T = Z,$$

where

$$C = (c_{ij})_{1,1}^{n\ \tilde{n}}, \quad Z = (z_{ij})_{1,1}^{n\ \tilde{n}}, \quad B = (B_i(x_j))_{1,1}^{n\ n} \quad \text{and} \quad \tilde{B} = (\tilde{B}_i(\tilde{x}_j))_{1,1}^{\tilde{n}\ \tilde{n}}.$$

The matrix B is $2m-1$-banded and nonsingular (c.f. [6]). Similarly, $\tilde{B}$ is $2\tilde{m}-1$-banded and nonsingular. Thus (2.4) has a unique solution for any given data $\{z_{ij}\}$.

The system (2.4) can be rewritten as two linear systems:

$$(2.5) \qquad BU = Z$$

$$(2.6) \qquad \tilde{B}C^T = U^T.$$

To solve (2.5) it suffices to find the LU decomposition of B, then back-solve $\tilde{n}$ times to find the $\tilde{n}$ columns of U. Equation (2.6) can then be solved by decomposing $\tilde{B}$ and performing n back substitutions. FORTRAN subroutines for generating B and $\tilde{B}$ and decomposing them as well as for solving the related systems can be found in [7].

It is known (cf. [9]), that the natural interpolating spline is the unique solution of the variational problem

$$(2.7) \qquad\qquad \underset{f \,\varepsilon\, U}{\text{minimize } J(f)}$$

where

$$(2.8) \qquad J(f) = \int_{x_1}^{x_n} \int_{\tilde{x}_1}^{\tilde{x}_{\tilde{n}}} (f^{(m,\tilde{m})}(x,y))^2 dx\, dy$$

$$+ \sum_{i=1}^{n} \int_{\tilde{x}_1}^{\tilde{x}_{\tilde{n}}} [f^{(0,\tilde{m})}(x_i,y)]^2 dy + \sum_{j=1}^{\tilde{n}} \int_{x_1}^{x_n} [f^{(m,0)}(x,\tilde{x}_j)]^2 dx,$$

$$(2.9) \qquad U = \{f \,\varepsilon\, L_2^{m,\tilde{m}}(H) : f(x_i,\tilde{x}_j) = z_{i,j},\; i = 1,\ldots,n,\; j = 1,\ldots,\tilde{n}\},$$

$$(2.10) \qquad L_2^{m,\tilde{m}}(H) = \{f : J(f) \text{ exists and is finite}\},$$

and H denotes the rectangle $\{(x,y) : x_1 \leq x \leq x_n,\; \tilde{x}_1 \leq y \leq \tilde{x}_{\tilde{n}}\}$.

§3. Bivariate Natural Smoothing

In this section we relax the interpolation conditions and consider a smoothing problem in which the goodness of fit is balanced against a measure of smoothness similar to $J(f)$ in the previous section. To measure goodness of fit, let

$$(3.1) \qquad E(f) = \sum_{i=1}^{n} \sum_{j=1}^{\tilde{n}} w_i \tilde{w}_j [f(x_i,\tilde{x}_j) - z_{ij}]^2$$

where $\{w_i\}_1^n$ and $\{\tilde{w}_j\}_1^{\tilde{n}}$ are prescribed positive weights. To measure the smoothness of f, let

168

$$(3.2) \qquad J_2(f) = \int_{x_1}^{x_n} \int_{\tilde{x}_1}^{\tilde{x}_{\tilde{n}}} [f^{(m,\tilde{m})}(x,y)]^2 \, dxdy$$

$$(3.3) \qquad J_1(f) = \sum_{j=1}^{\tilde{n}} \tilde{w}_j \int_{x_1}^{x_n} [f^{(m,0)}(x,\tilde{x}_j)]^2 \, dx + \sum_{i=1}^{n} w_i \int_{\tilde{x}_1}^{\tilde{x}_{\tilde{n}}} [f^{(0,\tilde{m})}(x_i,y)]^2 \, dy.$$

The smoothing problem of interest in this section is the following:

$$(3.4) \qquad \underset{f \varepsilon L_2^{m,\tilde{m}}(H)}{\text{minimize}} \quad J_2(f) + w\, J_1(f) + w^2 E(f),$$

where w is a positive smoothing parameter balancing goodness of fit against smoothness. We give a constructive solution to problem (3.4) in Theorem 3.3 below. First we need two lemmas. It is known (cf. [6]) that for each $1 \leq \nu \leq m$. there exist $\{\beta_{i\nu}\}_{i=1}^{n}$ and a polynomial p_j of degree $m - 1$ such that

$$(3.5) \qquad B_\nu(x) = p_\nu(x) + \sum_{i=1}^{n} \beta_{i\nu}(x-x_i)_+^{2m-1} \ .$$

For each $1 \leq \mu \leq \tilde{n}$, let $\{\tilde{\beta}_{j\mu}\}_{j=1}^{\tilde{n}}$ be the analogous coefficients in the one-sided expansion of $\tilde{B}_\mu$.

<u>Lemma 3.1.</u> Let s be a tensor-product natural spline as in (2.2) with coefficients $\{c_{ij}\}$. Then for any $f \in L_2^{m,\tilde{m}}(H)$,

$$(3.6) \qquad \int_{x_1}^{x_n} \int_{\tilde{x}_1}^{\tilde{x}_{\tilde{n}}} f^{(m,\tilde{m})}(x,y) \, s^{(m,\tilde{m})}(x,y) \, dxdy$$

$$= (-1)^{m+\tilde{m}}(2m-1)!(2\tilde{m}-1)! \sum_{i=1}^{n} \sum_{j=1}^{\tilde{n}} f(x_i,\tilde{x}_j) \sum_{\nu=1}^{n} \sum_{\mu=1}^{\tilde{n}} c_{\nu\mu} \beta_{i\nu} \tilde{\beta}_{j\mu}.$$

<u>Proof:</u> Integrating by parts $\tilde{m}$-1 times with respect to the y-variable, we obtain

$$I := \int_{x_1}^{x_n} \int_{\tilde{x}_1}^{\tilde{x}_{\tilde{n}}} f^{(m,\tilde{m})}(x,y) s^{(m,\tilde{m})}(x,y) dx dy$$

$$= (-1)^{\tilde{m}-1} \int_{x_1}^{x_n} \int_{\tilde{x}_1}^{\tilde{x}_{\tilde{n}}} f^{(m,1)}(x,y) s^{(m,2\tilde{m}-1)}(x,y) dx dy.$$

The boundary terms drop out because s is a natural spline and hence

$$s^{(0,j)}(\tilde{x}_1) = s^{(0,j)}(\tilde{x}_{\tilde{n}}) = 0, \quad j = \tilde{m},\ldots,2\tilde{m}-2.$$

Integrating m-1 times by parts on the x-variable, we get

$$I = (-1)^{m+\tilde{m}} \int_{x_1}^{x_n} \int_{\tilde{x}_1}^{\tilde{x}_{\tilde{n}}} f^{(1,1)}(x,y) s^{(2m-1,2\tilde{m}-1)}(x,y) dx dy.$$

Since $s^{(2m-1,2\tilde{m}-1)}(x,y)$ is constant on each subrectangle $[x_i,x_{i+1}) \otimes [\tilde{x}_j,\tilde{x}_{j+1})$, we find

$$I \, (-1)^{m+\tilde{m}} = \sum_{i=1}^{n-1} \sum_{j=1}^{\tilde{n}-1} \int_{x_i}^{x_{i+1}} \int_{\tilde{x}_j}^{\tilde{x}_{j+1}} f^{(1,1)}(x,y) s^{(2m-1,2\tilde{m}-1)}(x_i+,\tilde{x}_j+) dx dy$$

$$= \sum_{i=1}^{n-1} \sum_{j=1}^{\tilde{n}-1} s^{(2m-1,2\tilde{m}-1)}(x_i+,\tilde{x}_j+) \, f(x,y) \, \Big|_{x_i}^{x_{i+1}} \Big|_{\tilde{x}_j}^{\tilde{x}_{j+1}}.$$

Substituting

$$s^{(2m-1,2\tilde{m}-1)}(x_i+,\tilde{x}_j+) = \sum_{\nu=1}^{n} \sum_{\mu=1}^{\tilde{n}} c_{\nu\mu} \, B_\nu^{(2m-1)}(x_i+) \, \tilde{B}_\mu^{(2\tilde{m}-1)}(x_i+),$$

using the facts that (cf. (3.5))

$$B_\nu^{(2m-1)}(x_i+) = (2m-1)! \sum_{p=1}^{i} \beta_{p\nu},$$

$$\tilde{B}_\mu^{(2\tilde{m}-1)}(\tilde{x}_j+) = (2\tilde{m}-1)! \sum_{q=1}^{j} \tilde{\beta}_{q\mu},$$

and rearranging, we arrive at (3.6). ∎

__Lemma 3.2.__ Under the same hypotheses as Lemma 3.1,

$$(3.7) \qquad \int_{x_1}^{x_n} f^{(m,0)}(x,y)\, s^{(m,0)}(x,y)\,dx$$

$$= (-1)^m (2m-1)! \sum_{i=1}^{n} f(x_i,y) \sum_{\nu=1}^{n} \sum_{\mu=1}^{\tilde{n}} c_{\nu\mu} \beta_{i\nu} \tilde{B}_\mu(y)$$

and

$$(3.8) \qquad \int_{\tilde{x}_1}^{\tilde{x}_{\tilde{n}}} f^{(0,\tilde{m})}(x,y) s^{(0,\tilde{m})}(x,y)\,dy =$$

$$= (-1)^{\tilde{m}} (2\tilde{m}-1)! \sum_{j=1}^{\tilde{n}} f(x,\tilde{x}_j) \sum_{\nu=1}^{n} \sum_{\mu=1}^{\tilde{n}} c_{\nu\mu} B_\nu(x) \tilde{\beta}_{j\mu}.$$

__Proof:__ It suffices to prove (3.7)--the proof of (3.8) is similar.
Integrating, m times by parts we obtain

$$\int_{x_1}^{x_n} f^{(m,0)}(x,y) s^{(m,0)}(x,y)\,dx = (-1)^{m-1}\int_{x_1}^{x_n} f^{(1,0)}(x,y) s^{(2m-1,0)}(x,y)\,dx$$

$$= (-1)^{m-1} \sum_{i=1}^{n-1} \int_{x_i}^{x_{i+1}} f^{(1,0)}(x,y) s^{(2m-1,0)}(x_i+,y)\,dx$$

$$= (-1)^{m-1} \sum_{i=1}^{n-1} [f(x_{i+1},y)-f(x_i,y)] s^{(2m-1,0)}(x_i+,y)$$

$$= (-1)^m \sum_{i=1}^{n} f(x_i,y)\, \text{jump}[s^{(2m-1,0)}(\cdot,y)]_{x_i}.$$

Now, by (3.5),

$$\text{jump}[s^{(2m-1,0)}(\cdot,y)]_{x_i} = \sum_{\nu=1}^{n} \sum_{\mu=1}^{\tilde{n}} c_{\nu\mu}\, \beta_{i\nu}\, \tilde{B}_\mu(y),$$

and the result follows. ∎

We are now ready to solve the smoothing problem (3.4).

__Theorem 3.3.__ For any given data and any smoothing parameter $w > 0$, there is a unique solution of problem (3.4). Moreover, this solution is a tensor-natural spline as in (2.2), and the coefficients $\{c_{ij}\}$ are uniquely determined by the linear system

$$(3.9) \qquad (B + w^{-1} E) \; C \; (\tilde{B} + w^{-1} \tilde{E})^{T} = Z,$$

where B, $\tilde{B}$, C, and Z are as in (2.4), and where (cf. [6]) $E = (E_{ij})$ and $\tilde{E} = (\tilde{E}_{ij})$ with

$$(3.10) \qquad E_{ij} = (-1)^{m}(2m-1)! \; w_{i}^{-1} \; \beta_{ij}$$

$$\tilde{E}_{ij} = (-1)^{\tilde{m}} (2\tilde{m}-1)! \; \tilde{w}_{i}^{-1} \; \tilde{\beta}_{ij}.$$

The matrices E and $\tilde{E}$ have band widths $2m+1$ and $2\tilde{m}+1$ respecively.

__Proof:__ It is convenient to introduce the following functional defined for ϕ, ψ, $\in L_2^{m,\tilde{m}}(H)$:

$$(3.11) \qquad J(\phi,\psi) = \int_{x_1}^{x_n}\int_{\tilde{x}_1}^{\tilde{x}_{\tilde{n}}} \psi^{(m,\tilde{m})}(x,y) \; \phi^{(m,\tilde{m})}(x,y)dxdy$$

$$+ w \sum_{j=1}^{n} w_j \int_{\tilde{x}_1}^{\tilde{x}_{\tilde{n}}} \psi^{(0,\tilde{m})}(x_j,y) \; \phi^{(0,\tilde{m})}(x_j,y)dy$$

$$+ w \sum_{j=1}^{\tilde{n}} \tilde{w}_j \int_{x_1}^{x_n} \psi^{(m,0)}(x,\tilde{x}_j) \; \phi^{(m,0)}(x,\tilde{x}_j)dx$$

$$+ w^2 \sum_{i=1}^{n} \sum_{j=1}^{\tilde{n}} w_i\tilde{w}_j \; \phi(x_i,\tilde{x}_j)[\psi(x_i,\tilde{x}_j)-z_{ij}].$$

Now if s is a spline of the form (2.2) with coefficients satisfying (3.9), then using Lemmas 3.1 and 3.2 to replace the integrals, it is not hard to see that

$$J(s,\psi) = 0, \text{ all } \psi \in L_2^{m,\tilde{m}}(H).$$

We can now show that the system (3.9) is nonsingular. Suppose c_H is a solution of (3.9) with $Z \equiv 0$, and let s_H be the corresponding spline. We show that $s_H \equiv 0$. Indeed, by (3.11), $J(s_H, s_H) = 0$, and thus $s_H^{(m,\tilde{m})}(x,y) = 0$ a.e. on H. This implies that

$$(3.12) \qquad s_H(x,y) = \sum_{\nu=1}^{m} x^{\nu-1}\phi_\nu(y) + \sum_{\mu=1}^{\tilde{m}} y^{\mu-1}\psi_\mu(x),$$

where ϕ_ν and ψ_μ are functions in $L_2^{\tilde{m}}[\tilde{x}_1, \tilde{x}_{\tilde{n}}]$ and $L_2^{m}[x_1, x_n]$, respectively. But $J(s_H, s_H) = 0$ also implies that

$$(3.13) \qquad \int_{x_1}^{x_n} [s_H^{(m,0)}(x, \tilde{x}_j)]^2 dx = 0, \quad j = 1,\ldots,\tilde{n}$$

$$(3.14) \qquad \int_{\tilde{x}_1}^{\tilde{x}_{\tilde{n}}} [s_H^{(0,\tilde{m})}(x_i, y)]^2 dy = 0, \quad i = 1,\ldots,n$$

and

$$(3.15) \qquad s_H(x_i, \tilde{x}_j) = 0, \quad i = 1,\ldots,n \quad \text{and} \quad j = 1,\ldots,\tilde{n}.$$

Substituting (3.12) in (3.13), we conclude that

$$\int_{x_1}^{x_n} [\sum_{\mu=1}^{m} \tilde{x}_j^{\mu-1}\psi_\mu^{(m)}(x)]^2 dx = 0, \quad j = 1,\ldots,\tilde{n}.$$

This asserts that the polynomial

$$P(y) = \sum_{\mu=1}^{\tilde{m}} (\int_{x_1}^{x_n} [\psi_\mu^{(m)}(x)]^2 dx)y^{2\mu-2} + \ldots$$

vanishes at the $\tilde{n} \geq 2\tilde{m}$ points $\tilde{x}_1 < \ldots < \tilde{x}_{\tilde{n}}$. This implies each of the coefficients is zero; i.e., $\int_{x_1}^{x_n} [\psi_\mu^{(m)}(x)]^2 dx = 0$, and thus that ψ_μ is a polynomial of degree $m-1$ at most, $\mu = 1,\ldots,\tilde{m}$. Arguing similarly for the ϕ_ν, we conclude that

$$s_H(x,y) = \sum_{\nu=1}^{m} \sum_{\mu=1}^{\tilde{m}} e_{\nu\mu} \, x^{\nu-1} y^{\mu-1} \ .$$

But then (3.15) implies $s_H \equiv 0$, and the proof that (3.9) is nonsingular is complete.

It remains to show that s with coefficients determined by (3.9) solves the minimization problem (3.4). Let

$$\sigma_w(f) = J_2(f) + w \, J_1(f) + w^2 E(f).$$

Then for $\phi \in L_2^{m,\tilde{m}}(H)$,

$$(3.16) \qquad \sigma_w(\phi) = \sigma_w(s) + 2 \, J(s,\phi-s)$$

$$+ w^2 \sum_{i=1}^{n} \sum_{j=1}^{\tilde{n}} w_i \tilde{w}_j [\phi(x_i,\tilde{x}_j)-s(x_i,\tilde{x}_j)]^2$$

$$+ \int_{x_1}^{x_n} \int_{\tilde{x}_1}^{\tilde{x}_{\tilde{n}}} [\phi^{(m,\tilde{m})}(x,y)-s^{(m,\tilde{m})}(x,y)]^2 dxdy$$

$$+ w \sum_{i=1}^{n} w_i \int_{\tilde{x}_1}^{\tilde{x}_{\tilde{n}}} [\phi^{(0,\tilde{m})}(x_i,y)-s^{(0,\tilde{m})}(x_i,y)]^2 dy$$

$$+ w \sum_{j=1}^{\tilde{n}} \tilde{w}_j \int_{x_1}^{x_n} [\phi^{(m,0)}(x,\tilde{x}_j)-s^{(m,0)}(x,\tilde{x}_j)]^2 dx.$$

Since $J(s,\phi-s) = 0$ by (3.11), it follows that $\sigma_w(\phi) \geq \sigma_w(s)$, and equality can hold iff each of the other terms in (3.16) is zero. Arguing as we did above for s_H, we conclude that $\sigma_w(\phi) = \sigma_w(s)$ iff $\phi \equiv s$. $\equiv$

The system (3.9) can be efficiently solved by splitting it into two systems as we did in the interpolation case (cf. (2.5)-(2.6)). In particular, we can solve (3.9) by solving

$$(3.17) \qquad\qquad (B+w^{-1}E) \, U = Z$$

$$(3.18) \qquad\qquad (\tilde{B}+w^{-1}\tilde{E})C^T = U^T.$$

As before, we can solve these systems by performing the LU-decompositions of $(B+w^{-1}E)$ and $(\tilde{B}+w^{-1}\tilde{E})$, followed by $\tilde{n}$ and n back solutions for (3.17) and (3.18), respectively. FORTRAN subroutines for generating the matrices B, E, $\tilde{B}$, $\tilde{E}$ and for solving the systems (3.17)-(3.18) can be found in [7].

We close this section with some remarks about the behavior of the spline s solving the smoothing problem (3.4) as w is varied. As $w \to 0$, problem (3.4) approaches the problem of minimizing the functional $J_2(f)$, and the spline s_w converges to the unique tensor polynomial $p \in P_{m,\tilde{m}}$ which fits the data best in the L_2-norm. Here

$$(3.19) \qquad P_{m,\tilde{m}} = \{p(x,y) = \sum_{i=1}^{m} \sum_{j=1}^{\tilde{m}} c_{ij} \, x^{i-1} y^{j-1}\}.$$

On the other hand, as $w \to \infty$, s_w converges to the tensor-natural interpolating spline discussed in §2. The behavior of the error $E(s_w)$ is also of interest. As $w \to 0$, $E(s_w)$ converges to $E(s_0)$, the error in the least square polynomial fit. As $w \to \infty$, $E(s_w)$ converges to zero. Clearly $E(s_w)$ is a continuous function of w.

§4. <u>Choosing the smoothing parameter w.</u>

The amount of smoothing produced by the method of §3 is controlled by the parameter w; the larger w is chosen, the tighter the fit. In practice it may be difficult to estimate a reasonable value for w without some trial and error. In this section we discuss an iterative method for selecting w to give a predetermined tightness of fit.

Let $0 < \varepsilon < E(0)$ be given. Because of the properties of $E(w)$ discussed at the end of the previous section, it is clear that there exists a unique $w^* > 0$ such that

$$(4.1) \qquad\qquad E(s_{w^*}) = \varepsilon$$

where s_{w^*} is the solution of problem (3.4) corresponding to w^*. Since (4.1) is a nonlinear equation, to solve it we use an iterative procedure.

Instead of solving (4.1) directly, it is equivalent to find $w*$ satisfying

$$(4.2) \qquad \frac{1}{F(w*)} = \frac{1}{\sqrt{\epsilon}} \ , \quad F(w) = \sqrt{E(s_w)} \ .$$

Now starting with an initial guess w^0, we can compute improved estimates by Newton's method which in this case reads

$$w^{i+1} = w^i + \frac{F^2(w^i) \ (\sqrt{\epsilon} - F(w^i))}{\sqrt{\epsilon} \ F(w^i) \ F'(w^i)} \ .$$

The quantity $F^2(w^i) = E(s_{w^i})$ can be computed directly once we have constructed the spline s_{w^i}. We now show how $F(w^i) \ F'(w^i)$ can be efficiently computed.

First we need a lemma on differentiating matrix expressions.

<u>Lemma 4.1.</u> Let B and E be $n \times n$ matrices, z an n-vector and w real. Then

$$(4.4) \qquad \frac{d}{dw} (B+w^{-1}E)^{-1}z = (B+w^{-1}E)^{-1}w^{-2}E(B+w^{-1}E)^{-1}z.$$

<u>Proof:</u> Let $g(w) = (B+w^{-1}E)^{-1}z$. Then $(B+w^{-1}E)g = z$. Differentiating with respect to w yields $(B+w^{-1}E)g' - w^{-2}Eg = 0$. Substituting g back into this expression gives (4.4). ∎

<u>Lemma 4.2.</u> Given $w > 0$, let s_w be the smoothing spline of Theorem 3.3. Let $F^2(w) = E(s_w)$. Then

$$(4.5) \qquad F(w)F'(w) = \sum_{i=1}^{n} \sum_{j=1}^{\tilde{n}} w_i \tilde{w}_j \ \Phi_{ij} \ \Phi_{ij}' \ ,$$

where

$$(4.6) \qquad \Phi = B(B+w^{-1}E)^{-1}z(\tilde{B}+w^{-1}\tilde{E})^{-T}\tilde{B}^T - z$$

and

176

$$(4.7) \qquad \Phi' = w^{-2}[B(B+w^{-1}E)^{-1}Z(\tilde{B}+w^{-1}\tilde{E})^{-T}\tilde{E}^T(\tilde{B}+w^{-1}\tilde{E})^{-1}\tilde{B}^T$$

$$+ B(B+w^{-1}E)^{-1}E(B+w^{-1}E)^{-1}Z(\tilde{B}+w^{-1}\tilde{E})^{-T}\tilde{B}^T].$$

<u>Proof</u>: By (3.9), $C_w = (B+w^{-1}E)^1 Z(\tilde{B}+\bar{w}^1 \tilde{E})^T$. Then $s_w(x_i,\tilde{x}_j)-z_{ij} = \Phi_{ij}$ all $1 \leq i \leq n$, $1 \leq j \leq \tilde{n}$, and thus

$$F^2(w) = \sum_{i=1}^{n} \sum_{j=1}^{\tilde{n}} w_i\tilde{w}_j \, \Phi_{ij}^2(w).$$

Then

$$2F(w)F'(w) = 2 \sum_{i=1}^{n} \sum_{j=1}^{\tilde{n}} w_i\tilde{w}_j \, \Phi_{ij} \, \Phi'_{ij} \, ,$$

where Φ and Φ' are given in (4.6) and (4.7), respectively. $\blacksquare$

While the expression (4.5) looks quite formidable, it can, in fact, be computed rather efficiently. To compute Φ_{ij} we simply evaluate the spline s_w at the point $(x_i,\tilde{x}_j)$ and subtract z_{ij}. We now consider computing Φ'_{ij}. The matrices $(B+w^{-1}E)$ and $(\tilde{B}+w^{-1}\tilde{E})$ have already been decomposed in order to solve (3.9). Using them we solve the equations

$$(4.8) \qquad (B+w^{-1}E) \, U = Z$$

and

$$(4.9) \qquad (\tilde{B}+w^{-1}\tilde{E}) \, V = Z^T.$$

Let

$$(4.10) \qquad R_1 = Z - BU$$

$$(4.11) \qquad R_2^T = Z^T - \tilde{B} \, V \, .$$

The quantities BU and $\tilde{B}V$ can be computed by evaluating splines with coefficients U and V, respectively. Now

$$(B+w^{-1}E)(B+w^{-1}E)^{-1}Z = Z$$

implies

$$w^{-1} E(B+w^{-1}E)^{-1}Z = Z - B(B+w^{-1}E)^{-1}Z$$
$$= Z - BU = R_1.$$

Similarly,

$$w^{-1}\tilde{E}(\tilde{B}+w^{-1}\tilde{E})^{-1} Z^T = Z^T - \tilde{B}V = R_2^T .$$

Next we solve

$$(4.12) \qquad (B+w^{-1}E) \; C_1 \; (\tilde{B}+w^{-1}\tilde{E})^T = R_1$$

and

$$(4.13) \qquad (B+w^{-1}E) \; C_2 \; (\tilde{B}+w^{-1}\tilde{E})^T = R_2 .$$

Then

$$\Phi' = w^{-1}B \; [C_1+C_2] \; \tilde{B}^T,$$

and this expression can be evaluated by computing the values of the spline with coefficients $C_1 + C_2$.

§5. Remarks

1. The methods presented here for interpolation and smoothing of data on a grid have been programmed in FORTRAN and tested on a wide variety of test data.

2. Tensor methods of the type discussed in this paper cannot be applied to the problem of interpolating and smoothing data which is given at points scattered in the plane. For a review of several methods for the scattered case see [14] and references therein.

3. In this paper we have dealt with natural splines. As is well-known, the use of natural splines results in some loss of accuracy near the boundary when fitting smooth functions. When derivative information is available at the grid points along the sides of the rectangle H, it is possible to use tensor splines which take account of this information. Methods of this type are discussed in [4].

References

1. Anselone, P. M. and Laurent, P. J.   A general method for the construction of interpolating and smoothing spline functions. Numer. Math. 12 (1968), 66-82.

2. Carasso, C. and Laurent, P. J.   On the numerical construction and the practical use of interpolating spline functions. Information Processing 68 (Proc. IFIP Congress, Edinburgh, 1968), A. J. H. Morell, ed., North Holland, Amsterdam, 1969, 86-89.

3. deBoor, Carl.   Efficient computer manipulation of tensor products, ACM Trans. Math. Software 5(1979), 173-182.

4. Hu, C. and Schumaker, L. L.   Complete spline smoothing, CAT Report #65, Texas A&M University, 1984.

5. Larkin, F. M.   Optimal estimation of bounded linear Functionals from noisy data.   IFIP Congress, 1971, TA-1-89-98.

6. Lyche, T. and Schumaker, L. L.   Computation of smoothing and interpolating natural splines via local bases, SIAM J. Numer. Anal. 10 (1973), 1027-1038.

7. Lyche, Tom, K. Sepehrnoori, and L.L. Schumaker, Fortran subroutines for computing smoothing and interpolating natural splines, Adv. Eng. Software 5(1983), 2 - 5.

8. Mehlum, Even.   A curve fitting method based on a variational criterion, BIT, 4(1964), 213-223.

9. Nielson, G., Smoothing and interpolating splines, PhD dissertation, Univ. of Utah, 1972.

10. Reinsch, C. H., Smoothing by spline functions.   Numer. Math., 10(1967), 177-183.

11. Reinsch, C.H., Smoothing by spline functions II, Numer.Math.16(1971), 451-454.

12. Schoenberg, I. J.   Spline functions and the problem of graduation. Proc. Nat. Acad. Sci. U.S.A., 52(1964), 947-950.

13. Schumaker, L. L.   Some algorithms for the computation of interpolating and approximating spline functions. Theory and Application of Spline Functions. T. N. E. Greville, ed., Academic Press, New York, 1969, 87-102.

14. Schumaker, L.L., Fitting surfaces to scattered data, in <u>Approximation Theory II</u>, G.G. Lorentz, C.K. Chui, and L.L. Schumaker, eds, Academic Press, New York, 1976, 203 - 268.

15. Woodford, C. H. An algorithm for data smoothing using spline functions, BIT $\underline{10}$(1971), 501-510.

Chui Li Hu
Larry L. Schumaker
Center for Approximation Theory
Texas A&M University
College Station, Texas 77843

Supported in part by contract NAS9-16664

International Series of
Numerical Mathematics, Vol. 74
© 1985 Birkhäuser Verlag Basel

BEST APPROXIMATION BY SPLINE FUNCTIONS :

THEORY AND NUMERICAL METHODS

Günther Nürnberger

Fakultät für Mathematik und Informatik
Universität Mannheim
6800 Mannheim, Federal Republic of Germany

ABSTRACT

A brief survey of best approximation by splines with fixed knots in various norms with special emphasis on numerical methods is given. In addition, an algorithm for computing (continuous) piecewise polynomials with free knots is discussed. Most of the results were proved in the last decade.

INTRODUCTION. A fundamental problem in applied mathematics is to

approximate a given function by functions of a simple form which

can be easily evaluated on a computer. Two function classes which

have these properties play an important role in numerical analy-

sis: the space

$$\Pi_m = \left\{ \sum_{i=0}^{m} a_i t^i : a_0, \ldots, a_m \ \text{real} \right\}$$

of polynomials of degree m, and, more general, the space

$$S_m(x_1, \ldots, x_k) = \{ s \in C^{m-1}[a,b] : s|_{[x_i, x_{i+1}]} \in \Pi_m , i=0, \ldots, k \}$$

of splines of degree m with k fixed knots $a=x_0<x_1<\ldots<x_k<x_{k+1}=b$.

The problem of best approximation with respect to a given norm $\|\cdot\|$ can be stated as follows. Determine for a given function $f \in C[a,b]$ a spline $s_f \in S_m(x_1,\ldots,x_k)$ such that

$$\| f - s_f \| = \min\{ \| f - s \| : s \in S_m(x_1,\ldots,x_k) \} \ .$$

The spline s_f is called best approximation of f from $S_m(x_1,\ldots,x_k)$.

In the middle of the 19 th century P.L. Chebyshev was the first who considered best approximation by polynomials systematically in connection with a mechanical problem. This domain is intensively treated in any classical book on approximation theory listed up in the references.

The first systematic study of best approximation by splines with fixed knots goes back to Rice [1967] and Schumaker [1968 a] . Since then, various problems concerning characterization, uniqueness, continuity and computation in best spline approximation were completely solved. But still several deep problems, in particular for splines with free knots and multivariate splines remained open up to the present.

In this manuscript a brief survey of the theory and numerical methods in best approximation by splines with fixed knots is given. Moreover, an algorithm for piecewise polynomials with free knots is described. On the other hand, we will not consider best approximation in the free knot and multivariate case. These domains are not fully developed at present.

Section 1 deals with characterization, uniqueness and continuity in uniform approximation. In section 2 a Remez type algorithm for computing spline approximations in the uniform norm is described. Section 3 is devoted to uniqueness and interpolation at canonical points in L_1-approximation. In section 4 uniqueness in one-sided L_1-approximation and the relationship to Gauß quadrature formulas are considered. Finally, in section 5 an algorithm for (continuous) piecewise polynomials with free knots is discussed. We close this section with a general remark on the relationship of interpolation and best approximation. A relatively simple method for computing spline approximations is the interpolation of functions which suffices for several practical purposes. On the other hand at present only a few results on the comparison of the error for Lagrange interpolation with splines of arbitrary knot location and the minimal deviation are available. For special functions interpolation may even fail numerically. The reader is refered to the book of de Boor [1978], where many numerical examples of interpolation are given. According to our experience the computation of best spline approximations with fixed knots and (continuous) piecewise polynomials with free knots is a practicable method on a modern computer. Since the fixed knots can be chosen arbitrarily, this may be important for developing algorithms for (differentiable) splines with free knots.

1. <u>UNIFORM APPROXIMATION</u>. The space of all r-times continuously differentiable real-valued functions on an interval [a,b] is denoted by $C^r[a,b]$.

The results in this paper also hold for splines with multiple knots, however for the sake of brevity we only consider the case of simple knots.

Let points $a = x_0 \leq x_1 \leq \ldots \leq x_k \leq x_{k+1} = b$ be given. For every integer $i \in \{1,\ldots,k\}$ let $1 \leq m_i \leq m$ be the number of knots which coincide with x_i (e.g. if $x_{i-1} < x_i = x_{i+1} < x_{i+2}$, then $m_i = 2$). We call

$$S_m(x_1,\ldots,x_k) = \{s \in C[a,b] : s\big|_{[x_i,x_{i+1}]} \in \Pi_m , i=0,\ldots,k,$$

and s has $m-m_i$ continuous derivatives at x_i, $i=1,\ldots,k\}$

the space of <u>splines of degree m with knots $x_1,\ldots,x_k$ of multiplicities $m_1,\ldots,m_k$</u>. If $m_1 = \ldots = m_k = 1$, then the space of splines with simple knots is obtained.

More general, the results in this paper, stated for simple knots, also hold for Chebyshev splines (for the definition see Schumaker [1981]).

Best approximation by splines can be considered in various norms. Let a function $h \in C[a,b]$ be given. The <u>uniform norm</u> of h is defined by

$$\| h \|_\infty = \sup\{|h(t)| : t \in [a,b]\}$$

and for $1 \leq p < \infty$ the <u>L_p-norm</u> is defined by

$$\| h \|_p = \left(\int_a^b |h(t)|^p dt \right)^{1/p}$$

Best approximations with respect to the uniform norm (respectively L_p-norm) are called <u>best uniform approximations</u> (respectively <u>best L_p-approximations</u>).

It is well-known and easy to verify that best approximations from a given finite-dimensional subspace always exist.

In the following we consider characterization and uniqueness of best spline approximations in the uniform norm, where alternation properties of the error play an important role.

Given $h \in C[a,b]$, we call points $a \leq t_1 < \ldots < t_r \leq b$ <u>alternating extreme points</u> of h, if there exists a sign $\sigma \in \{-1,1\}$ such that $\sigma(-1)^i h(t_i) = \| h \|_\infty, i=1,\ldots,r$. If I is a subinterval of $[a,b]$, then we denote by $A_I(h)$ the maximal number of alternating extreme points of h in I.

In the following results let a function $f \in C[a,b]$ and a spline $s_f \in S_m(x_1,\ldots,x_k)$ be given.

We begin with a characterization of best spline approximations which was independently proved in Rice [1967] and Schumaker [1968 a]. Extensions of this result were given in Sommer [1980 a] and Nürnberger, Schumaker, Sommer and Strauß [1984 b].

<u>THEOREM 1.1.</u> The following statements are equivalent:
(i) The spline s_f is a best uniform approximation of f from $S_m(x_1,\ldots,x_k)$.
(ii) There exists an interval $[x_p,x_{p+q}] \subset [a,b]$, $q \geq 1$, such that $A_{[x_p,x_{p+q}]}(f-s_f) \geq q+m+1$.

If $\{s_1,\ldots,s_{k+m+1}\}$ is a basis of $S_m(x_1,\ldots,x_k)$, then a subset $\{t_1,\ldots,t_{k+m+1}\}$ of $[a,b]$ is called <u>poised</u> with respect to $S_m(x_1,\ldots,x_k)$, if $\det(s_i(t_j))_{i,j=1}^{k+m+1} \neq 0$.

In the following let additional knots $x_{-m} < \ldots < x_{-1} < a$ and $b < x_{k+2} < \ldots < x_{k+m+1}$ be given.

In contrast to the classical case of polynomials, best uniform spline approximations are not always unique. The reason is that sets may not be poised. The following theorem of Schoenberg & Whitney [1953] characterizes poised sets. Extensions of this results were given in Nürnberger, Schumaker, Sommer & Strauß [1983], [1984 a].

__THEOREM 1.2.__ A set $T = \{t_1,\ldots,t_{k+m+1}\}$ with $a \le t_1 < \ldots < t_{k+m+1} \le b$ is poised with respect to $S_m(x_1,\ldots,x_k)$ if and only if every interval $(x_i,x_{i+m+j}) \subset (x_{-m},x_{k+m+1})$, $j \ge 1$, contains at least j points of T.

The notion of strong uniqueness goes back to Newman & Shapiro [1963] and plays an important role in the computation of best approximations (compare section 2). A spline is called __strongly unique best approximation__ of f from $S_m(x_1,\ldots,x_k)$, if there exists a constant $K_f > 0$ such that for all $s \in S_m(x_1,\ldots,x_k)$,

$$\| f - s \|_\infty \ge \| f - s_f \|_\infty + K_f \| s - s_f \|_\infty \ .$$

The __strong unicity constant__ $K(f)$ of f is defined to be the maximum of all such numbers K_f.

Schaback [1978] gave a sufficient condition for strong uniqueness. The following characterization proved in Nürnberger [1982 a] was deduced from a general theorem for weak Chebyshev spaces. Extensions of this result were given in Nürnberger, Schumaker, Sommer and Strauß [1984 b].

__THEOREM 1.3.__ The following statements are equivalent:
(i) The spline s_f is a strongly unique best uniform approximation of f from $S_m(x_1,\ldots,x_k)$.
(ii) For every interval $(x_i,x_{i+m+j}) \subset (x_{-m},x_{k+m+1})$, $j \ge 1$, we have
$A_{(x_i,x_{i+m+j})}(f - s_f) \ge j+1 \ .$

A function $h \in C[a,b]$ is called <u>flat of order m from the left</u> (respectively <u>from the right</u>) at a given point $t_0 \in (a,b)$, if

$$\liminf_{\substack{t \to t_0 \\ t < t_0}} \frac{|h(t_0) - h(t)|}{|t_0 - t|^m} = 0$$

(respectively

$$\liminf_{\substack{t \to t_0 \\ t > t_0}} \frac{|h(t_0) - h(t)|}{|t_0 - t|^m} = 0 \;) \; .$$

In general, unique and strongly unique best approximations are not the same. Building on earlier work of Rice [1967], Schumaker [1968 a] and Strauß [1975 b], the following characterization of uniqueness was proved in Nürnberger & Singer [1982]. For extensions of this result see Nürnberger, Schumaker, Sommer and Strauß [1984 b].

<u>THEOREM 1.4.</u> The following statements are equivalent:

(i) The spline s_f is a unique best uniform approximation of f from $S_m(x_1, \ldots, x_k)$.

(ii) The following three conditions hold:

(a) For every interval $[x_i, x_{i+m+j}] \subset [x_{-m}, x_{k+m+1}]$, $j \geq 1$, we have $A_{[x_i, x_{i+m+j}]}(f - s_f) \geq j+1$.

(b) If $A_{(x_i, x_{i+m+j}]}(f - s_f) = j$ (respectively $A_{[x_i, x_{i+m+j})}(f - s_f) = j$), then $f - s_f$ is flat of order m from the right at x_i (respectively from the left at x_{i+m+j}).

(c) If $A_{(x_i, x_{i+m+j})}(f - s_f) = j$, then $f - s_f$ is flat of order m from the right at x_i or from the left at x_{i+m+j}.

Since in practical computations functions are only given up to some error, this leads to a question of stability: what are the functions from the interior (briefly denoted by int) of

$$SU(S_m(x_1,\ldots,x_k)) = \{f \in C[a,b] : f \text{ has a strongly unique best}$$
$$\text{uniform approximation from } S_m(x_1,\ldots,x_k)\}$$

and

$$U(S_m(x_1,\ldots,x_k)) = \{f \in C[a,b] : f \text{ has a unique best uniform}$$
$$\text{approximation from } S_m(x_1,\ldots,x_k)\} .$$

The next theorem proved in Nürnberger [1983 a] characterizes those functions. Related results on optimization were obtained in Nürnberger [1985].

THEOREM 1.5. If $s_f \in S_m(x_1,\ldots,x_k)$ is a best uniform approximation of $f \in C[a,b]$, then the following statements are equivalent:

(i) $f \in \text{int } SU(S_m(x_1,\ldots,x_k))$.

(ii) $f \in \text{int } U(S_m(x_1,\ldots,x_k))$.

(iii) $A_{[a,b]}(f - s_f) \geq k+m+2$ and for every $[x_p,x_{p+q}] \subsetneqq [a,b]$. $q \geq 1$, we have $A_{[x_p,x_{p+q}]}(f - s_f) < q+m+1$.

A subspace $G = \text{span}\{g_1,\ldots,g_n\}$ of $C[a,b]$ is called <u>weak Chebyshev</u>, if for every set $\{t_1,\ldots,t_n\}$ with $a \leq t_1 < \ldots < t_n \leq b$, $\det(g_i(t_j))_{i,j=1}^n$ is of the same sign. It is well-known that the space $S_m(x_1,\ldots,x_k)$ has this property (see e.g. Schumaker [1981]). We call

$$K(S_m(x_1,\ldots,x_k)) = \{f \in C[a,b] : \text{span}(S_m(x_1,\ldots,x_k) \cup \{f\}) \text{ is weak}$$
$$\text{Chebyshev}\}$$

the <u>convexity cone</u> of $S_m(x_1,\ldots,x_k)$.

It follows from results of Micchelli [1977] and Zwick [1984 b] that if $f \in C^{m+1}[a,b]$, then $f \in K(S_m(x_1,\ldots,x_k))$ if and only if there exists a sign $\sigma \in \{-1,1\}$ such that

$$\sigma(-1)^i f^{(m+1)}(t) \geq 0 \quad , \quad t \in [x_i,x_{i+1}] \quad , \quad i=0,\ldots,k \quad .$$

By using the above theorems Zwick [1984 a] proved the following uniqueness result for functions from the convexity cone.

__THEOREM 1.6.__ If $m \geq 2$, then $K(S_m(x_1,\ldots,x_k)) \cap C^1[a,b]$ is a subset of int $SU(S_m(x_1,\ldots,x_k))$.

A general formula for computing the strong unicity constant for splines was proved in Nürnberger [1982 b] which we formulate in a special case. Extensions were obtained in Nürnberger [1983 b], where numerical examples are given. Earlier results on the strong unicity constant for polynomials were given in Cline [1973], Henry & Roulier [1978], Schmidt [1980] and Blatt [1982].

__THEOREM 1.7.__ Let $s_f \in S_m(x_1,\ldots,x_k)$ be a strongly unique best uniform approximation of $f \in C[a,b]$ such that $f - s_f$ has exactly $k+m+2$ extreme points $a \leq t_1 < \ldots < t_{k+m+2} \leq b$. For each $j \in \{1,\ldots,m+k+2\}$ let $s_j \in S_m(x_1,\ldots,x_k)$ be the unique function with

$$s_j(t_i) = \text{sgn}(f(t_i) - s_f(t_i)) \quad , \quad i=1,\ldots,k+m+2 \quad , \quad i \neq j \quad .$$

Then we have

$$K(f) = \min\{1/\|s_j\|_\infty : j=1,\ldots,k+m+2\} \quad .$$

The mapping $P_{S_m(x_1,\ldots,x_k)}$ which associates to each $f \in C[a,b]$ the set of its best uniform approximations from $S_m(x_1,\ldots,x_k)$ is called __metric projection__.

Since $P_{S_m(x_1,\ldots,x_k)}$ is a set-valued mapping we have to say what we mean by continuity. Three different continuity concepts are considered. The metric projection $P_{S_m(x_1,\ldots,x_k)}$ is called <u>upper semicontinuous</u> (respectively <u>lower semicontinuous</u>), if the set $\{f \in C[a,b] : P_{S_m(x_1,\ldots,x_k)}(f) \cap A \neq \emptyset\}$ is closed (respectively open) for every closed (respectively open) subset A of $S_m(x_1,\ldots,x_k)$. A continuous mapping $F : C[a,b] \rightarrow S_m(x_1,\ldots,x_k)$ is called <u>continuous selection</u> for $P_{S_m(x_1,\ldots,x_k)}$, if

$$F(f) \in P_{S_m(x_1,\ldots,x_k)}(f) \quad \text{for all } f \in C[a,b].$$

By a result of Singer [1970] the metric projection onto any finite-dimensional subspace is upper semicontinuous.

The next theorem follows from a result on the lower semicontinuity of the metric projection for arbitrary finite-dimensional subspaces due to Blatter, Morris & Wulbert [1968] (see also Brosowski & Wegmann [1973]).

<u>THEOREM 1.8.</u> The metric projection $P_{S_m(x_1,\ldots,x_k)}$ is not lower semicontinuous.

The following result on continuous selections for the spline metric projection was proved in Nürnberger & Sommer [1978 b]. In a series of papers these authors gave a complete characterization of continuous selections for the metric projection onto arbitrary finite-dimensional subspaces (see the survey of Nürnberger & Sommer [1985], where also remarks on the relationship of continuous selections and the convergence of algorithms are given). For additional results on continuous selections see the survey in Deutsch [1983].

$\underline{\text{THEOREM 1.9.}}$ There exists a continuous selection for $P_{S_m(x_1,\ldots,x_k)}$ if and only if $k \leq m+1$.

Blatt, Nürnberger & Sommer [1981] showed that the selection F constructed in the proof of Theorem 1.9 for $k \leq m+1$ is even $\underline{\text{point-wise-Lipschitz-continuous}}$ (i.e. for each $f \in C[a,b]$ there exists a constant $C_f > 0$ such that for all $\tilde{f} \in C[a,b]$,

$$\| F(f) - F(\tilde{f}) \|_\infty \leq C_f \| f - \tilde{f} \|_\infty).$$

Blatter & Schumaker [1982], [1983] proved that continuous selections for $P_{S_m(x_1,\ldots,x_k)}$ are never unique.

We close this section with a theorem on the approximation power of splines. This subject is intensively treated in Schumaker [1981], where many historical notes are given (see also the references). The $\underline{\text{modulus of continuity}}$ of a function $h \in C[a,b]$ is defined by

$$\omega(h;\delta) = \sup\{ |h(t_1) - h(t_2)| : t_1,t_2 \in [a,b], \ |t_1 - t_2| \leq \delta \},$$

where $\delta > 0$.

$\underline{\text{THEOREM 1.10.}}$ Let $m \geq 1$ and $j \in \{0,\ldots,m\}$ be given. Then there exists a constant $K > 0$ (depending only on m and j) such that for every $f \in C^j[a,b]$ and every set of knots $\{x_1,\ldots,x_k\}$ we have

$$d(f,S_m(x_1,\ldots,x_k)) \leq K\delta^j \omega(f^{(j)};\delta) \quad ,$$

where $\delta = \max\{x_{i+1} - x_i : i=0,\ldots,k\}$. (Note that

$$\omega(f^{(j)};\delta) \leq \delta \| f^{(j+1)} \|_\infty,$$

if $f \in C^{j+1}[a,b]$).

$\underline{\text{2. REMEZ TYPE ALGORITHM.}}$ In this section a brief description of a Remez type algorithm for computing best uniform spline appro-

ximations which was developed in Nürnberger & Sommer [1983 a] ,
[1983 b] is given. Already Schumaker [1969 a] observed that the
idea of the classical Remez algorithm for Haar spaces can also
be used for spline spaces. Esch & Eastman [1969] used an optimization approach for computing spline approximations. An algorithm
for strict spline approximations was developed in Strauß [1984 b],
[1984 c] .

We first need some results on nonzero determinants. Let
$\{g_1, \ldots, g_N\}$ be a basis of $S_m(x_1, \ldots, x_k)$, where $N = k + m + 1$. If
$M = \{t_1, \ldots, t_{N+1}\}$ is a subset of $[a,b]$, then we set

$$D(M) = \begin{vmatrix} g_1(t_1) & \cdots & g_N(t_1) & (-1)^1 \\ \vdots & & \vdots & \vdots \\ g_1(t_{N+1}) & \cdots & g_N(t_{N+1}) & (-1)^{N+1} \end{vmatrix} .$$

LEMMA 2.1. For a set $M = \{t_1, \ldots, t_{N+1}\}$ with $a \le t_1 < \ldots < t_{N+1} \le b$
the following conditions are equivalent:

(i) $D(M) \ne 0$.

(ii) Every interval $(x_i, x_{i+m+j}) \subset (x_{-m}, x_{k+m+1})$, $j \ge 1$, contains at
least j points from M.

REMARK 2.2. By using Lemma 2.1 Nürnberger & Sommer [1983 a]
showed that for each set M with $D(M) \ne 0$ there exists a unique
maximal interval $I_M = [x_p, x_{p+q}] \subset [a,b]$, $q \ge 1$, such that every
interval $(x_i, x_{i+m+j}) \subset (x_{p-m}, x_{p+q+m})$, $j \ge 1$, contains at least $j + 1$
points from $M \cap I_M$. Moreover, if a point from $M \cap I_M$ is replaced
by an arbitrary point in $[a,b]$, then $D(\tilde{M}) \ne 0$ for the resulting
set $\tilde{M}$.

We now briefly describe the algorithm. Let a function $f \in C[a,b] \smallsetminus S_m(x_1,\ldots,x_k)$ be given. We choose a sufficiently small $\varepsilon > 0$ and set

$$I_\varepsilon = [a,b] \smallsetminus \bigcup_{i=1}^{k} \left[(x_i - \varepsilon, x_i) \cup (x_i, x_i + \varepsilon) \right] .$$

We only work on I_ε in order to get a stable algorithm. Then we select a set $M_1 = \{t_{1,1}, \ldots, t_{N+1,1}\}$ with $a \leq t_{1,1} < \ldots < t_{N+1,1} \leq b$ and $D(M_1) \neq 0$. Since $D(M_1) \neq 0$, there exists a unique $s_1 \in S_m(x_1,\ldots,x_k)$ and a unique $\lambda_1 \in \mathbb{R}$ such that

$$(-1)^i (f(t_{i,1}) - s_1(t_{i,1})) = \lambda_1 , \quad i=1,\ldots,N+1 .$$

For $n \geq 1$ we proceed by induction as follows. We choose a point $t_n \in I_\varepsilon$ with $|f(t_n) - s_n(t_n)| \approx \||(f - s_n)|_{I_\varepsilon}\|_\infty$ and replace a point from M_n by t_n such that a new set $M_{n+1} = \{t_{1,n}, \ldots, t_{N+1,n}\}$ with $a \leq t_{1,n} < \ldots < t_{N+1,n} \leq b$ and $D(M_{m+1}) \neq 0$ is obtained. (It can be shown by easily constructed examples that the exchange of the classical Remez algorithm in general does not yield this property.)

Our rule of exchange is described as follows. Let I_{M_n} be the unique interval from Remark 2.2 and let $t_{p_n,n}$ (respectively $t_{q_n,n}$) be the first (respectively last) point from $M_n \cap I_{M_n}$. Moreover, we set $t_{0,n} = -\infty$ and $t_{N+2,n} = \infty$. Then there exists an integer $j \in \{0,\ldots,N+1\}$ such that $t_{j,n} < t_n < t_{j+1,n}$.

CASE 1. $j \in \{1,\ldots,N\}$.

If $\mathrm{sgn}(f(t_{j,n}) - s_n(t_{j,n})) = \mathrm{sgn}(f(t_n) - s_n(t_n))$ (respectively $\mathrm{sgn}(f(t_{j+1,n}) - s_n(t_{j+1,n})) = \mathrm{sgn}(f(t_n) - s_n(t_n))$), then we replace $t_{j,n}$ (respectively $t_{j+1,n}$) by t_n, if $D(M_{n+1}) \neq 0$ for the resulting set M_{n+1}. Otherwise, we replace $t_{p_n,n}$ (respectively $t_{q_n,n}$)

by t_n, if $t_n > t_{q_n,n}$ (respectively $t_n < t_{p_n,n}$). It follows from Remark 2.2 that $D(M_{n+1}) \neq 0$.

CASE 2. $j = 0$.

If $\text{sgn}(f(t_{1,n}) - s_n(t_{1,n})) = \text{sgn}(f(t_n) - s_n(t_n))$, then we replace $t_{1,n}$ by t_n. Otherwise, we replace $t_{q_n,n}$ by t_n. It follows from Remark 2.2 that $D(M_{n+1}) \neq 0$.

CASE 3. $j = N+1$.

If $\text{sgn}(f(t_{N+1,n}) - s_n(t_{N+1,n})) = \text{sgn}(f(t_n) - s_n(t_n))$, then we replace $t_{N+1,n}$ by t_n. Otherwise, we replace $t_{p_n,n}$ by t_n. It follows from Remark 2.2 that $D(M_{n+1}) \neq 0$.

Since in all cases $D(M_{n+1}) \neq 0$, there exists a unique $s_{n+1} \in S_m(x_1,\ldots,x_k)$ and a unique $\lambda_{n+1} \in \mathbb{R}$ such that

$$(-1)^i (f(t_{i,n+1}) - s_{n+1}(t_{i,n+1})) = \lambda_{n+1} , \quad i=1,\ldots,N+1 .$$

Moreover, the following rule should be used. It can be shown that after finitely many steps $|\lambda_n| \approx \| (f - s_n)|_{I_\varepsilon} \|_\infty$. From this point on the almost extreme point t_n of $f - s_n$ should be only taken from small neighborhoods of the points $t_{i,n}$, $i=1,\ldots,N+1$, such that the position of the resulting points with respect to the knot intervals is not changed.

The following convergence result holds. Since the above sequences depend on I_ε we set $(\lambda_n^{(\varepsilon)}) = (\lambda_n)$ and $(s_n^{(\varepsilon)}) = (s_n)$.

THEOREM 2.3. (i) $\lambda_n^{(\varepsilon)} \uparrow d(f, S_m(x_1,\ldots,x_k))$ for $n \to \infty$.
(ii) $s_n^{(\varepsilon)} \to s_f^{(\varepsilon)}$ for $n \to \infty$, where $s_f^{(\varepsilon)} \in S_m(x_1,\ldots,x_k)$ is a best uniform approximation of f on I_ε.

(iii) $\| f - s^{(\varepsilon)} \|_\infty \to d(f, S_m(x_1, \ldots, x_k))$ for $\varepsilon \to 0$.

(iv) If f has a strongly unique best uniform approximation $s_f \in S_m(x_1, \ldots, x_k)$, then $s_n^{(\varepsilon)} \to s_f$ for $n \to \infty$.

This result shows that, although we work on I_ε (with fixed $\varepsilon > 0$), for practical purposes the algorithm actually yields a best approximation of f from $S_m(x_1, \ldots, x_k)$ on the whole interval $[a,b]$. Nürnberger [1983 a] observed that the convergence is quadratic, if, roughly speaking, $f \in \text{int } SU(S_m(x_1, \ldots, x_k))$ and the error $f - s_f$ has exactly N+1 extreme points.

In general, to increase the speed of convergence it is desirable to replace simultaneously all points from M_n by almost extreme points of $f - s_n$.

We stop the algorithm, if

$$\| f - s_n \|_\infty - d(f, S_m(x_1, \ldots, x_k)) \leq \| f - s_n \|_\infty - \lambda_n \leq \varepsilon_0 \ ,$$

where ε_0 is the desired accuracy. Moreover, if f has a strongly unique best approximation s_f, then we get the following error estimation:

$$\| s_n - s_f \|_\infty \leq \frac{1}{K(f)} \Big(\| f - s_n \|_\infty - \| f - s_f \|_\infty \Big)$$

$$\leq \frac{1}{K(f)} \Big(\| f - s_n \|_\infty - \lambda_n \Big) \leq \frac{\varepsilon_0}{K(f)} \ ,$$

where $K(f)$ is the strong unicity constant of f.

When we tested the algorithm by several numerical examples we found out that most of the standard functions are from $\text{int } SU(S_m(x_1, \ldots, x_k))$.

3. $\underline{L_1\text{-APPROXIMATION.}}$ In this section we consider uniqueness and determination of best spline approximations by interpolation in

the L_1-norm.

We first note that it is well-known and easy to verify that for $1 < p < \infty$ best L_p-approximations from an arbitrary finite-dimensional subspace are always unique (see e.g. Singer [1970]).

Independently, Galkin [1974] and Strauß [1975 a] proved the following uniqueness theorem for L_1-approximations. Extensions of this result were given in Carroll & Braess [1974], Sommer [1979], [1983 a], Strauß [1981], Nürnberger, Schumaker, Sommer & Strauß [1984 b] and Kroó [1985].

<u>THEOREM 3.1.</u> For every $f \in C[a,b]$ there exists a unique best L_1-approximation from $S_m(x_1,\ldots,x_k)$.

In the following we show that best L_1-approximations for functions from the convexity cone can be obtained by interpolation at certain points. If $1 \le r \le k+m+1$, then we call points $a < t_1 < \ldots < t_r < b$ <u>canonical points</u> of $S_m(x_1,\ldots,x_k)$, if

$$\sum_{i=1}^{r} (-1)^i \int_{t_i}^{t_{i+1}} s(t)\,dt = 0 \quad \text{for all } s \in S_m(x_1,\ldots,x_k),$$

where $t_0 = a$ and $t_{r+1} = b$.

The next theorem on uniqueness and poisedness of canonical points was proved in Micchelli [1977]. Extensions of this result were obtained in Sommer [1979] and Nürnberger, Schumaker, Sommer & Strauß [1984 b].

<u>THEOREM 3.2.</u> For the space $S_m(x_1,\ldots,x_k)$ there exists a unique set of $k+m+1$ canonical points which is poised with respect to $S_m(x_1,\ldots,x_k)$.

The following theorem on the determination of best L_1-approximations by interpolation is due to Micchelli [1977]. Extensions of this result were proved in Sommer [1979] and Nürnberger, Schumaker, Sommer & Strauß [1984 b].

__THEOREM 3.3.__ For every $f \in K(S_m(x_1,\ldots,x_k))$ the best L_1-approximation $s_f \in S_m(x_1,\ldots,x_k)$ is uniquely determined by interpolation at the set of canonical points $\{t_1,\ldots,t_{k+m+1}\}$ of $S_m(x_1,\ldots,x_k)$, i.e.

$$s_f(t_i) = f(t_i) \quad , \quad i=1,\ldots,k+m+1.$$

__4. ONE-SIDED L_1-APPROXIMATION AND GAUSS QUADRATURE FORMULAS.__ In this section we consider uniqueness of best one-sided L_1-approximation from spline spaces and the relationship to interpolation and quadrature formulas.

Given $f \in C[a,b]$, we call $s_f \in S_m(x_1,\ldots,x_k)$ with $s_f \le f$ a __best one-sided L_1-approximation__ of f, if

$$\| f - s_f \|_1 = \min\{ \| f - s \|_1 : s \in S_m(x_1,\ldots,x_k) , s \le f\} \quad .$$

We begin with a non-unicity theorem due to Pinkus [1976] (see also DeVore [1968] and Strauß [1982]). Non-unicity results for arbitrary finite-dimensional subspaces were independently proved in Pinkus & Totik [1984] and Nürnberger [1984 a] , [1984 b] . The last author gave a characterization of unicity spaces by using optimization techniques.

__THEOREM 4.1.__ There exist functions in $C[a,b]$ which have more than one best one-sided L_1-approximation from $S_m(x_1,\ldots,x_k)$.

In contrast to Theorem 4.1 the following unicity theorem due to Pinkus [1976] holds. Extensions of this result were given in Sommer & Strauß [1981], Strauß [1982] and Nürnberger, Schumaker, Sommer & Strauß [1984 b].

THEOREM 4.2. If $m \geq 2$, then for every function in $C^1[a,b]$ there exists a unique best one-sided L_1-approximation from $S_m(x_1, \ldots, x_k)$.

A classical result of C.F. Gauß says that for Π_{2m+1} there exists a quadrature formula of $m+1$ points and positive weights which is exact for Π_{2m+1}. We will see that an analogous result holds for $S_m(x_1, \ldots, x_k)$.

For a subset $T = \{t_1, \ldots, t_r\}$ of $[a,b]$ the <u>index</u> of T, denoted by $I(T)$, is the number of points in T counting each point in $\{a,b\}$ as $1/2$. A quadrature formula $Q(f) = \sum_{i=1}^{r} a_i f(t_i)$, where $a \leq t_1 < \ldots < t_r \leq b$ and $a_1, \ldots, a_r > 0$ is called <u>Gauß quadrature formula</u> for $S_m(x_1, \ldots, x_k)$, if $I(\{t_1, \ldots, t_r\}) = (m+k+1)/2$ and $\int_a^b s(t)dt = Q(s)$ for all $s \in S_m(x_1, \ldots, x_k)$. We only consider the case when $t_r < b$. Analogous results hold for $t_r = b$.

Micchelli & Pinkus [1977] proved the following uniqueness theorem for Gauß quadrature formulas. For related results on quadrature formulas for Haar spaces and spline spaces see Schoenberg [1965], [1966], [1969], [1973], Karlin & Studden [1966], DeVore [1968], Karlin [1969 a] , [1971], Sommer & Strauß [1981], Strauß [1982], [1984 a] and Nürnberger, Schumaker, Sommer & Strauß [1984 b].

THEOREM 4.3. If $m \geq 2$, then there exists a unique Gauß quadrature formula for $S_m(x_1, \ldots, x_k)$.

198

The next theorem on the determination of best one-sided L_1-approximations by interpolation is due to Micchelli & Pinkus [1977]. Extensions of this result were given in Strauß [1982].

<u>THEOREM 4.4.</u> If $m \geq 2$, then for every $f \in K(S_m(x_1,\ldots,x_k)) \cap C^1[a,b]$ the best one-sided L_1-approximation $s_f \in S_m(x_1,\ldots,x_k)$ is uniquely determined by interpolation at the points $a \leq t_1 < \ldots < t_r < b$ of the Gauß quadrature formula of $S_m(x_1,\ldots,x_k)$, i.e.

$$s_f(t_i) = f(t_i) \;,\; i=1,\ldots,r \quad \text{and} \quad s_f'(t_i) = f'(t_i) \;,\; t_i \neq a \;,\; i=1,\ldots,r.$$

<u>5. ALGORITHM FOR PIECEWISE POLYNOMIALS.</u> An algorithm for computing segment approximations in various norms was developed in Nürnberger, Sommer & Strauß [1984]. We give a brief description of this method for best piecewise polynomials in the uniform norm.

We call
$$PP_{m,k} = \{s : [a,b] \to \mathbb{R} : \text{there exist knots } x_1 < \ldots < x_k \text{ in}$$
$$(a,b) \text{ such that } s\big|_{(x_i,x_{i+1})} \in \Pi_m \;,\; i=0,\ldots,k\}$$

the set of <u>piecewise polynomials</u> of degree m with k free knots

Let a function $f \in C[a,b]$ be given. A piecewise polynomial $s_f \in PP_{m,k}$ is called <u>best uniform approximation</u> of f, if $\| f - s_f \|_\infty = \min\{\| f - s \|_\infty : s \in PP_{m,k}\}$. The set of knots corresponding to s_f is called <u>optimal</u>. For a subinterval I of [a,b] we set $d(f,\Pi_m,I) = \min \{\|(f - p)\big|_I\|_\infty : p \in \Pi_m\}$. A set of knots $\{x_1,\ldots,x_k\}$ is called <u>leveled</u>, if

$$d(f,\Pi_m,[x_{i-1},x_i]) = d(f,\Pi_m,[x_i,x_{i+1}]) \;,\; i=1,\ldots,k \;.$$

The next result due to Lawson [1964] (see Meinardus [1967]) is used in the algorithm.

__THEOREM 5.1.__ (i) For every set of knots $\{x_1,\ldots,x_k\}$ we have

$$\min_{0\leq i\leq k} d(f,\Pi_m,[x_i,x_{i+1}]) \leq d(f,PP_{m,k}) \leq \max_{0\leq i\leq k} d(f,\Pi_m,[x_i,x_{i+1}]) \ .$$

(ii) There exists an optimal set of knots which is leveled. Conversely, every beveled set of knots is optimal.

We now briefly describe the algorithm. In the following we compute for every $n \geq 1$ a set of knots $\{x_{1,n},\ldots,x_{k,n}\}$ and set $d_{i,n} = d(f,\Pi_m,[x_{i,n},x_{i+1,n}])$, $i=0,\ldots,k$. In the first step we choose a set of knots $\{x_{1,1},\ldots,x_{k,1}\}$ and compute the values $d_{0,1},\ldots,d_{k,1} > 0$. This can be done by the classical Remez algorithm (see e.g. Meinardus [1967]).

We set $a_1 = \min\{d_{i,1} : i=0,\ldots,k\}$, $b_1 = \max\{d_{i,1} : i=0,\ldots,k\}$, $a_1 = 10^{\alpha_1}$ and $b_1 = 10^{\beta_1}$.

Then we proceed by induction as follows. For $n \geq 1$ we set $\delta_{n+1} = (\alpha_n + \beta_n)/2$ and $d_{n+1} = 10^{\delta_{n+1}}$. Then there exists a set of knots $\{x_{1,n+1},\ldots,x_{k,n+1}\}$ such that $d_{i,n+1} = d_{n+1}$, $i=0,\ldots,k-1$. Running from $i=1$ to $i=k$ we compute such a set of knots and set $c_{n+1} = d_{k,n+1}$. Then we set $\tilde{a}_{n+1} = \min\{d_{n+1},c_{n+1}\}$, $\tilde{b}_{n+1} = \max\{d_{n+1},c_{n+1}\}$, $a_{n+1} = \max\{a_n,\tilde{a}_{n+1}\}$, $b_{n+1} = \min\{b_n,\tilde{b}_{n+1}\}$, $a_{n+1} = 10^{\alpha_{n+1}}$ and $b_{n+1} = 10^{\beta_{n+1}}$.

A simple method to compute $x_{1,n+1}$ is to choose points y_1 and y_2 such that $\tilde{d}_1 = d(f,\Pi_m,[a,y_1]) < d_{n+1}$ and $\tilde{d}_2 = d(f,\Pi_m,[a,y_2]) > d_{n+1}$. Let $\tilde{d}_1 = 10^{\tilde{\delta}_1}$ and $\tilde{d}_2 = 10^{\tilde{\delta}_2}$. Then we use the regula falsi method for exponents $\tilde{\delta}_1$, $\tilde{\delta}_2$ and δ_{n+1} by setting

$$y_3 = y_1 + (y_2 - y_1) \frac{\delta_{n+1} - \tilde{\delta}_1}{\tilde{\delta}_2 - \tilde{\delta}_1} \ .$$

In this way a sequence (y_p) is obtained which converges to $x_{1,n+1}$.

Of course, we stop after very few steps. The other knots $x_{2,n+1}, \ldots, x_{k,n+1}$ are computed analogously. (A parabola method is described in Nürnberger, Sommer & Strauß [1984]).

It is shown in Nürnberger, Sommer & Strauß [1984] that for $d(f, PP_{m,k}) = 10^{\delta}$ we have

$$|\delta - \delta_n| \le \frac{1}{2}|\alpha_{n-1} - \beta_{n-1}| \le \ldots \le \frac{1}{2^n}|\alpha_0 - \beta_0|$$

for all n and thus

$$\lim_{n \to \infty} d_n = d(f, PP_{m,k}) \quad .$$

The corresponding knot sequences converge to a leveled set of knots, if this set is unique.

In practice it suffices to compute the values $d(f, \Pi_m, I)$ for the appearing knot intervals I only approximately by interpolation at Chebyshev points (for details see Nürnberger, Sommer & Strauß [1984]). By shifting the optimal polynomial pieces a nearly best continuous spline approximation is obtained. Moreover, for approximation on a finite set the algorithm yields a nearly optimal set of knots. Numerical examples are given in Nürnberger, Sommer & Strauß [1984].

Further different types of algorithms for segment approximation in various norms were developed in Lawson [1964], Pavlidis & Maika [1974], Karon [1978], Kioustelidis [1980] and McLaughlin & Zacharski [1980].

In this paper we have not discussed best approximation by splines with free knots, by rational splines and by splines in several variables. In these domains several deep problems are unsolved at present.

The reader interested in approximation by splines with free knots is refered to Arndt [1974], Augsburger [1967], Barrar & Loeb [1970], [1976 a] , [1976 b] , [1978], Barrow & Smith [1978], Barrow, Chui, Smith & Ward [1978], Bojanov [1979], de Boor [1963], [1969], [1972], [1974], Braess [1974 a] , [1974 b] , Braess & Dyn [1983], Braess & Nürnberger [1981], Burchard [1974], Chui, Smith & Ward [1977], Cromme [1976], Jetter [1978], Jetter & Lange [1978], Johnson [1960], Karlin [1969 b] , [1976 a] , [1976 b] , Loeb [1980], Malcolm [1977], Meinardus [1966 a] , [1966 b] , [1967], Nürnberger, Schumaker, Sommer & Strauß [1984 c] , Powell [1968], Rice [1969], Schaback [1978], Schoenberg [1969], Schumaker [1968 b], [1969 b] , [1979], Strauß [1979], [1984 a] , Taylor [1979] and others .

Approximation by rational splines was investigated in Braess & Werner [1974], Schaback [1973] and Werner [1974], [1979], [1980].

A survey on multivariate splines is given in de Boor [1982] and Dahmen & Micchelli [1983] and general multivariate methods are considered in the survey of Cheney [1983].

REFERENCES

Ahlberg J.H., Nilson E.N. and Walsh J.L.
   (1967) The theory of splines and their applications,
           Academic Press, New York.

Arndt H.
   (1974) On uniqueness of best spline approximations with free
           knots, J. Approx. Theory 11, 118-125.

Augsburger W.
   (1967)   Segmentapproximation in L_p-Räumen, Dissertation, TH
            Clausthal.

Barrar R.B. and Loeb H.L.
   (1970)   Existence of best spline approximations with free knots,
            J. Math. Anal. Appl. 31, 383-390.
   (1976a) Multiple zeros and applications to optimal linear func-
            tionals, Numer. Math. 25, 257-262.
   (1976b) On a non-linear characterization problem for mono-
            splines, J. Approx. Theory 18, 220-240.
   (1978)   On monosplines with odd multiplicities of least norm,
            J. Analyse Math. 33, 12-38.

Barrow D.L. and Smith P.W.
   (1978)   Asymptotic properties of best $L_2[0,1]$ approximation
            by splines with variable knots, Quart. Appl. Math.
            36, 293-304.

Barrow D.L., Chui C.K., Smith P.W. and Ward J.D.
   (1978)   Unicity of best L_2-approximation by second order
            splines with variable knots, Math. Comp. 32, 1131-1141.

Blatt H.-P.
   (1982)   Strenge Eindeutigkeitskonstanten und Fehlerabschätzun-
            gen bei linearer Tschebyscheff-Approximation, in
            Numerische Methoden der Approximationstheorie, Collatz
            L., Meinardus G. and Werner H. eds., ISNM 59, Birk-
            häuser-Verlag, Basel, 9-25.

Blatt H.-P., Nürnberger G. and Sommer M.
   (1981)   A characterization of pointwise-Lipschitz-continuous
            selections for the metric projection, Numer. Funct.
            Anal. and Optimiz. 4, 101-122.

Blatter J. and Schumaker L.L.
   (1982)   The set of continuous selections of a metric projec-
            tion in C(X), J. Approx. Theory 36, 141-155.
   (1983)   Continuous selections and maximal alternators for
            spline approximation, J. Approx. Theory 38, 71-80.

Blatter J., Morris P.D. and Wulbert D.E.
   (1968)   Continuity of the set valued metric projection, Math.
            Annalen 178, 12-24.

Böhmer K.
   (1974)   Spline-Funktionen, Teubner-Verlag, Stuttgart.

Bojanov B.D.
   (1979)   Uniqueness of the monosplines of least deviation, in
            Numerical Integration, Hämmerlin G. ed., ISNM 15,
            Birkhäuser-Verlag, Basel, 67-97.

de Boor C.
(1963) Best approximation properties of spline functions of
        odd degree, J. Math. Mech. 12, 747-750.
(1968) On uniform approximation by splines, J. Approx. Theory
        1, 219-235.
(1969) On the approximation by γ polynomials, in Approximation
        with special emphasis on spline functions, Schoenberg
        I.J. ed., Academic Press, New York, 157-183.
(1972) Good approximation by splines with variable knots, in
        Spline functions and approximation theory, Meir A. and
        Sharma A. eds., Birkhäuser-Verlag, Basel, 57-72.
(1974) Good approximation by splines with variable knots II, in
        Conference on the numerical solution of differential
        equations, Lecture Notes in Mathematics 363, Springer-
        Verlag, Berlin, 12-20.
(1978) A practical guide to splines, Springer-Verlag, New York.
(1982) Topics in multivariate approximation theory, in Topics
        in Numerical Analysis, Lecture Notes in Mathematics 965,
        Springer-Verlag, Berlin, 40-78.

de Boor C. and Fix G.J.
(1973) Spline approximation by quasi-interpolants, J. Approx.
        Theory 8, 19-45.

Braess D.
(1974a) Chebyshev approximation by spline functions with free
        knots, Numer. Math. 17, 357-366.
(1974b) On the nonuniqueness of monosplines with least L_2-norm,
        J. Approx. Theory 12, 91-93.

Braess D. and Dyn N.
(1982) On the uniqueness of monosplines and perfect splines
        of least L_1- and L_2-norm, J. d'Anal. Math. 41, 217-233.

Braess D. and Nürnberger G.
(1981) Nonuniqueness of best L_p-approximation for generalized
        convex functions by splines with free knots, Numer.
        Funct. Anal. and Optimiz. 4, 199-209.

Braess D. and Werner H.
(1974) Tschebyscheff-Approximation mit einer Klasse rationaler
        Splinefunktionen II, J. Approx. Theory 10, 379-399.

Brosowski B. and Wegmann R.
(1973) On the lower semicontinuity of the set- valued metric
        projection, J. Approx. Theory 8, 84-100.

Brosowski B., Deutsch F. and Nürnberger G.
(1980) Parametric approximation, J. Approx. Theory 29, 261-277.

Burchard H.G.
(1974) Splines with optimal knots are better, J. Applicable
        Anal. 3, 309-319.

Butzer P.L. and Berens H.
(1967) Semi-groups of operators and approximation, Springer-
        Verlag, Berlin.

Carroll M.P. and Braess D.
  (1974) On uniqueness of L_1-approximation for certain families
          of spline functions, J. Approx. Theory 12, 362-364.

Cheney E.W.
  (1966) Introduction to approximation theory, McGraw Hill, New
          York.
  (1983) The best approximation of multivariate functions by
          combinations of univariate ones, in Approximation
          Theory IV, Chui C.K., Schumaker L.L. and Ward J.D. eds.,
          Academic Press, New York, 1-26.

Chui C.K., Smith P.W. and Ward J.D.
  (1977) On the smoothness of best L_2 approximants from non-
          linear spline manifolds, Math. Comp. 31, 17-23.

Cline A.K.
  (1973) Lipschitz conditions on uniform approximation operators,
          J. Approx, Theory 8, 160-172.

Collatz L. and Krabs W.
  (1973) Approximationstheorie, Teubner-Verlag, Stuttgart.

Cromme L.J.
  (1976) Eine Klasse von Verfahren zur Ermittlung bester nicht-
          linearer Tschebyscheff-Approximationen, Numer. Math.
          25, 447-459.

Dahmen W. and Micchelli C.A.
  (1983) Recent progress in multivariate splines, in Approxima-
          tion Theory IV, Chui C.K., Schumaker L.L. and Ward J.D.
          eds., Academic Press, New York, 27-121.

Demko S. and Varga R.S.
  (1974) Extended L_p error bounds for spline and L-spline in-
          terpolation, J. Approx. Theory 12, 242-264.

DeVore R.A.
  (1968) One-sided approximation of functions, J. Approx. Theo-
          ry 1, 11-25.
  (1972) The approximation of continuous functions by positive
          linear operators, Lecture Notes in Mathematics 293,
          Springer-Verlag, Berlin.

DeVore R.A. and Scherer K.
  (1976) A constructive theory for approximation by splines
          with an arbitrary sequence of knot sets, in Approxi-
          mation Theory, Schaback R. and Scherer K. eds., Lec-
          ture Notes in Mathematics 556, Springer-Verlag, Berlin,
          167-183.

Deutsch F.
  (1983) A survey of metric selections, Contemporary Mathematics
          18, 49-71.

Deutsch F., Nürnberger G. and Singer I.
(1980) Weak Chebyshev subspaces and alternation, Pacific J.
         Math. 89, 9-31.

Esch R.E. and Eastman W.L.
(1969) Computational mehtods for best spline approximation,
         J. Approx. Theory 2, 85-96.

Galkin R.V.
(1974) The uniqueness of the element of best mean approxima-
         tion to a continuous function using splines with
         fixed nodes, Math. Notes 15, 3-8.

Hedstrom G.W. and Varga R.S.
(1971) Application of Besov spaces to spline approximation,
         J. Approx. Theory 4, 295-327.

Henry M.S. and Roulier J.A.
(1978) Lipschitz and strong unicity constants for changing
         dimension, J. Approx. Theory 22, 85-94.

Jetter K.
(1978) L_1-Approximation verallgemeinerter konvexer Funktionen
         durch Splines mit freien Knoten, Math. Z. 164, 53-66.

Jetter K. and Lange G.
(1978) Die Eindeutigkeit L_2-optimaler Monosplines, Math. Z.
         158, 23-34.

Johnson R.S.
(1960) On monosplines of least deviation, Trans. Amer. Math.
         Soc. 96, 458-477.

Jones R.C. and Karlovitz L.A.
(1970) Equioscillation under nonuniqueness in the approxima-
         tion of continuous functions, J. Approx. Theory 3,
         138-145.

Karlin S.
(1968) Total positivity, Stanford, California.
(1969a) Best quadrature formulas and interpolation by splines
         satisfying boundary conditions, in Approximation
         with special emphasis on spline functions, Schoenberg
         I.J. ed., Academic Press, New York, 447-466.
(1969b) The fundamental theorem of algebra for monosplines
         satisfying certain boundary conditions and applica-
         tions to optimal quadrature formulas, in Approxima-
         tion with special emphasis on spline functions,
         Schoenberg I.J. ed., Academic Press, New York, 467-484.
(1971) Best quadrature formulas and splines, J. Approx. Theory
         4, 59-90.
(1976a) On a class of best nonlinear approximation problem and
         extended monosplines, in Studies in spline functions
         and approximation theory, Karlin S., Micchelli C.A.,

            Pinkus A. and Schoenberg I.J. eds., Academic Press,
            New York, 19-66.
  (1976b) A global improvement theorem for polynomial monosplines,
            in Studies in spline functions and approximation theo-
            ry, Karlin S., Micchelli C.A., Pinkus A. and Schoen-
            berg I.J. eds., Academic Press, New York, 67-82.

Karlin S. and Studden W.J.
  (1966)   Tchebycheff systems: with applications in analysis and
            statistics, Interscience, New York.

Karon J.M.
  (1978)   Computing improved Chebyshev approximations by the
            continuation method I: Description of an algorithm,
            SIAM J. Numer. Anal. 15, 1269-1288.

Kioustelidis J.B.
  (1980)   Optimal segmented approximations, Computing 24, 1-8.

Kroó A.
  (1985)   Some theorems on best L_1-approximation of continuous
            functions, Acta Math. Hungar., to appear.

Lawson C.L.
  (1964)   Characteristic properties of the segmented rational
            minimax approximation problem, Numer. Math. 6, 293-301.

Loeb H.
  (1980)   The monospline of least norm and related problems, in
            Approximation Theory III, Cheney E.W. ed., Academic
            Press, New York, 21-39.

Lorentz G.G.
  (1966)   Approximation of functions, Holt, Rinehart and Winston,
            New York.

Lyche T. and Schumaker L.L.
  (1975)   Local spline approximation methods, J. Approx. Theory
            15, 294-325

Malcom M.A.
  (1977)   On the computation of nonlinear spline functions,
            SIAM J. Numer. Anal. 14, 254-282.

McLaughlin H.W. and Zacharski J.J.
  (1980)   Segmented approximation, in Approximation Theory III,
            Cheney E.W. ed., Academic Press, New York, 647-754.

Meinardus G.
  (1966a) Über ein Monotonieprinzip bei linearen Approximationen,
            ZAMM 46, 227-238.
  (1966b) Zur Segmentapproximation mit Polynomen, ZAMM 46,
            239-246.

(1967)   Approximation of functions: theory and numerical me-
          thods, Springer-Verlag, Berlin.

Meinardus G. and Nürnberger G.
  (1985) Approximation theory and numerical methods for delay
          differential equations, in Delay equations, approxi-
          mation and application, Meinardus G. and Nürnberger G.
          eds., ISNM, Birkhäuser-Verlag, Basel, to appear.

Müller M.W.
  (1978) Approximationstheorie, Akademische Verlagsgesellschaft,
          Wiesbaden.

Micchelli C.A.
  (1977) Best L^1-approximation by weak Chebyshev systems and
          the uniqueness of interpolating perfect splines,
          J. Approx. Theory 19, 1-14.

Micchelli C.A. and Pinkus A.
  (1977) Moment theory for weak Chebyshev systems with appli-
          cations to monosplines, quadrature formulae and best
          one-sided L_1-approximation by spline functions with
          fixed knots, SIAM J. Math. Anal. 8, 206-230.

Newman D.J. and Shapiro H.S.
  (1963) Some theorems on Chebyshev approximation, Duke Math. J.
          30, 673-684.

Nürnberger G.
  (1980) Nonexistence of continuous selections of the metric
          projection and weak Chebyshev systems, SIAM J. Math.
          Anal. 11, 460-467.
  (1982a) A local version of Haar's theorem in approximation
          theory, Numer. Funct. Anal. and Optimiz. 5, 21-46.
  (1982b) Strong univity constants for spline functions, Numer.
          Funct. Anal. and Optimiz. 5, 319-347.
  (1983a) Strong unicity of best approximations: a numerical
          aspect,Numer. Funct. Anal. and Optimiz. 6, 399-421.
  (1983b) Strong unicity constants for finite-dimensional sub-
          spaces, in Approximation Theory IV, Chui C.K., Schu-
          maker L.L. and Ward D. eds., Academic Press, New York,
          643-648.
  (1984a) Unicity in one-sided L_1-approximation and quadrature
          formulae, J. Approx. Theory, to appear.
  (1984b) Global unicity in semi-infinite optimization, preprint.
  (1985) Unicity in semi-infinite optimization, in Parametric
          optimization and approximation, Brosowski B. and
          Deutsch F. eds., Birkhäuser-Verlag, Basel, to appear.

Nürnberger G. and Singer I.
  (1982) Uniqueness and strong uniqueness of best approximations
          by spline subspaces and other subspaces, J. Math. Anal.
          Appl. 90, 171-184.

Nürnberger G. and Sommer M.
 (1978a) Weak Chebyshev subspaces and continuous selections for
         the metric projection, Trans. Amer. Math. Soc. 238,
         129-138.
 (1978b) Characterization of continuous selections of the me-
         tric projection for spline functions, J. Approx. Theo-
         ry 22, 320-330.
 (1983a) Alternation for best spline approximations, Numer.
         Math. 41, 207-221.
 (1983b) A Remez type algorithm for spline functions, Numer.
         Math. 41, 117-146.
 (1985) Continuous selections in Chebyshev approximation, in
         Parametric optimization and approximation, Brosowski B.
         and Deutsch F. eds., Birkhäuser-Verlag, Basel, to
         appear.

Nürnberger G., Sommer M. and Strauß H.
 (1984) An algorithm for segment approximation, preprint.

Nürnberger G., Schumaker L.L., Sommer M. and Strauß H.
 (1983) Interpolation by generalized splines, Numer. Math. 42,
         195-212.
 (1984a) Generalized Tchebycheffian splines, SIAM J. Math. Anal.
         15, 790-804.
 (1984b) Approximation by generalized splines, preprint.
 (1984c) Approximation by splines with free knots, preprint.

Pavlidis T. and Maika A.P.
 (1974) Uniform piecewise polynomial approximation with vari-
         able joints, J. Approx. Theory 12, 61-69.

Pinkus A.
 (1976) One-sided L_1-approximation by splines with fixed knots,
         J. Approx. Theory 18, 130-135.

Pinkus A. and Totik V.
 (1984) One-sided L_1-approximation, preprint.

Popov V.A.
 (1975) On approximation of absolutely continuous functions by
         splines, Mathematica (Cluj) 8, 1299-1301.

Powell M.J.
 (1968) On best L_2 spline approximations, in Numerische Mathe-
         matik, Differentialgleichungen,Approximationstheorie,
         Collatz L., Meinardus G. and Unger H. eds., ISNM 9,
         Birkhäuser-Verlag, Basel, 317-337.
 (1981) Approximation theory and methods, Cambridge Univer-
         sity Press.

Prenter P.M.
 (1975) Splines and variational methods, John Wiley & Sons,
         New York.

Rice J.R.
(1964)   The approximation of functions I, Addison & Wesley,
         Reading, Massachusetts.
(1967)   Characterization of Chebyshev approximation by splines,
         SIAM J. Numer. Anal. 4, 557-567.
(1969)   The approximation of functions II, Addison & Wesley,
         Reading, Massachusetts.

Rivlin T.J.
(1969)   An introduction to approximation of functions, Blais-
         dell, Massachusetts.

Sard A.
(1963)   Linear approximation, Amer. Math. Soc., Providence, R.I.

Schaback R.
(1973)   Spezielle rationale Splinefunktionen, J. Approx. Theo-
         ry 7, 281-292.
(1978)   On alternation numbers in nonlinear Chebyshev appro-
         ximation, J. Approx. Thoery 23, 379-391.

Scherer K.
(1974)   Stetigkeitsmoduli und beste Approximation durch poly-
         nomiale Splines, in Spline Funktionen, Böhmer K.,
         Meinardus G. and Schempp W. eds., Bibliographisches
         Institut, Mannheim, 289-302.

Schmidt D.
(1980)   A characterization of strong unicity constants, in
         Approximation Theory III, Cheney E.W. ed., Academic
         Press, New York, 805-810.

Schoenberg I.J.
(1965)   On monosplines of least deviation and quadrature for-
         mulae, SIAM J. Numer. Anal. 2, 144-170.
(1966)   On monosplines of least square deviation and best
         quadrature formulae II, SIAM J. Numer. Anal. 3, 321-328.
(1969)   Monosplines and quadrature formulae, in Theory and
         applications of spline functions, Greville T.N.E. ed.,
         Academic Press, New York, 157-207.
(1973)   Cardinal spline interpolation, CBMS 12, SIAM, Phila-
         delphia.

Schoenberg I.J. and Whitney A.
(1953)   On Polya frequency functions III. The positivity of
         translation determinants with application to the inter-
         polation problem by spline curves, Trans. Amer. Math.
         Soc. 74, 246-259.

Schönhage A.
(1971)   Approximationstheorie, Walter de Gruyter & Co., Berlin.

Schultz M.H.
(1973)   Spline Analysis, Prentice Hall, Englewood Cliffs, New
         Jersey.

Schumaker L.L.
  (1968a) Uniform approximation by Tchebycheffian spline func-
          tions, J. Math. Mech. 18, 369-378.
  (1968b) Uniform approximation by Chebyshev spline functions II:
          free knots, SIAM J. Numer. Anal. 5, 647-656.
  (1969a) Some algorithms for the computation of interpolating
          and approximationg spline functions, in Theory and
          applications of spline functions, Greville T.N.E. ed.,
          Academic Press, New York, 87-102.
  (1969b) On the smoothness of best spline approximations, J.
          Approx, Theory 2, 410-418.
  (1979) L_2 approximation by splines with free knots, in Appro-
          ximation in Theorie und Praxis, Meinardus G. ed.,
          Bibliographisches Institut, Mannheim, 157-182.
  (1981) Spline functions: basic theory, Wiley-Interscience,
          New York.

Shapiro H.S.
  (1971) Topics in approximation theory, Lecture Notes in
          Mathematics 187, Springer-Verlag, Berlin.

Sharma A. and Meir A.
  (1966) Degree of approximation by spline interpolation, J.
          Math. Mech. 15, 759-768.

Singer I.
  (1970) Best approximation in normed linear spaces by elements
          of linear subspaces, Springer-Verlag, Berlin.

Sommer M.
  (1979) L_1-approximation by weak Chebyshev spaces, in Approxi-
          mation in Theorie und Praxis, Meinardus G. ed., Bib-
          liographisches Institut, Mannheim, 85-102.
  (1980a) Characterization of continuous selections for the
          metric projection for generalized splines, SIAM J.
          Math. Anal. 11, 23-40.
  (1980b) Nonexistence of continuous selections of the metric
          projection for a class of weak Chebyshev spaces, Trans.
          Amer. Math. Soc. 260, 403-409.
  (1982) Characterization of continuous selections of the me-
          tric projection for a class of weak Chebyshev spaces,
          SIAM J. Math. Anal. 13, 280-294.
  (1983a) Weak Chebyshev spaces and best L_1-approximation, J.
          Approx. Theory 39, 54-71.
  (1983b) Continuous selections and convergence of best L_p-
          approximations in subspaces of spline functions, Nu-
          mer. Funct. Anal. and Optimiz. 6, 213-234.

Sommer M. and Strauß H.
  (1977) Eigenschaften von schwach tschebyscheffschen Räumen,
          J. Approx. Theory 21, 257-268.
  (1981) Unicity of best one-sided L_1-approximations for cer-
          tain classes of spline functions, Numer. Funct. Anal.
          and Optimiz. 4, 413-435.

Strang G. and Fix G.
  (1973) An analysis of the finite element method, Prentice
            Hall, Englewood Cliffs, N.J.

Strauß H.
  (1975a) L_1-Approximation mit Splinefunktionen, in Numerische
           Methoden der Approximationstheorie, Collatz L., Mei-
           nardus G. and Werner H. eds., ISNM 26, Birkhäuser-
           Verlag, Basel, 151-162.
  (1975b) Eindeutigkeit bei der gleichmäßigen Approximation mit
           Tchebyscheffschen Splinefunktionen, J. Approx. Theory
           15, 78-82.
  (1979) Optimale Quadraturformeln und Perfektsplines, J. Approx.
           Theory 27, 203-226.
  (1981) Eindeutigkeit in der L_1-Approximation, Math. Z. 176,
           63-74.
  (1982) Unicity in best one-sided L_1-approximation, Numer.
           Math. 40, 229-243.
  (1984a) Monotonicity of quadrature formulae of Gauß type and
           comparison theorems for monosplines, Numer. Math. 44,
           337-347.
  (1984b) Characterization of strict approximations in subspaces
           of spline functions, J. Approx. Theory 41, 309-328.
  (1984c) An algorithm for the computation of strict approxima-
           tions in subspaces of spline functions, J. Approx.
           Theory 41, 329-344.

Swartz B. and Varga R.S.
  (1972) Error bounds for spline and L-spline interpolation,
           J. Approx. Theory 6, 6-49.

Taylor G.D.
  (1979) Data fitting: some adaptive methods, in Approximation
           in Theorie und Praxis, Meinardus G. ed., Bibliogra-
           phisches Institut Mannheim, 291-304.

Varga R.S.
  (1971) Functional analysis and approximation theory in nume-
           rical analysis, CBMS 3, SIAM, Philadelphia.

Watson G.A.
  (1980) Approximation theory and numerical methods, Wiley-
           Interscience, Chichester.

Werner H.
  (1974) Tschebyscheff-Approximation mit einer Klasse rationa-
           ler Splinefunktionen, J. Approx. Theory 10, 74-92.
  (1979) An introduction to non-linear splines, in Polynomial
           and spline approximation, Sahney B.N. ed., D. Reidel
           Publishing Company, Boston, 247-306.
  (1980) The development of nonlinear splines and their appli-
           cations, in Approximation Theory III, Cheney E.W. ed.,
           Academic Press, New York, 125-150.

Zielke R.
(1979)   Discontinuous Cebysev systems, Lecture Notes in Mathematics 707, Springer-Verlag, Berlin.

Zwick D.
(1984a) Strong uniqueness of best spline approximation for a class of piecewise n-convex functions, preprint.
(1984b) The generalized convexity cone of splines with multiple knots, preprint.

International Series of
Numerical Mathematics, Vol. 74
© 1985 Birkhäuser Verlag Basel

RECONSTRUCTION AND APPROXIMATION OF
FUNCTIONS FROM SAMPLES

Q.I. RAHMAN   and   G. SCHMEISSER

Dépt. de Mathématiques   Mathematisches Institut

et de Statistique   Universität Erlangen-Nürnberg

Université de Montréal   Bismarckstraße 1 1/2

Montréal, H3C 3J7   D-8520 Erlangen

We begin by recalling some methods which are used numerically for the reconstruction and approximation of functions from samples. Then we discuss a modified approach which has many interesting aspects. Most of the results mentioned here were first presented at a Conference on Approximation Theory held in Oberwolfach from November 16 to 21, 1981 but, due to prolonged illness of the first named author, could not be submitted for publication so far.

1. The Sampling Theorem and its Limitations.

In many branches of electrical engineering a problem arises which may be stated as follows: *Is it possible to reconstruct or at least to approximate sufficiently well a signal* f *from values (samples) taken at equally spaced points* $n\pi/\tau$ $(\tau > 0;\ n \in \mathbb{Z})$?

Before we give an answer let us introduce the following two NOTATIONS. By UCB($\mathbb{R}$) we denote the class of all functions which are *uniformly continuous* and *bounded* on the real line. Furthermore, if $f \in L(\mathbb{R})$ we write $f^{\wedge}$ for its *Fourier transform* defined by

214

$$f^{\wedge}(y) := \frac{1}{\sqrt{2\pi}} \int_{-\infty}^{\infty} f(x)e^{-ixy}dx \ .$$

As an answer to the above question we mention the SAMPLING THEOREM (Whittaker, Shannon, Kotel'nikov).

If

$$(1.1) \qquad f \in UCB(\mathbb{R}) \cap L(\mathbb{R})$$

and

$$(1.2) \qquad f^{\wedge}(y) = 0 \quad for \quad |y| \geq \tau \ ,$$

then

$$(1.3) \quad f(x) = \sum_{n=-\infty}^{\infty} f(\frac{n\pi}{\tau}) \frac{\sin(\tau x - n\pi)}{\tau x - n\pi} =: C_{\tau}[f](x)$$

and the series converges absolutely and uniformly on $\mathbb{R}$.

The assumption (1.1) implies $f \in L^2(\mathbb{R})$ which means in the applications that the signal f is of finite energy. By the additional assumption (1.2) the signal only contains frequencies less than τ . Therefore the engineers say that *for a signal* f *of finite energy which is band-limited by* τ *the representation* (1.3) *holds*. The infinite series in (1.3) is often called *Whittaker's cardinal series*.

Among the assumptions of the Sampling Theorem the one in (1.2) is considered to be quite strong in engineering since it already excludes duration-limited signals. In fact, either a function itself or its Fourier transform can have finite support but not both. However, if we relax (1.2) we can no longer *reconstruct* a signal from its samples but only *approximate* it. The following statement holds.

EXTENDED SAMPLING THEOREM (Brown, Boas, Butzer & Splettstoesser). *If in the Sampling Theorem the condition* (1.2) *is relaxed to* $f^{\wedge} \in L(\mathbb{R})$, *then*

$$(1.4) \qquad f(x) = \lim_{\tau \to \infty} C_{\tau}[f](x)$$

uniformly in $x \in \mathbb{R}$.

For more information about the above two theorems and Whittaker's cardinal series we refer to the survey articles of P.L. Butzer [4], A.J. Jerri [8], F. Stenger [14] and the literature quoted there. The operator C_τ has many interesting properties but the following considerations raise certain questions:

(a) *Evaluation* . The interpolant $C_\tau[f](x)$ is an infinite series. In its terms the part not depending on f decreases quite slowly when n tends to infinity so that efficient numerical evaluation of $C_\tau[f](x)$ is possible only if f decreases rapidly. It is desirable that the fundamental functions of the interpolation process decrease as fast as possible.

(b) *Reconstruction* . By the Paley-Wiener theorem (see [1, chap. 6.8]) a function satisfying the assumptions of the Sampling Theorem is the restriction to $\mathbb{R}$ of an entire function of exponential type τ (which we again denote by f), i.e. for every $\varepsilon > 0$

$$\left| f(z) \right| = \mathcal{O}(e^{(\tau+\varepsilon)|z|})$$

as z tends to infinity in the complex plane. By a theorem of F. Carlson (see [1, (9.2.1)]) all entire functions of exponential type less than τ (not only those with $f \in L(\mathbb{R})$) are uniquely determined by their values $f(n\pi/\tau)$ ($n \in \mathbb{Z}$). Therefore the reconstruction of f from samples on the real line should be possible under a considerably weaker restriction on the growth of f than the one represented by (1.1).

(c) *Approximation* . The speed of convergence in (1.4) may be slow. A quantitative study shows that C_τ shares some of the bad properties known from Lagrange interpolation by polynomials. In particular it misses by a factor $\log \tau$ the order of best approximation by entire functions of exponential type τ .

2. The Discovery of Splines.

In 1946 Schoenberg published a paper [12] which is generally believed to be the origin for the introduction of splines in mathematics. The motivation for this research was nothing but the problem raised under (a).

Restricting himself to integer nodes ($\tau = \pi$) Schoenberg pointed out that the series in (1.3) is a special case of an interpolation series

$$(2.1) \qquad \sum_{n = -\infty}^{\infty} f(n)\, L(\,x - n)\, ,$$

where L is a *fundamental function* of Lagrange interpolation, i.e.

$$L(n) = \begin{cases} 1 & \text{for } n = 0 \\ 0 & \text{for all other integers } n\,. \end{cases}$$

He proposed to replace the function L used in (1.3), namely

$$(2.2) \qquad L(x) := \frac{\sin \pi x}{\pi x}\, ,$$

by one of faster "damping". Amongst his options he favoured what is now called a B-spline. However, except for the simplest case, the hat function, a B-spline is not a fundamental function of Lagrange interpolation; a fundamental function has to be formed from a B-spline and multiples of its translates. The resulting *cardinal spline* L is no longer of finite support but it decreases at least exponentially (see [13]). This is a fairly good solution of the evaluation problem (a).

3. The Window Function Method.

The engineers took note of Schoenberg's research but they often mix it with an idea which seems to be more in line with their way of thinking: the *window function method* (see e.g. [10]). In its simplest form a window function Φ is a characteristic function of a compact interval containing the origin. To use a window in the series (2.1) means to multiply L by Φ. Notice that even for completely arbitrary Φ the resulting function $\tilde{L} := \Phi L$ is again a fundamental function of Lagrange interpolation provided only that $\Phi(0) = 1$. Of course, using a characteristic function Φ is a very crude method (equivalent to truncation of (2.1)) since $\tilde{L}$ will not even be continuous although we enjoy its being of finite support. A more elegant way often pursued these days is to use a B-spline window. Then the corresponding interpolant (2.1) becomes differentiable up to a certain order and for

every x the series (2.1) has got only a finite number of non-zero terms.
This is an excellent solution of the evaluation problem (a).

However, there remains something to be desired in view of the reconstruction problem (b). Neither Schoenberg's cardinal spline interpolation nor
the method of B-spline windows can represent entire functions of exponential
type except for some very trivial ones.

We are not aware of studies to the approximation problem (c) in the
case of B-spline windows.Schoenberg in his cardinal spline interpolation [12],
[13] assumes that he is given an infinite sequence of data which he has to
smooth and does not ask for approximation of functions.

4. A Bandlimited Pseudo-Window.

We seek to make progress in respect of the problems under (b) and
(c) without losing much of what has been obtained for (a) by the methods described in Sections 2 and 3 . It is not difficult to see that for the reconstruction problem (b) a bandlimited function Φ would be the right tool.
In fact (see also [6]), let Φ be of exponential type $\delta < 1$ with $\Phi(0)=1$
and suppose that f is any given entire function of exponential type σ .
Then, for fixed y , the product

$$F_y : \quad x \mapsto f(x)\,\Phi(\tau(y - x))$$

is of exponential type $\sigma + \tau\delta$. Now, if Φ brings F_y "down" to a function
belonging to $L(\mathbb{R})$ we may apply the Sampling Theorem to F_y with
$\tau > \sigma/(1-\delta)$ and obtain

$$f(x)\,\Phi(\tau(y - x)) \;=\; \sum_{n=-\infty}^{\infty} f\!\left(\frac{n\pi}{\tau}\right)\Phi\!\left(\tau\!\left(y - \frac{n\pi}{\tau}\right)\right)\frac{\sin(\tau x - n\pi)}{\tau x - n\pi} \quad .$$

Hence, putting y = x we arrive at the reconstruction formula

$$f(x) \;=\; \sum_{n=-\infty}^{\infty} f\!\left(\frac{n\pi}{\tau}\right) L(\tau x - n\pi)$$

with

$$L(x) := \Phi(x)\,\frac{\sin x}{x} \quad .$$

Unfortunately an entire function Φ cannot be a window function in the proper sense since it is not of finite support. But what does finite support mean numerically ? If an entire function decreases so rapidly on $\mathbb{R}$ that outside a finite interval its modulus is less than the accuracy of a computer it must numerically be accepted as a function of finite support. Now the question arises how fast an entire function Φ of exponential type with $\Phi(0) = 1$ can decrease on the real line. As an answer it is known [9] that Φ can behave like

$$(4.1) \qquad \left| \Phi(x) \right| = \mathcal{O}(e^{-w(|x|)}) \qquad \text{as } x \to \pm\infty$$

where w is a positive function which, besides certain regularity conditions, satisfies

$$(4.2) \qquad \int_1^\infty \frac{w(x)}{x^2}\, dx < \infty \; .$$

This latter condition is also necessary. In [7] it was shown that the desired function Φ with the property (4.1) can be found in the form

$$(4.3) \qquad \Phi(x) = \prod_{n=1}^{\infty} \frac{\sin a_n x}{a_n x}$$

where the sequence (a_n) is determined by the given w satisfying (4.2) and some mild side conditions. In particular, given $\alpha > 1$ and $\delta > 0$ we constructed in [7] as an example an entire function $\Phi(\alpha, \delta, \,.\,)$ of exponential type δ such that

$$\left| \Phi(\alpha, \delta, x) \right| = \mathcal{O}\left(\exp(-|x| / (\log |x|)^\alpha) \right) \qquad \text{as } x \to \pm\infty \; .$$

We will restrict ourselves to this rate of decrease since by (4.2) we cannot do much better anyway.

Setting

$$(4.4) \qquad \Lambda(z) := \Lambda(\alpha, \delta, z) := \Phi\left(\frac{1+\alpha}{2}, \frac{\delta}{2}, z \right) \frac{\sin z}{z}$$

our modified cardinal series becomes

$$(4.5) \quad \mathscr{C}_\tau [f](z) := \mathscr{C}_{\tau,\alpha,\delta} [f](z) := \sum_{n=-\infty}^{\infty} f\left(\frac{n\pi}{t}\right) \Lambda(\tau z - n\pi).$$

5. Properties of the Modified Cardinal Series.

NOTATIONS. In what follows we shall always assume that $\alpha > 1$ and $\delta \in (0,1)$ are fixed. Accordingly we may simply write $\mathscr{C}_\tau$ instead of $\mathscr{C}_{\tau,\alpha,\delta}$ for the operator defined in (4.5). By c_1, c_2, c_3 etc. we denote appropriate positive constants which may depend only on α and δ, whereas by $\gamma_1, \gamma_2, \gamma_3$ etc. we denote positive constants which may in addition depend on everything in consideration (like the function f , the intervals I and I_ϵ etc.) *except* for τ, n, x and y . Next we set

$$w_\alpha(x) := |x| / (\log(2 + |x|))^\alpha$$

and denote by K_α the class of all real or complex valued functions f which are defined and continuous on $\mathbb{R}$ such that

$$(5.1) \qquad |f(x)| = \mathcal{O}(\exp w_\alpha(x)) \qquad \text{as} \quad x \to \pm \infty .$$

Finally let E_τ denote the class of all entire functions of exponential type τ .

5.1. Applicability and Interpolation Property.

THEOREM 1. *For every* $f \in K_\alpha$ *the operator* $\mathscr{C}_\tau$ *is applicable. The series* $\mathscr{C}_\tau [f](z)$ *converges absolutely and uniformly on every compact subset of* $\mathbb{C}$. *The function* $\mathscr{C}_\tau[f]$ *is an entire function of exponential type* $(1 + \delta) \tau$ *which interpolates* f *in the points* $n\pi/\tau$ $(n \in \mathbb{Z})$, *i.e.*

$$\mathscr{C}_\tau [f]\left(\frac{n\pi}{\tau}\right) = f\left(\frac{n\pi}{\tau}\right) .$$

Proof. For the fundamental function Λ in (4.4) the following estimate holds [11, Lemma 1 and its proof] :

$$(5.2) \quad |\Lambda(x + iy)| \leq c_1 \exp((1 + \delta)|y| - w_\beta(x)), \quad \beta := \frac{1 + 3\alpha}{4} .$$

This is used to obtain the convergence properties. Next, assuming without loss of generality that $x > 0$ and splitting the summation in (4.5) into

220

$$\sum_{n=-\infty}^{\infty} \ldots = \sum_{n=-\infty}^{0} \ldots + \sum_{n=1}^{\lfloor 2\tau x/\pi \rfloor -1} \ldots + \sum_{n=\lfloor 2\tau x/\pi \rfloor}^{\infty} \ldots$$

we find with the help of (5.2)

$$| \mathscr{C}_\tau \lceil f \rceil (x + iy) | \leq c_2 \; e^{(1+\delta)\tau |y|} \; \mathcal{O}(x \exp (c_3 \, w_\alpha(x)))$$

as $x \to \infty$ where the $\mathcal{O}$ term does not depend on y . This shows that $\mathscr{C}_\tau \lceil f \rceil$ is of exponential type $(1 + \delta)\tau$. The interpolation property is obvious.

Under an additional assumption on f we can deduce a representation for the interpolation error.

THEOREM 2. *Let* $f \in K_\alpha$ *and suppose that its variation on* $\lceil -x, x \rceil$ *does not grow faster than* $\mathcal{O}(\exp(w_\alpha(x)))$ *as* $x \to \pm \infty$. *Then for* $x \in \mathbb{R}$

$$(5.3) \qquad f(x) \; - \; \mathscr{C}_\tau \lfloor f \rfloor (x) =$$

$$\lim_{K \to \infty} \; \frac{\tau}{\pi} \sum_{k=-K}^{K} (e^{i2k\tau x} - 1) \int_{-\infty}^{\infty} f(u) \, \Lambda(\tau(x-u)) e^{-i2k\tau u} du \; .$$

Proof. We follow basically an idea of Boas $\lfloor 2 \rfloor$. Under the above assumptions

$$u \mapsto f(u) \cdot \Lambda(\tau \cdot (x - u)) \qquad (x \text{ fixed})$$

is a function of bounded variation belonging to $L(\mathbb{R})$.
Hence by Poisson's summation formula

$$(5.4) \qquad \mathscr{C}_\tau \lfloor f \rfloor (x) := \sum_{n=-\infty}^{\infty} f\left(\frac{n\pi}{\tau}\right) \Lambda\left(\tau\left(x - \frac{n\pi}{\tau}\right)\right) =$$

$$= \frac{\tau}{\pi} \sum_{k=-\infty}^{\infty} \int_{-\infty}^{\infty} f(u) \, \Lambda(\tau(x-u)) \, e^{-i2k\tau u} \, du \; .$$

Likewise the function

$$y \;\mapsto\; f(y)\,\Phi(\,\tau(x-y)) \qquad\qquad (\ x \ \text{ fixed }\ )$$

with $\Phi := \Phi(\frac{1+\alpha}{2},\frac{\delta}{2},\cdot)$ from (4.4) is of bounded variation and belongs to $L(\mathbb{R})$. It can therefore be recovered from its Fourier transform by taking the Cauchy principal value in the formal inversion formula. Thus for $y = x$ we obtain

$$f(x) = \lim_{K \to \infty} \frac{1}{2\pi} \int_{-(2K+1)\tau}^{(2K+1)\tau} \int_{-\infty}^{+\infty} f(u)\,\Phi(\,\tau(x-u))e^{iv(x-u)}\,du\,dv \;.$$

Splitting the outer integral into

$$\int_{-(2K+1)\tau}^{(2K+1)\tau} \cdots \;=\; \sum_{k=-K}^{K} \int_{(2k-1)\tau}^{(2k+1)\tau} \cdots$$

we may exchange the order of integration and obtain

$$(5.5)\quad f(x) = \lim_{K \to \infty} \frac{\tau}{\pi} \sum_{k=-K}^{K} e^{i2k\tau x} \int_{-\infty}^{+\infty} f(u)\,\Lambda(\tau(x-u))e^{-i2k\tau u}\,du \;.$$

Now (5.3) follows from (5.4) and (5.5) .

5.2. Reconstruction Power.

THEOREM 3. *For every* $f \in K_{\alpha} \cap E_{(1-\delta)\tau}$ *the identity*

$$(5.6)\qquad\qquad f = \mathscr{C}_{\tau}[f]$$

holds. Furthermore the fundamental function Λ *is a reproducing kernel such that*

$$(5.7) \qquad f(z) \;=\; \frac{\tau}{\pi} \int_{-\infty}^{\infty} f(t) \cdot \Lambda(\,\tau(z - t))dt$$

for all $\quad f \in K_{\alpha} \cap E_{(1-\delta)\tau}\,.$

Proof . The proof of (5.6) was already indicated in Section 4 . It also follows immediately from Theorem 2 by noting that $f\,\Lambda(\tau(x - .))$ is of exponential type less than 2τ so that for $k \neq 0$ the integrals in (5.3) vanish.

Finally (see [14, p. 172]), for $g \in L^2(\mathbb{R}) \cap E_{\tau}$

$$g(z) \;=\; \frac{\tau}{\pi} \int_{-\infty}^{\infty} g(t) \cdot \frac{\sin(\tau(z-t))}{\tau(z-t)} \; dt \;.$$

Applying this to the admissible function $z \mapsto f(z)\,\Phi(\tau(y - z))$ and setting $y = z$ we obtain (5.7).

REMARK. The identity (5.6) holds already under the weaker assumption that f instead of belonging to K_{α} satisfies the growth condition (5.1) at the nodes, i.e.

$$\left| f\left(\frac{n\pi}{\tau} \right) \right| \;=\; \mathcal{O}\,(\exp w_{\alpha}(n)) \qquad\qquad \text{as} \quad n \to \pm\,\infty\,.$$

(see [11, Lemma 2].).

5.3. Approximation Power.

NOTATIONS. Let $\|\,.\,\|_{\mathbb{R}}$ and $\|\,.\,\|_{[a,b]}$ be the supremum norm on $\mathbb{R}$ and $[a, b]$, respectively. We denote by $\omega(f;.)$ the modulus of continuity of f and by $A_{\tau}(f)$ the error of best approximation to f by entire functions of exponential type τ with respect to $\|\,.\,\|_{\mathbb{R}}$.

LEMMA. *The so-called Lebesgue function* L_{τ} *of* $\mathscr{C}_{\tau}$, *defined by*

$$L_{\tau}(x) \;:=\; \sum_{n=-\infty}^{\infty} |\,\Lambda(\,\tau x - n\pi\,)|,$$

is bounded on $\mathbb{R}$ *by a constant* c_4 .

Proof. Obviously $L_\tau(x)$ has period π/τ. Hence

$$\sup_{x \in \mathbb{R}} L_\tau(x) = \max_{0 \leq x \leq \pi/\tau} \sum_{n=-\infty}^{\infty} |\Lambda(\tau x - n\pi)|$$

$$= \max_{0 \leq t \leq \pi} \sum_{n=-\infty}^{\infty} |\Lambda(t - n\pi)| .$$

Recalling (5.2) we obtain the desired result.

The following theorems show the excellent properties of $\mathcal{C}_\tau$ with respect to the approximation problem (c).

THEOREM 4 . *For every* $f \in UCB(\mathbb{R})$

$$(5.8) \qquad \| f - \mathcal{C}_\tau[f] \|_{\mathbb{R}} \leq (1 + c_4) A_{(1-\delta)\tau}(f) .$$

Proof. Let g be an entire function of exponential type $(1-\delta)\tau$ establishing a best approximation to f in the norm $\| \cdot \|_{\mathbb{R}}$. Of course $g \in K_\alpha$ and so by Theorem 3 $g = \mathcal{C}_\tau[g]$. Hence using the preceding lemma

$$\| f - \mathcal{C}_\tau[f] \|_{\mathbb{R}} \leq \| f - g \|_{\mathbb{R}} + \| \mathcal{C}_\tau[f] - \mathcal{C}_\tau[g] \|_{\mathbb{R}}$$

$$\leq (1 + c_4) \| f - g \|_{\mathbb{R}} = (1 + c_4) A_{(1-\delta)\tau}(f) .$$

Various estimates of $A_\sigma(f)$ in terms of the regularity properties of f are known (see [15, p. 257-261]). They can be used to deduce from Theorem 4 other error bounds which may be more convenient. The dependance on δ can often be absorbed by the constants in the estimates. As typical examples we mention

COROLLARY 1. *If* $f \in UCB(\mathbb{R})$ *then*

$$(5.9) \qquad \| f - \mathcal{C}_\tau[f] \|_{\mathbb{R}} \leq c_5 \, \omega(f; \tfrac{1}{\tau}) .$$

If in addition $f^{(k)}$ *exists and belongs to* $UCB(\mathbb{R})$, *then*

$$(5.10) \qquad \| f - \mathcal{C}_\tau[f] \|_{\mathbb{R}} \leq c_6 \, \tau^{-k} \, \omega(f^{(k)}; \tfrac{1}{\tau}) .$$

The operator $\mathcal{C}_\tau$ has good approximation properties even for those functions $f \in K_\alpha$ which are not uniformly continuous and bounded as the following theorem shows.

THEOREM 5. *If* $f \in K_\alpha$, *then*

$$(5.11) \qquad \lim_{\tau \to \infty} \mathcal{C}_\tau [f] (x) = f(x)$$

for all $x \in \mathbb{R}$. *More precisely, if* $I = [a,b]$ *is any compact interval and* $I_\varepsilon := [a - \varepsilon, b + \varepsilon]$, *then for every* $\varepsilon > 0$

$$(5.12) \qquad \| f - \mathcal{C}_\tau [f] \|_I \leq c_5 \, \omega_{I_\varepsilon} (f; \tfrac{1}{\tau}) + Q_\varepsilon(\tau) \quad \text{where} \quad \omega_{I_\varepsilon}(f ; \cdot)$$

denotes the modulus of continuity of the restriction of f *to* I_ε *and*

$$(5.13) \qquad \lim_{\tau \to \infty} \frac{\log | Q_\varepsilon(\tau) | \; (\log \varepsilon\tau)^\alpha}{\varepsilon\tau} = - \infty .$$

Proof. It is sufficient to prove (5.12) and (5.13).
Set

$$f^*(x) := \begin{cases} f(a - \varepsilon) & \text{if } x \leq a - \varepsilon \\ f(x) & \text{if } x \in I_\varepsilon \\ f(b + \varepsilon) & \text{if } x \geq b + \varepsilon . \end{cases}$$

Then $f^* \in UCB(\mathbb{R})$ and

$$\| f - \mathcal{C}_\tau [f] \|_I = \| f^* - \mathcal{C}_\tau [f] \|_I \leq$$

$$\leq \| f^* - \mathcal{C}_\tau [f^*] \|_I + \| \mathcal{C}_\tau [f^* - f] \|_I$$

$$\leq \| f^* - \mathcal{C}_\tau [f^*] \|_\mathbb{R} + \| \mathcal{C}_\tau [f^* - f] \|_I .$$

By Corollary 1

$$\| f^* - \mathcal{C}_\tau [f^*] \|_\mathbb{R} \leq c_5 \, \omega(f^*; \tfrac{1}{\tau}) \leq c_5 \, \omega_{I_\varepsilon} (f, \tfrac{1}{\tau}) .$$

Hence it remains to verify (5.13) with

$$Q_\varepsilon(\tau) := \| \mathcal{C}_\tau [f^* - f] \|_I .$$

Denote by $N := N(\tau)$ the set of all integers n such that $n\pi/\tau \notin I_\varepsilon$. Then obviously

$$(5.14) \qquad |\mathscr{C}_\tau[f^* - f](x)| \leq \sum_{n \in N} (M + |f(\tfrac{n\pi}{\tau})|) \cdot |\Lambda(\tau x - n\pi)|,$$

where

$$M := \max(|f(a - \varepsilon)|, |f(b + \varepsilon)|).$$

Now, for fixed $x \in I$ there exists a mapping ϕ from N into the set of non-negative integers such that

$$\varepsilon + \phi(n)\,\pi/\tau \leq |x - n\pi/\tau| < \varepsilon + (\phi(n) + 1)\pi/\tau.$$

From (5.2) we find for $n \in N$

$$(5.15) \qquad |\Lambda(\tau x - n\pi)| \leq c_1 \exp(-w_\beta(\varepsilon\tau + \phi(n)\,\pi)).$$

Since for $x \in I$

$$|n\pi/\tau| \leq |x| + |x - n\pi/\tau| \leq \gamma_1 + \varepsilon + (\phi(n) + 1)\pi/\tau$$

we conclude using that $f \in K_\alpha$

$$(5.16) \qquad M + |f(\tfrac{n\pi}{\tau})| \leq \gamma_2 \exp(\gamma_3 w_\alpha(\varepsilon + \phi(n)\pi/\tau)).$$

Now (5.15) and (5.16) show that for sufficiently large τ

$$(5.17) \qquad (M + |f(\tfrac{n\pi}{\tau})|)\,|\Lambda(\tau x - n\pi)|$$

$$\leq \gamma_4 \exp(-\gamma_5 w_\beta(\varepsilon\tau + \phi(n)\pi)).$$

We also note that for a given integer $\nu \geq 0$ there are at most two $n \in N$ such that $\phi(n) = \nu$. Therefore (5.14) and (5.17) yield

$$Q_\varepsilon(\tau) \leq 2\gamma_4 \sum_{\nu=0}^{\infty} \exp(-\gamma_5 w_\beta(\varepsilon\tau + \nu\pi)).$$

Using the integral comparison criterion and setting

$$v(x) := \int_x^{\infty} \exp(-\gamma_5 w_\beta(\xi))\,d\xi$$

we find for large τ

$$Q_\epsilon(\tau) \leq 2\gamma_4 \left(-v'(\epsilon\tau) + \frac{1}{\pi} \cdot v(\epsilon\tau) \right).$$

By l'Hospital's rule

$$\lim_{\tau \to \infty} \frac{v(\epsilon\tau)}{\tau v'(\epsilon\tau)} = 0 \ .$$

Hence

$$Q_\epsilon(\tau) = o(\tau v'(\epsilon\tau)) \qquad\qquad \text{as} \quad \tau \to \infty$$

which gives (5.13).

It is well known that if $\omega(f; t) = o(t)$ as $t \to 0$, then f must be a constant. Hence, in general, $Q_\epsilon(\tau)$ decreases much faster than $\omega_{I_\epsilon}(f; 1/\tau)$ as $\tau \to \infty$. Therefore we may state

COROLLARY 2 . *If in the situation of Theorem 5 the restriction of* f *to* I_ϵ *is not a constant, then*

$$\| f - \mathscr{C}_\tau[f] \|_I \leq \gamma_6 \, \omega_{I_\epsilon}\left(f; \frac{1}{\tau} \right) \ .$$

If a function f is holomorphic in a strip

$$S_d := \{ z : |\operatorname{Im} z| \leq d \}$$

it is known (see [15, p. 308])that the error of the best approximation by entire functions of exponential type τ converges geometrically as $\tau \to \infty$. By Theorem 4 we get the same result for the error of the interpolation by $\mathscr{C}_\tau[f]$. More quantitative results can be deduced from the following

THEOREM 6 . *Let* f *be holomorphic in the strip* S_d , *continuous on the closure of* S_d *and suppose that* $f(\cdot + iy) \in K_\alpha$ *with* (5.1) *holding uniformly for* $y \in [-d, d]$. *Then for* $z = x + iy \in S_d$

$$(5.18) \qquad f(z) - \mathcal{C}_\tau[f](z) = \frac{\sin\tau z}{2\pi i}\left(-\int_{-\infty}^{\infty} \frac{f(t + id)\ \Phi(\tau(z-t-id))}{(t+id-z)\ \sin(\tau(t + id))}\ dt\right.$$

$$\left. + \int_{-\infty}^{\infty} \frac{f(t-id)\ \Phi(\tau(z - t + id))}{(t - id - z)\ \sin(\tau(t - id))}\ dt\right)$$

where $\ \Phi := \phi(\frac{1 + \alpha}{2}, \frac{\delta}{2}, \cdot)\ $ *is the function in* $\ (4.4)$.

Proof. Let us assume that f is holomorphic in the closure of S_d since otherwise we may first prove (5.18) for $d' < d$ and extend the result by a limiting process. Denote by $\mathcal{L}_N$ the positively oriented boundary of the rectangle with vertices at $\pm(N + 1/2)\pi/\tau \pm id$. Then by the residue theorem

$$(5.19) \qquad f(z) - \sum_{n=-N}^{N} f(n\pi/\tau)\ \Lambda(\tau z - n\pi)$$

$$= \frac{\sin\tau z}{2\pi i} \int_{\mathcal{L}_N} \frac{f(\xi)\ \Phi(\tau(z - \xi))}{(\xi - z)\ \sin\tau\xi}\ d\xi .$$

Analoguously to (5.2) we can estimate

$$(5.20) \qquad |\Phi(z)| \leq c_7 \exp(\delta|y| - w_\beta(x)), \quad \beta := \frac{1 + 3\alpha}{4} .$$

Hence in the integral in (5.19) the contributions of the line segments parallel to the imaginary axis vanish as $N \to \infty$ and we obtain (5.18).

COROLLARY 3. *Under the assumptions of Theorem* 6

$$(5.21) \qquad |f(z) - \mathcal{C}_\tau[f](z)| \leq$$

$$\frac{\gamma_7}{\tau(d - |y|)} \cdot \frac{\cosh\tau y}{\sinh\tau d} \cdot \exp(\delta\tau(d + |y|) + w_\alpha(x)) .$$

In particular, for $\lambda := (1 - \delta)d - (1-\delta)\cdot|y| > 0$ *we have point-wise*

228

$$(5.22) \qquad |f(z) - \mathcal{C}_\tau[f](z)| = \mathcal{O}(\tfrac{1}{\tau} e^{-\lambda\tau}) \quad as \quad \tau \to \infty \ .$$

REMARK. The proof will show that if f is in addition bounded on S_d , then $w_\alpha(x)$ can be dropped on the right hand side of (5.21).

Proof of Corollary 3. Using the estimates

$$|\sinh y| \le |\sin(x + iy)| \le \cosh y$$

we deduce from (5.18)

(5.23)

$$|f(z) - \mathcal{C}_\tau[f](z)| \le \frac{1}{\pi} \ \frac{\cosh \tau y}{\sinh \tau d} \cdot \frac{1}{d - |y|} \cdot M(x,y,\tau,d)$$

where

$$M(x,y,\tau,d) := \max_{\varepsilon = \pm i} \int_{-\infty}^{\infty} |f(t + \varepsilon d)\ \Phi(\tau(z - t - \varepsilon d))|\ dt \ .$$

By our assumptions on f

$$|f(x + iy)| \le \gamma_7 \exp w_\alpha(x) \qquad\qquad for \ |y| \le d \ .$$

Hence together with (5.20)

$$(5.24) \quad M(x,y,\tau,d) \le c_7\gamma_7 \exp(\delta\tau(d + |y|)) \int_{-\infty}^{\infty} \exp(w_\alpha(t) - w_\beta(\tau(x-t)))\ dt.$$

Let us finally estimate the integral. For this purpose we first note that

$$w_\alpha(u + x) \le w_\alpha(|u| + |x|) + c_8 \le w_\alpha(u) + w_\alpha(x) + c_8$$

and for large τ

$$w_\alpha(u) - w_\beta(\tau u) \le -\frac{1}{2} w_\beta(\tau u) \ .$$

Therefore, substituting $u := t - x$ we obtain

$$(5.25) \quad \int_{-\infty}^{\infty} \exp(w_\alpha(t) - w_\beta(\tau(x-t)))dt = \int_{-\infty}^{\infty} \exp(w_\alpha(u+x) - w_\beta(\tau u))\, du$$

$$\leq \exp(w_\alpha(x) + c_8) \int_{-\infty}^{\infty} \exp\left(-\frac{1}{2} w_\beta(\tau u)\right) du$$

$$= \frac{2}{\tau} \exp(w_\alpha(x) + c_8) \int_{0}^{\infty} \exp\left(-\frac{1}{2} w_\beta(v)\right) dv \ .$$

The last integral depends only on α . Combining (5.23), (5.24) and (5.25) we obtain (5.21) which immediately implies (5.22).

5.4. Quadrature.

Let ω be a non-negative continuous weight function defined on $\mathbb{R}$ such that

$$\int_{-\infty}^{\infty} \omega(x) \, | f(x) | \, dx$$

exists for every function f of exponential type. An example of such a weight function is $\omega(x) = \exp(-x^2)$. In the situation of Theorem 3 we may multiply both sides of (5.6) by ω and integrate along $\mathbb{R}$. It is not difficult to see that on the right hand side the order of summation and integration may be exchanged. This way we obtain

THEOREM 7 . *Let ω be a weight function as specified above. Then there exists a quadrature formula*

$$(5.26) \quad \int_{-\infty}^{\infty} \omega(x) \, f(x) \, dx = \sum_{n=-\infty}^{\infty} A_{\tau n} \cdot f\left(\frac{n\pi}{\tau}\right) + R_\tau[f]$$

which is applicable to every $f \in K_\alpha$. The infinite series in (5.26) converges absolutely. Moreover $R_\tau[f] = 0$ for all $f \in K_\alpha \cap E_{(1-\delta)\tau}$.

REMARKS. 1. Notice that in particular $R_\tau[f] = 0$ whenever f is a polynomial (of arbitrary degree!). A corresponding result does not exist for quadrature on finite intervals. This has to do with density properties of the set of polynomials.

2. If f is holomorphic in a strip S_d we may use Theorem 6 to obtain a representation of $R_\tau[f]$.

For further comments see [6] .

5.5. Numerical Aspects.

The function Φ which plays a crucial role in the definition of the operator $\mathcal{C}_\tau$ is not easy to handle since it is given by an infinite product. It is therefore important to mention that we may truncate the product (4.3) thus replacing $\Phi(x)$ by

$$\Phi_N(x) := \prod_{n=1}^{N} \frac{\sin a_n x}{a_n x} \qquad (N \geq 1) .$$

In this case our results remain true except that K_α has to be replaced by $P_{N,\varepsilon}$, the class of all functions defined and continuous on $\mathbb{R}$ such that

$$|f(x)| = \mathcal{O}(|x|^{N-\varepsilon}) \qquad (\varepsilon > 0)$$

as $x \to \pm \infty$. As a consequence only Theorem 2 , (5.13) and (5.21) have to be modified. This way we obtain a sequence of operators $\mathcal{C}_{\tau,N}$ applying to larger classes of functions with increasing N . The empty product (N=0 and $\Phi_N \equiv 1$) leads to Whittaker's operator C_τ . Therefore we can gradually improve from C_τ up to $\mathcal{C}_\tau$ by increasing N . The main disadvantage of small N is that we lose in the evaluation problem (a). *The larger* N *gets, the faster the series* $\mathcal{C}_{\tau,N}[f](x)$ *converges, the less samples we need for approximating* $\mathcal{C}_{\tau,N}[f](x)$ *with a prescribed accuracy.*

Instead of using a truncated product (4.3) we may as well take just *one* sine function and factor out its zeros of small modulus. The result can be written in terms of the gamma function. This leads us to

$$\Phi_n : z \mapsto \frac{((n-1)!)^2}{\Gamma(n + z\delta/\pi)\, \Gamma(n - z\delta/\pi)}$$

which is an entire function of exponential type δ such that $\Phi_n(0) = 1$ and

$$| \Phi_n(x) | = \mathcal{O}(|x|^{-2n+1}) \qquad \text{as } x \to \pm \infty.$$

6. A Twodimensional Sampling Problem.

Boas [3] proved that an entire harmonic function u of exponential type less than π is uniquely determined by its values on the lattice points n, n+i (n $\in \mathbb{Z}$). He asked for a reconstruction formula. Ching & Chui [5] found a partial answer, somewhat analoguous to the Sampling Theorem. It requires that u and u (. + i) belong to $L^2(\mathbb{R})$. Involving the pseudo-window $\Phi = \Phi((1 + \alpha)/2, \delta/2, .)$ via its Fourier transform we constructed an entire harmonic function H of exponential type $\pi + \delta$ such that the following theorem holds (see [11]).

THEOREM 8. *If* u *is an entire harmonic function of exponential type* $\pi - \delta$ *such that*

$$\left. \begin{array}{l} | \, u(n) \, | \\ \\ | \, u(n + i) \, | \end{array} \right\} = \mathcal{O} (\exp w_\alpha(n)) \qquad \qquad as \qquad n \to \underline{+} \, \infty \; ,$$

then

$$u(z) = \sum_{n = -\infty}^{\infty} u(n) \cdot H(z-n) + u(n+i) \cdot H(\overline{z}-n+i).$$

The series converges absolutely and uniformly on every compact subset of $\mathbb{C}$.

REFERENCES

1. R.P. BOAS: Entire functions. Academic Press, New York 1954.

2. R.P. BOAS: Summation formulas and band-limited signals. Tôhoku Math. J.
24 (1972), 121-125.

3. R.P. BOAS: A uniqueness theorem for harmonic functions. J. Approximation
Theory 5 (1972), 425-427.

4. P.L. BUTZER: The Shannon sampling theorem and some of its generalizations.
An overview. In: B. Sendov et al. (eds.): "Constructive Function
Theory 81". Publ. House Bulg. Acad. Science, Sofia 1983,
p. 258-273.

5. C.-H. CHING and C.K. CHUI: A representation formula for harmonic functions.
Proc. Amer. Math. Soc. 39 (1973), 349-352.

6. R. GERVAIS, Q.I. RAHMAN and G. SCHMEISSER: A quadrature formula of infinite
order. In: G. Hämmerlin (ed.): "Numerical Integration". ISNM 57,
Birkhäuser, Basel 1982, p. 107-118.

7. R. GERVAIS, Q.I. RAHMAN and G. SCHMEISSER: A bandlimited function simula-
ting a duration-limited one. In: P.L. Butzer et al. (eds.):
"Anniversary Volume on Approximation Theory and Functional Analysis". ISNM 65, Birkhäuser, Basel 1984, p. 355-362.

8. A.J. JERRI: The Shannon sampling theorem - its various extensions and
applications: A tutorial review. Proc. IEEE 65(1977), 1565-1596.

9. N. LEVINSON: Gap and Density Theorems. Amer. Math. Soc. Colloq. Publ. 26,
New York 1940.

10. A.V. OPPENHEIM and R.W.SCHAFER: Digital Signal Processing. Prentice-Hall
Inc., Englewood Cliffs, New Jersey 1975.

11. Q.I. RAHMAN and G. SCHMEISSER: Representation of entire harmonic functions
from given values. J. Math. Anal. Appl. (to appear).

12. I.J. SCHOENBERG: Contributions to the problem of approximation of equidistant data by analytic functions. Quart. Appl. Math. 4(1946), 45-99 and 112-141.

13. I. J. SCHOENBERG: Cardinal Spline Interpolation. Regional Conference Series in Applied Mathematics 12, SIAM, Philadelphia 1973.

14. F. STENGER: Numerical methods based on Whittaker cardinal, or sinc functions. SIAM Review 23 (1981), 165-224.

15. A.F. TIMAN: Theory of Approximation of Functions of a Real Variable. Pergamon Press, Oxford 1963.

International Series of
Numerical Mathematics, Vol. 74
© 1985 Birkhäuser Verlag Basel

TIME DELAY AND DOPPLER FREQUENCY SHIFT
IN RADAR/SONAR DETECTION,
WITH APPLICATION TO FOURIER OPTICS

Walter Schempp

Lehrstuhl für Mathematik I

Universität Siegen

Siegen

Dedicated to Prof. Dr. Karl Zeller
on the occasion of his 60th birthday

This paper presents a solution of the radar synthesis problem and the invariance problem for radar ambiguity surfaces over the symplectic time-frequency plane via harmonic analysis on the real Heisenberg nilpotent group. In the same vein, the paper is concerned with the phase anomaly of Fourier and microwave optics. As a mathematical by-product, an identity for Laguerre functions of different orders pops up. Some of its special cases, to wit, some theta identities are explicitly calculated.

1. Introduction

Let us start with a concise explanation of the basic mathematical principles of signal detection in order to point out the fundamental rôle played by the time delay and the Doppler frequency shift in radar detection of moving targets.

The purpose of radar (=**ra**dio **d**etection **and r**anging) and sonar systems is basically to detect the presence of a target, and to extract information of interest such as **range** d, radial **velocity** v relative to the transmitter, etc., about the target. The figure below shows an elementary form of a radar system using a common antenna for both transmission and reception, which is achieved by means of a duplexer.

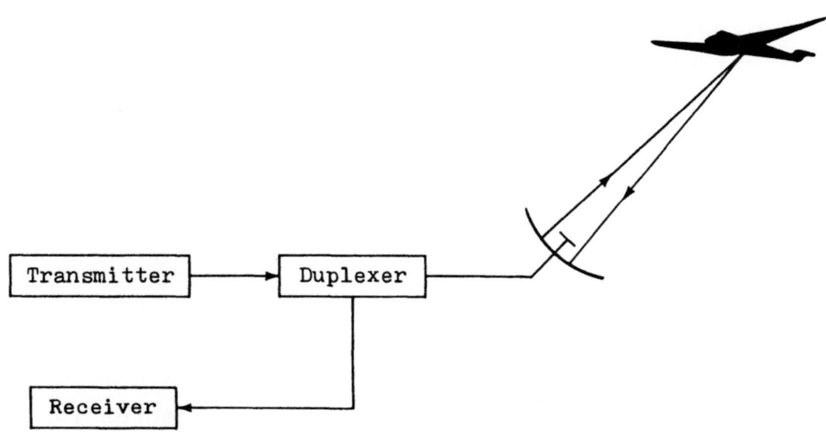

In the transmission mode, the antenna radiates periodically a narrow beam of electromagnetic energy of a few centimeters' wavelength in the form of trains ("blocks") of coherent radar pulses of large amplitude and brief duration and the same carrier frequency ω . If a remote target lies in the path of the propagating beam, a portion of the transmitted signal energy is reflected by the target and picked up by the antenna (operating in the reception mode). The echo signal is then processed in the receiver to detect the presence of the target and estimate the parameters for tracking the target.

The signals received in radar and in most communication systems consist of a high-frequency carrier modulated in amplitude (or phase) by functions f of time that vary much more slowly than the cycles of the carrier. In radar the parameters chiefly serving to distinguish or resolve two echo signals are their

$$\text{arrival times x}$$

and the

$$\text{Doppler shifts y}$$

of their carrier frequencies from a common reference frequency within the spectral band. The transmitted frequency ω is a natural choice for the reference frequency of a single train of coherent pulses.

The structure of the receiver and its performance depend upon the symmetrized auto-correlation or radar auto-ambiguity function $H(f;x,y)$ associated with the envelope f of the transmitted signal pulse

$$t \longrightarrow f(t)e^{2\pi i \omega t}.$$

It follows $d = \frac{1}{2}cx$ and $v = \frac{1}{2}c\frac{y}{\omega}$ where c denotes the velocity of electromagnetic radiation. We always assume that f belongs to the Schwartz space $\mathscr{S}(\mathbf{R})$ of infinitely differentiable complex-valued functions on the real line $\mathbf{R}$ that are rapidly decreasing at infinity. We will consider $\mathscr{S}(\mathbf{R})$ as an (everywhere dense) vector subspace of the standard complex Hilbert space $L^2(\mathbf{R})$. The energy of the signal is then given by the integral (squared L^2-norm)

$$\|f\|^2 = \int_{\mathbf{R}} |f(t)|^2 \, dt.$$

Thus the L^2-norm on the complex prehilbert space $\mathscr{S}(\mathbf{R})$ may be considered as the signal energy norm.

The **radar auto-ambiguity function** which plays a central rôle in the mathematical theory of radar and sonar systems (Ville [26], Woodward [30], Wilcox [29]) takes the form

$$H(f;x,y) = \int_{\mathbf{R}} f(t+\tfrac{1}{2}x)\overline{f}(t-\tfrac{1}{2}x)e^{2\pi i y t} \, dt.$$

While the radar auto-ambiguity function is complex-valued, the **Wigner distribution function** used in quantum mechanics (see Wigner [28] and, for instance, Balazs-Jennings [3] and Hillery-O'Connell-Scully-Wigner [11])

$$P(f;q,p) = \int_{\mathbf{R}} f(q+\tfrac{1}{2}t)\overline{f}(q-\tfrac{1}{2}t)e^{2\pi i p t} \, dt$$

is real-valued but not always positive on the position-momentum plane $\mathbf{R} \oplus \mathbf{R}$. The functions $H(f;.,.)$ and $P(f;.,.)$ are related through a double Fourier transform. The symmetrized cross-correlation or **radar cross-ambiguity function** $H(f,g;.,.)$ associated with $f \in \mathscr{S}(\mathbf{R})$ and $g \in \mathscr{S}(\mathbf{R})$ is similarly defined via the prescription

$$H(f,g;x,y) = \int_{\mathbf{R}} f(t+\tfrac{1}{2}x)\overline{g}(t-\tfrac{1}{2}x)e^{2\pi iyt}dt$$

and is of importance in communication theory.

The first problem to be solved is the **radar synthesis problem** (see, for instance, Wilcox [29]). It asks for an intrinsic characterization of those functions $F \in \mathscr{S}(\mathbf{R} \oplus \mathbf{R})$ on the time-frequency plane $\mathbf{R} \oplus \mathbf{R}$ (or "information plane" in the sense of Gabor [7]) for which there exists a complex-valued envelope $f \in \mathscr{S}(\mathbf{R})$ satisfying the identity

$$F = H(f;.,.).$$

In other words, the problem is to find necessary and sufficient conditions for a given complex-valued smooth function F in the two Fourier dual variables $x \in \mathbf{R}$ (separation in time) and $y \in \mathbf{R}$ (separation in frequency) such that F can be realized as a radar auto-ambiguity function $(x,y) \longrightarrow H(f;x,y)$ with respect to a complex-valued smooth waveform $t \longrightarrow f(t)$ in one (time) variable $t \in \mathbf{R}$. In Section 2 we will establish a solution of the radar synthesis problem by means of harmonic analysis on the real Heisenberg nilpotent group $\widetilde{A}(\mathbf{R})$. Ultimately, the solution is based on the analogy between non-relativistic quantum mechanics and signal theory which has been emphasized in the classical studie of Ville [26] and particularly in the fundamental work of Gabor [7]. However, their point of view is less formal than the one adopted in the present paper.

The image $\mathscr{F} = H(f;\mathbf{R},\mathbf{R})$ of the time frequency plane $\mathbf{R} \oplus \mathbf{R}$ under the radar auto-ambiguity function $H(f;.,.)$ is called to be the **radar ambiguity surface** over the time-frequency plane ge-

nerated by the complex envelope $f \in \mathscr{S}(\mathbf{R})$. For every signal the
radar ambiguity surface is peaked at the origin (0,0) of the
time-frequency plane $\mathbf{R} \oplus \mathbf{R}$ so that certainly not all the functions $F \in \mathscr{S}(\mathbf{R} \oplus \mathbf{R})$ can be realized as radar auto-ambiguity functions with respect to a suitable waveform $f \in \mathscr{S}(\mathbf{R})$. A second signal arriving with separations x in time and y in frequency that
lie under this central peak will be difficult to distinguish from
the first signal. For many types of signals the radar ambiguity
surface exhibits additional peaks elsewhere over the time-
frequency plane. These sidelobes may conceal weak signals with
arrival times and carrier frequencies far from those of the first
signal. In a measurement of the arrival time and frequency of a
single signal, the subsidiary peaks may lead to gross errors in
the result. The taller the sidelobes of the radar ambiguity surface, the greater the probability of such errors in time and Doppler frequency shift. It is desirable, therefore, for the central
peak of the radar ambiguity surface to be narrow, and for there
to be as few and as low sidelobes as possible.

By changing the waveform f of a radar signal of given energy
it is possible to change the accuracy of the range and relative
radial velocity measurements in such a manner that an increase of
the range accuracy results in a decrease of the velocity accuracy, and vice versa. The preceding statement constitutes the es-
sence of the **radar uncertainty principle.** If the waveform $f \in \mathscr{S}(\mathbf{R})$
is normalized such that $\|f\| = 1$ holds, the radar uncertainty
principle can be expressed in terms of the radar auto-ambiguity
function by the formula

$$\int_{\mathbf{R} \oplus \mathbf{R}} |H(f;x,y)|^2 \, dxdy = 1.$$

It parallels the Heisenberg uncertainty principle of quantum
mechanics according to which not all physical quantities observed
in any realizable experiment (even in principle only) can be determined with an arbitrarily high accuracy. Even under ideal ex-

perimental conditions, an increase in the measurement accuracy of
one quantity can be achieved only at the expense of decreasing
the measurement accuracy on another "canonically conjugate" quantity. The position coordinate q and its momentum p is one example
of two such canonically conjugate quantities: It is impossible to
determine simultaneously the position q and momentum p of a non-
relativistic quantum-mechanical particle. If $f \in \mathcal{S}(\mathbf{R})$ denotes a
normalized state vector, the identity

$$\int_{\mathbf{R} \oplus \mathbf{R}} |P(f;q,p)|^2 \, dq dp = 1$$

is an expression for the Heisenberg uncertainty principle in
terms of the Wigner distribution function.

As everyone knows, to fully understand any mathematical system one has to understand the transformations of the system and
especially those transformations of the system that leave some
particular aspect of the system invariant. In case of the mathematical theory of radar/sonar systems, a close investigation of
the radar uncertainty principle leads to a study of the geometry
of the radar ambiguity surfaces $\mathscr{F} = H(f;\mathbf{R},\mathbf{R})$ over the time-frequency plane by means of their energy preserving linear **automor-
phisms.** By such an automorphism of $\mathscr{F}$ we will understand a unitary operator S: $L^2(\mathbf{R}) \longrightarrow L^2(\mathbf{R})$ that maps the vector subspace
$\mathcal{S}(\mathbf{R})$ onto itself such that for all waveforms $f \in \mathcal{S}(\mathbf{R})$ and for
each pair $(x,y) \in \mathbf{R} \oplus \mathbf{R}$ there exists a pair $(x',y') \in \mathbf{R} \oplus \mathbf{R}$ depending on S that satisfies the identity

$$H(f;x,y) = H(S(f);x',y').$$

The second problem to be solved is the **invariance problem** for
radar ambiguity surfaces over the time-frequency plane of calculating explicitly their energy preserving linear automorphisms.
A solution of the invariance problem will be given in Section 3
by means of the linear oscillator representation of the meta-
plectic group $Mp(1,\mathbf{R}) = Sp(1,\mathbf{R}) \times \{+1,-1\}$. The result enables us

to determine, for instance, the radially symmetric, i.e.,
SO(2,**R**)-invariant radar ambiguity surfaces over the time-
frequency plane **R** $\oplus$ **R** and leads in a natural way to the Laguerre
functions. The process of cutting down to the compact Heisenberg
nilmanifold then yields a new identity for the Laguerre functions. Some consequences of this identity will be studied in Section 4 infra. Section 5 is concerned with a treatment of cardinal
spline interpolation and digital signal processing from the view
point of harmonic analysis on the compact Heisenberg nilmanifold.
Finally, Section 6 deals briefly with an application of the con-
cept of ambiguity surface to Fourier optics.

Acknowledgements. The author is very grateful to Professor Edwin
Hewitt (Seattle, WA) for inviting him to lecture on the subject
at the Department of Mathematics of the University of Washington
and for his friendship over the years. Moreover, the author's
thanks are due to Professor Kurt Bernardo Wolf (Mexico City) for
his hospitality at the Instituto de Investigaciones en Matema-
ticas Aplicadas y en Sistemas (IIMAS) of the Universidad Nacional
Autonoma de Mexico (UNAM).

2. The Radar Synthesis Problem

D. Gabor (Nobel prize 1971) has pointed out in his classical
paper [6] that there are two fundamentally distinct approaches to
the description of nature: that of time and that of frequency.
Both approaches are combined by the notion of **time-frequency** (or
information) **plane** which is of basic importance in information
theory. In the following we look upon the time-frequency plane as
the two-dimensional real vector space **R** $\oplus$ **R** of all pairs v=(x,y).
Define the standard symplectic (=non-degenerate antisymmetric bi-
linear) form B on **R** $\oplus$ **R** via the prescription

$$B(v_1,v_2) = \det \begin{pmatrix} x_1 & y_1 \\ x_2 & y_2 \end{pmatrix}.$$

It is well known that B may be identified with an element of the real vector space of exterior forms, $\Lambda^2(\mathbf{R} \times \mathbf{R}^*)$. A complex-valued function on $\mathbf{R} \oplus \mathbf{R}$ is called to be of **positive type** on the two-dimensional real **symplectic** vector space $(\mathbf{R} \oplus \mathbf{R};B)$ if for **all** finite sequences of vectors $(v_j)_{1 \leq j \leq N}$ in $\mathbf{R} \oplus \mathbf{R}$ the matrix

$$(e^{-\pi i B(v_j, v_k)} F(v_j - v_k))_{\substack{1 \leq j \leq N \\ 1 \leq k \leq N}}$$

is a positive definite Hermitian matrix. It should be observed that the notion of positive definiteness on the symplectic plane $(\mathbf{R} \oplus \mathbf{R};B)$ is essentially different from the notion of positive definiteness on the Euclidean plane $\mathbf{R} \oplus \mathbf{R}$.

A complex-valued function F on $\mathbf{R} \oplus \mathbf{R}$ is called to be of **pure positive type** on the symplectic vector space $(\mathbf{R} \oplus \mathbf{R};B)$ if each decomposition $F = F_1 + F_2$ of F into a sum of functions F_1 and F_2 of positive type on $(\mathbf{R} \oplus \mathbf{R};B)$ implies that F_1 and F_2 are proportional to F (cf. Souriau [25]).

The **real Heisenberg group** $\widetilde{A}(\mathbf{R})$ is the three-dimensional real Lie group with underlying manifold $(\mathbf{R} \oplus \mathbf{R}) \times \mathbf{R}$ and multiplication given by

$$(v_1, z_1) \cdot (v_2, z_2) = (v_1 + v_2, z_1 + z_2 + \tfrac{1}{2} B(v_1, v_2)).$$

Another realization of $\widetilde{A}(\mathbf{R})$ is given by the unipotent matrices

$$\begin{pmatrix} 1 & x & z \\ 0 & 1 & y \\ 0 & 0 & 1 \end{pmatrix}$$

with real entries x, y, z. See, for instance, Auslander [1]. The center $\widetilde{C} = \{(0,0,z) \mid z \in \mathbf{R}\}$ of $\widetilde{A}(\mathbf{R})$ is isomorphic to the additive group $\mathbf{R}$ and the descending central series as well as the derived series of $\widetilde{A}(\mathbf{R})$ are given by the filtration

$$\widetilde{A}(\mathbf{R}) \hookleftarrow \widetilde{C} \hookleftarrow \{1\} \ .$$

Thus $\widetilde{A}(\mathbf{R})$ is a connected, simply connected, 2-step nilpotent, three-dimensional, real Lie group with one-dimensional center and this property characterizes $\widetilde{A}(\mathbf{R})$ up to isomorphy.

Let us describe the topologically irreducible continuous unitary linear representations of $\widetilde{A}(\mathbf{R})$. Their knowledge goes back to v. Neumann [14]. Since $\widetilde{A}(\mathbf{R})$ is a connected solvable locally compact topological group all its finite dimensional topologically irreducible continuous unitary linear representations are necessarily one-dimensional. Since $\widetilde{A}(\mathbf{R})$ is simply connected, it forms a monomial real Lie group. Let $\boldsymbol{w}$ denote its Lie algebra, the three-dimensional **real Heisenberg nilpotent Lie algebra**. The Mackey machine or, adopting a more geometric point of view, the Kirillov coadjoint orbit picture in the dual vector space $\boldsymbol{w}^{*}$ of $\boldsymbol{w}$ show that the one-dimensional continuous unitary linear representations of $\widetilde{A}(\mathbf{R})$ can be parametrized by the set of pairs $(\alpha,\beta) \in \mathbf{R} \oplus \mathbf{R}$. The infinite dimensional topologically irreducible continuous unitary linear representations of $\widetilde{A}(\mathbf{R})$, however, form up to isomorphy a family (U_{λ}) where the parameter λ runs through the set $\mathbf{R}^{*} = \mathbf{R} - \{0\}$ of non-zero real numbers. As mentioned previously, this description of the unitary dual of $\widetilde{A}(\mathbf{R})$ goes back to the classical paper of v. Neumann [14]. It should be noticed that in quantum mechanics λ stands for Planck's constant. The prototype among the generic topologically irreducible continuous unitary linear representations of $\widetilde{A}(\mathbf{R})$ is the **linear Schrödinger representation** $U := U_1$ acting on the complex Hilbert space $L^{2}(\mathbf{R})$. It should be observed that the Schwartz space $\mathcal{S}(\mathbf{R})$ is the vector subspace of $L^{2}(\mathbf{R})$ of $\mathscr{C}^{\infty}$-vectors for U on which it acts via the prescription

$$U(v,z)f(t) = e^{2\pi i\left(z+\frac{1}{2}xy+yt\right)}f(t+x) \quad (t \in \mathbf{R}).$$

Let the symbol $\otimes$ denote the dyadic tensor product with respect to the prehilbert space structure of $\mathcal{S}(\mathbf{R})$ and tr_{U} the trace functional on $\mathcal{S}(\mathbf{R})$ composed with U. It follows that the radar cross-ambiguity function $H(f,g;.,.)$ is equal to the restriction of the **coefficient function**

$$c_{U,f,g} = tr_U(f \otimes g)$$

of U onto the polarized cross-section of $\tilde{A}(\mathbf{R})$ to the center $\tilde{C}$ (cf. Howe [12]). Thus we have the identity

$$H(f,g;x,y) = c_{U,f,g}(x,y,0)$$

for all functions $f \in \mathscr{d}(\mathbf{R})$, $g \in \mathscr{d}(\mathbf{R})$ and all pairs $(x,y) \in \mathbf{R} \oplus \mathbf{R}$, where $c_{U,f,g}$ is a square integrable function on $\tilde{A}(\mathbf{R})$ modulo $\tilde{C}$ for all elements $f \in L^2(\mathbf{R})$, $g \in L^2(\mathbf{R})$ by the flatness of the coadjoint orbit in $\mathbf{w}^*$ associated with U under the Kirillov correspondence (cf. Moore-Wolf [13]). The square integrability of the linear Schrödinger representation U of $\tilde{A}(\mathbf{R})$ modulo $\tilde{C}$ implies the Moyal orthogonality conditions of its coefficient functions $c_{U,f,g}$ which in turn imply the radar uncertainty principle

$$\int_{\mathbf{R} \oplus \mathbf{R}} |H(f;x,y)|^2 \, dxdy = \|f\|^4$$

in terms of the radar auto-ambiguity function with respect to the waveform $f \in L^2(\mathbf{R})$.

The link between the radar ambiguity functions and nilpotent harmonic analysis described above is of central importance for our approach to the mathematical theory of radar/sonar systems and related problems. It formalizes the analogy between non-relativistic quantum mechanics and signal theory mentioned in Section 1 supra. Moreover, it furnishes by a suitable adaptation of the Gelfand-Naimark-Segal (GNS) reconstruction to the Heisenberg nilpotent group $\tilde{A}(\mathbf{R})$ and an application of Schur's lemma the following solution of the radar synthesis problem (cf. [20], [21]).

<u>Theorem 1.</u> Let the function $F \in \mathscr{d}(\mathbf{R} \oplus \mathbf{R})$ be given. There exists a signal envelope $f \in \mathscr{d}(\mathbf{R})$ such that

$$F = H(f; ., .)$$

holds if and only if F is of pure positive type on the symplectic time-frequency plane $(\mathbf{R} \oplus \mathbf{R}; B)$. In this case, the waveforms f which can be synthesized from F are determined uniquely up to a multiplicative complex constant of modulus 1.

The preceding result proves the efficiency of the nilpotent harmonic analysis approach to the mathematical theory of radar/sonar detection.

3. The Invariance Problem for Radar Ambiguity Surfaces

In his pioneering work on the metaplectic group, Weil [27] reformulated large parts of the theory of quadratic forms using the language of unitary linear group representations. We will follow this use by letting act the symplectic group $Sp(\mathbf{R} \oplus \mathbf{R}; B) = Sp(1, \mathbf{R}) = SL(2, \mathbf{R})$ in the natural way on the real Heisenberg nilpotent group $\widetilde{A}(\mathbf{R})$ leaving its center $\widetilde{C}$ pointwise fixed. Moreover, we let $\widetilde{\sigma} \longrightarrow T_{\widetilde{\sigma}}$ denote the linear oscillator representation of the metaplectic group $Mp(1, \mathbf{R}) = Sp(1, \mathbf{R}) \times \mathbf{Z}/2\mathbf{Z}$ which doubly covers $Sp(1, \mathbf{R})$ by means of the covering epimorphism $\widetilde{\sigma} \longrightarrow \sigma$ ("isomorphism up to a sign"). An application of Segal's metaplectic formula which on its part is based on the Stone-von Neumann uniqueness theorem [14] then establishes the following solution of the invariance problem for radar ambiguity surfaces over the time-frequency plane.

Theorem 2. Let the unitary operator

$$S: L^2(\mathbf{R}) \longrightarrow L^2(\mathbf{R})$$

be a linear automorphism of the radar ambiguity surface $\mathscr{F}$ over the symplectic time-frequency plane $(\mathbf{R} \oplus \mathbf{R}; B)$ - then there exists a unique transformation $\sigma \in Sp(1,\mathbf{R})$ and a complex number $\zeta_{\widetilde{\sigma}}$ of modulus $|\zeta_{\widetilde{\sigma}}| = 1$ such that

$$S = \zeta_{\widetilde{\sigma}} T_{\widetilde{\sigma}}$$

holds.

The intertwining operators $T_{\widetilde{\sigma}}$ are Fourier transforms with respect to suitable coordinates of the symplectic time-frequency plane $(\mathbf{R} \oplus \mathbf{R}; B)$. The symmetries of the radar ambiguity surface displayed below from different viewpoints are computed by the preceding theorem.

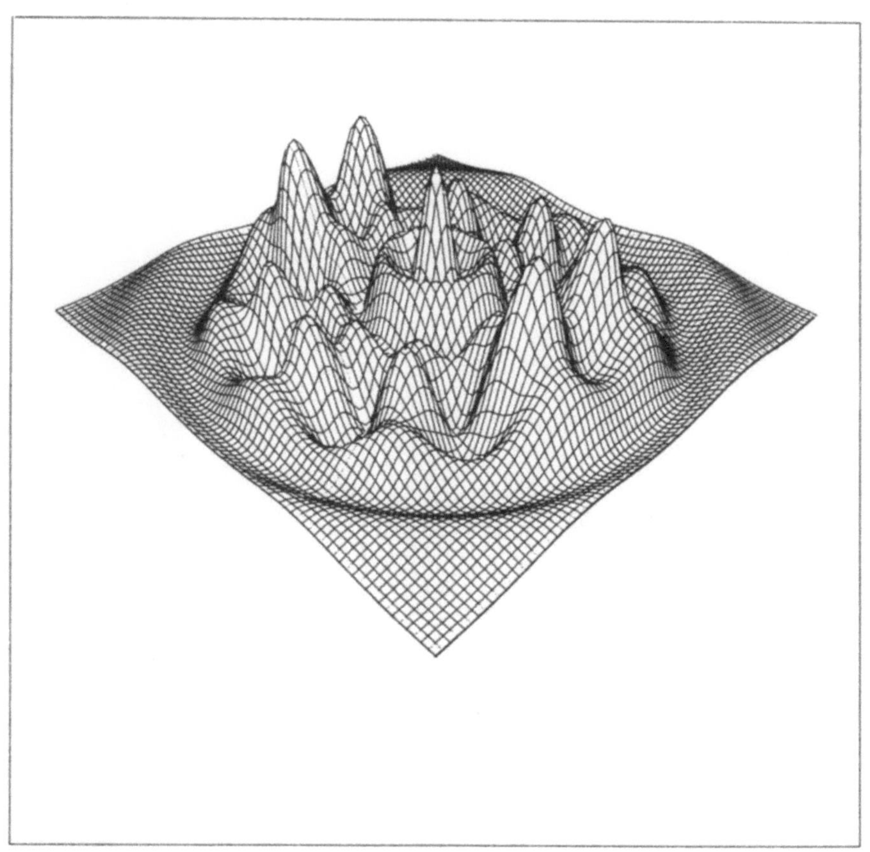

It is well known that $Sp(1,\mathbf{R})$ is generated by the matrices

$$\begin{pmatrix} a & 0 \\ 0 & a^{-1} \end{pmatrix} \quad \begin{pmatrix} 1 & 0 \\ u & 1 \end{pmatrix} \quad \begin{pmatrix} 0 & 1 \\ -1 & 0 \end{pmatrix}$$

where $a \neq 0$ and $u \neq 0$ denote real numbers. Based on a suitable polarization of $\widetilde{A}(\mathbf{R})$, an explicit computation of the operators $T_{\widetilde{\sigma}}$ for these matrices $\sigma \in Sp(1,\mathbf{R})$ gives rise to the following different form of the solution of the invariance problem for radar ambiguity surfaces over the time-frequency plane.

<u>Corollary 1.</u> The energy preserving linear automorphisms S of the radar ambiguity surface $\mathscr{F}$ = H(f;**R,R**) may be realized by finite sequences of time scalings, pointwise multiplications and convolutions on the time axis **R** of the waveform f∈$\mathscr{S}$(**R**) with the chirp signals

$$t \longrightarrow C_u \cdot e^{-\pi i u t^2} \qquad (C_u \in \mathbf{C}, \ |C_u|=1)$$

of real parameters u ≠ 0.

An explicit formula for the phase factor C_u of the chirp signals occurring in Corollary 1 will be given in Section 6 infra.

Let $(W_m)_{m\geq0}$ be the sequence of standardized Hermite functions (harmonic oscillator wave functions) and $(L_n^{(a)})_{n\geq0}$ the sequence of Laguerre functions of order $a>-1$. Then we obtain by the Bargmann-Fock-Segal model (or complex wave model, cf. Ogden-Vági [15]) of the linear Schrödinger representation U of $\widetilde{A}$(**R**) the following result.

<u>Corollary 2.</u> Let f ∈ $\mathscr{S}$(**R**) have energy-norm $\|f\|$ = 1. The radar ambiguity surface $\mathscr{F}$ = H(f;**R,R**) over the time-frequency plane **R ⊕ R** is SO(2,**R**)-invariant if and only if f = ζW_m for a certain integer m $\geq$ 0 and a complex number ζ of modulus $|\zeta|$ = 1. In this case the radar cross-ambiguity functions take the form

$$H(W_m,W_n;x,y) = \sqrt{\frac{n!}{m!}} \ (\sqrt{\pi}\,(x+iy))^{m-n} L_n^{(m-n)}(\pi\,(x^2+y^2)) \qquad (m\geq n\geq0)$$

for all pairs (x,y) ∈ **R ⊕ R**.

The figures displayed below show typical examples of radially symmetric radar ambiguity surfaces over the time-frequency plane **R ⊕ R**.

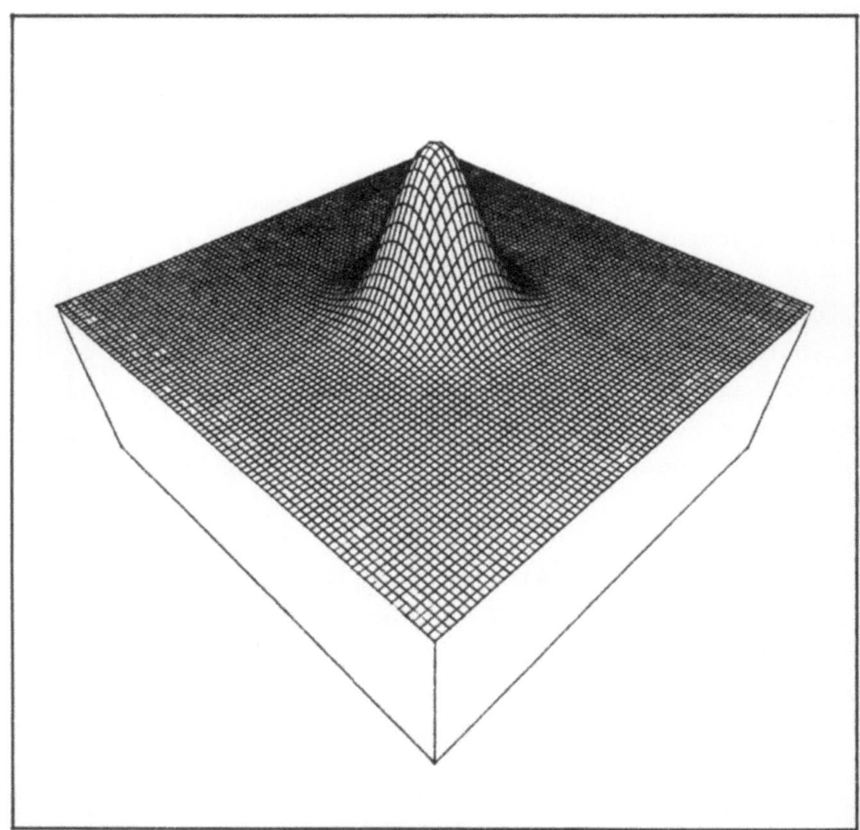

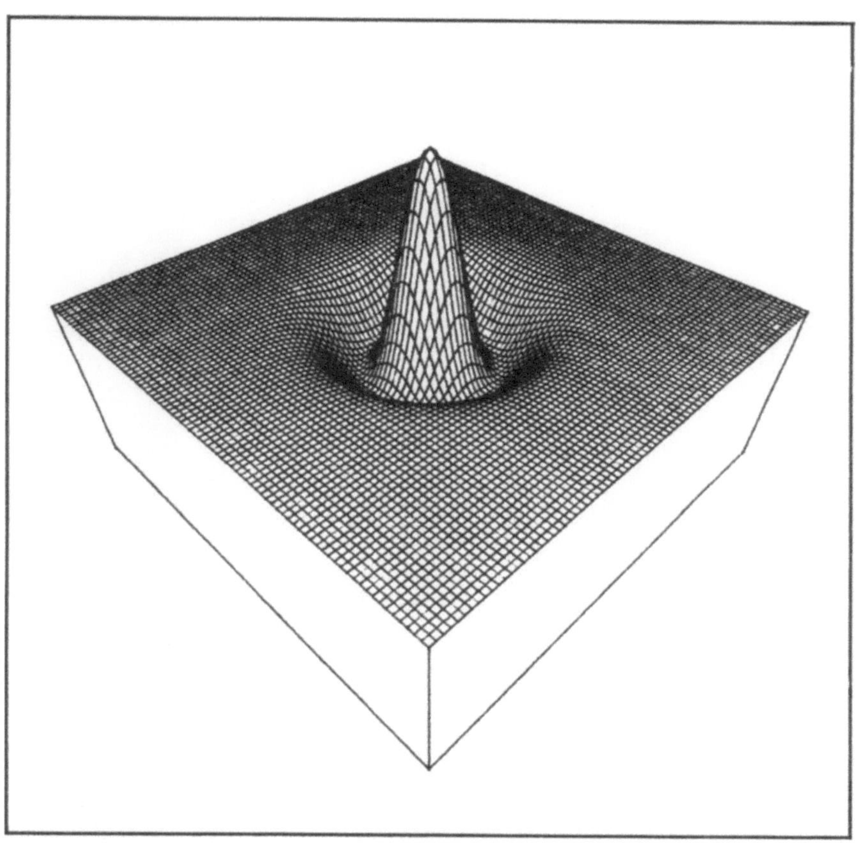

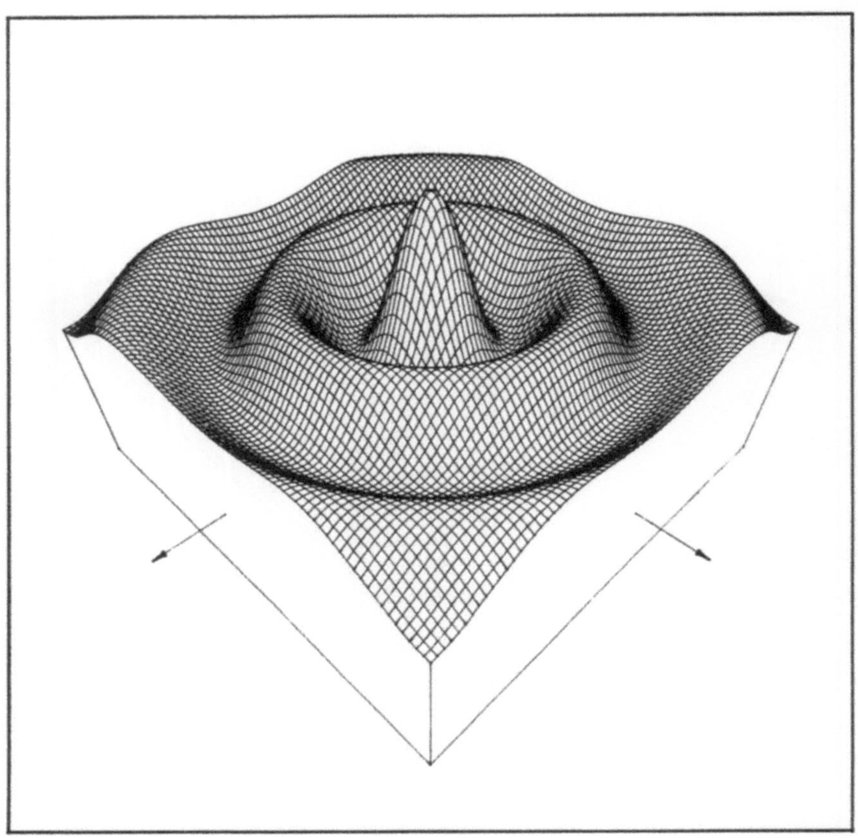

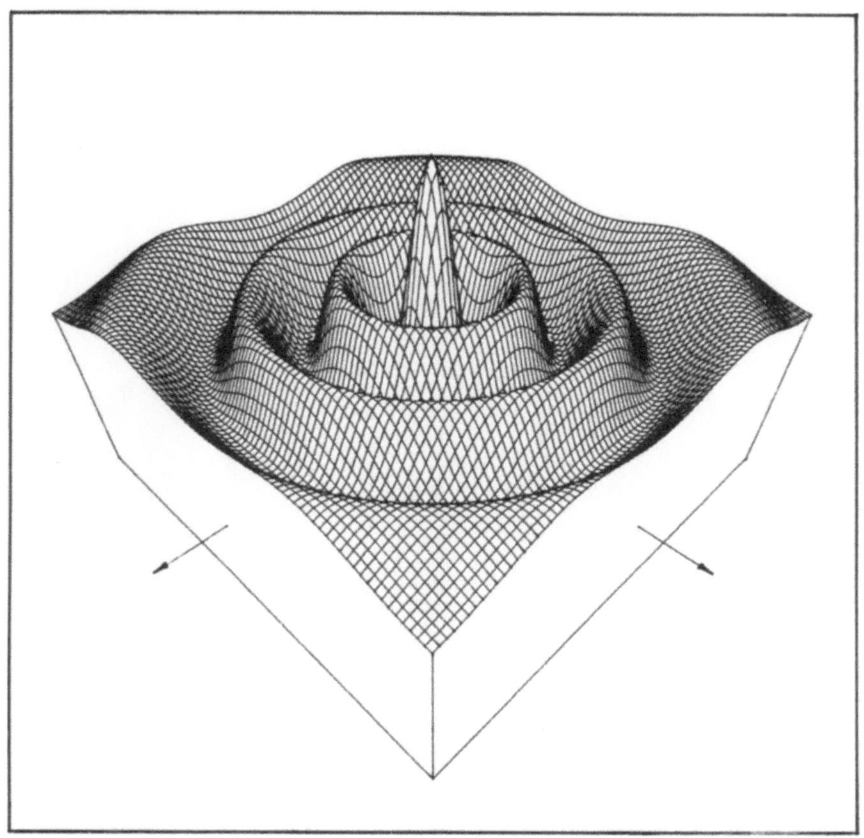

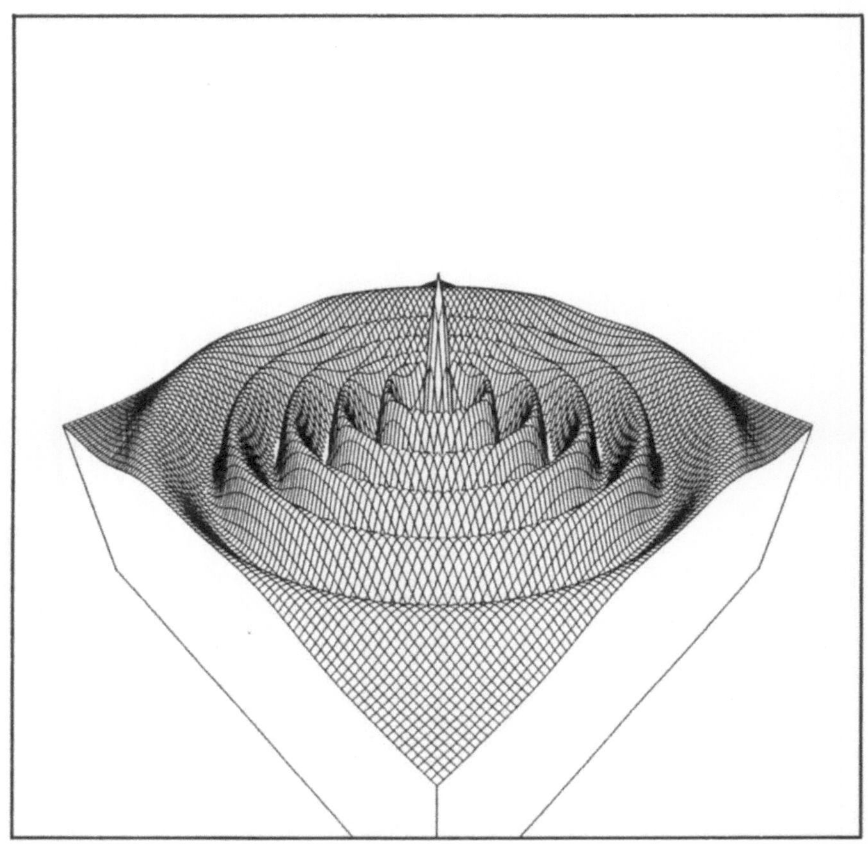

4. The Compact Heisenberg Nilmanifold

In $\widetilde{A}(\mathbf{R})$ we consider the arithmetic subgroup D defined by x, y, 2z $\in \mathbf{Z}$. The quotient $D \setminus \widetilde{A}(\mathbf{R})$ of right cosets modulo D forms a compact homogeneous manifold, the **Heisenberg nilmanifold.** It can be shown that $D \setminus \widetilde{A}(\mathbf{R})$ is a principal **circle bundle** over the two-dimensional compact torus group $\mathbf{T}^2$.

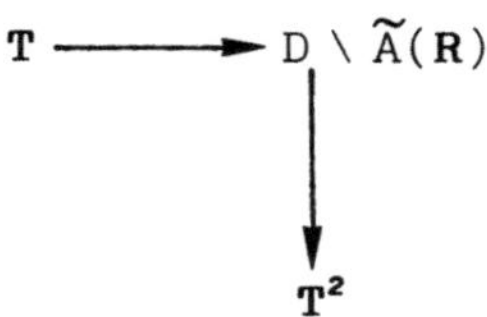

Construct the Weil-Brezin isomorphism, i.e., the periodization with a fundamental domain of $D \setminus \widetilde{A}(\mathbf{R})$ (cf. Auslander [1], [2]). It forms an intertwining operator between the lattice representation and the linear Schrödinger representation of the real Heisenberg nilpotent group $\widetilde{A}(\mathbf{R})$. Combined with the Plancherel theorem it yields the following identities for the radar auto-ambiguity functions (cf. [19], [22]).

<u>Theorem 3.</u> Suppose that m and n denote natural numbers $\geqq 0$ – then

$$\sum_{(\mu,\nu)\,\in\,\mathbf{Z}\,\times\,\mathbf{Z}} H(W_m;\mu,\nu).H(W_n;\mu,\nu) = \sum_{(\mu,\nu)\,\in\,\mathbf{Z}\,\times\,\mathbf{Z}} |H(W_m,W_n;\mu,\nu)|^2$$

holds.

<u>Corollary.</u> For all integers $m \geqq n \geqq 0$ the following identity for Laguerre functions of different orders holds:

$$\sum_{(\mu,\nu)\,\in\,\mathbf{Z}\,\times\,\mathbf{Z}} L_m^{(0)}(\pi(\mu^2+\nu^2)).L_n^{(0)}(\pi(\mu^2+\nu^2)) =$$

$$\frac{n!}{m!}\,\pi^{m-n} \sum_{(\mu,\nu)\in\mathbf{Z}\,\times\,\mathbf{Z}} (\mu^2+\nu^2)^{m-n}(L_n^{(m-n)}(\pi(\mu^2+\nu^2)))^2$$

In the case m=1, n=0 we get the identity

$$\pi \sum_{\mu \in \mathbf{Z}} \mu^2 e^{-\pi\mu^2} = \frac{1}{4} \sum_{\mu \in \mathbf{Z}} e^{-\pi\mu^2}$$

The case m=2, n=1 yields the identity

$$\pi^3 \sum_{\mu \in \mathbf{Z}} \mu^6 e^{-\pi\mu^2} = \frac{15}{32} \sum_{\mu \in \mathbf{Z}} (8\pi^2\mu^4 - 1) e^{-\pi\mu^2}$$

and m=3, n=2 yields

$$\pi^5 \sum_{\mu \in \mathbf{Z}} \mu^{10} e^{-\pi\mu^2} = \frac{45}{64} \sum_{\mu \in \mathbf{Z}} (16\pi^4\mu^8 - 140\pi^2\mu^4 + 21) e^{-\pi\mu^2}$$

The author is grateful to his student M. Schmidt for his help with these calculations. A direct proof of the preceding theta identities follows via the integral

$$\int_{\mathbf{R}} e^{-x^2} \cos ax \, dx = \sqrt{\pi}\, e^{-a^2/4}$$

and an application of the Poisson summation formula (see Grosjean [8]). The first theta identity above is also a consequence of the heat equation for the classical first order Jacobi theta functions (cf. Bellman [4]).

5. Cardinal Spline Interpolation

Let $m \geq 1$ be an integer and denote by $\Phi_m(P)$ the complex vector space of univariate **spline functions** of degree m-1 with knot set P. Thus $S \in \Phi_m(P)$ if and only if S is a (m-2)-times continuously differentiable complex-valued function on **R** and the restrictions of S to the subsequent intervals with end points in P are polynomials of degree $\leqq m-1$ with complex coefficients. In the case of spline knots set $P = \mathbf{Z}$ the **cardinal spline interpolation problem** for a given bi-infinite sequence $(y_n)_{n \in \mathbf{Z}} \in L^2(\mathbf{Z})$ reads as follows: Does there exist a cardinal spline function $S \in \Phi_m(\mathbf{Z})$ such that

$$S(n) = y_n \qquad (n \in \mathbf{Z})$$

holds?

An application of the Poisson-Weil factorization of the Fourier cotransform (cf. [17] and Auslander [2]) combined with an argument concerning the inversion of Toeplitz matrices shows that when m is even the cardinal spline interpolation problem admits a unique solution. However, when m is odd the knots of the splines must be displaced by 1/2 to ensure the existence of a unique solution of the cardinal spline interpolation problem (Subbotin-Schoenberg theorem; see Schoenberg [24]).

Let $PW(\mathbf{C})$ denote the Paley-Wiener space of all entire functions of exponential type at most π that are square integrable on the real line $\mathbf{R}$. In view of the Paley-Wiener theorem the Fourier transform is an isometric isomorphism of the separable complex Hilbert space $PW(\mathbf{C})$ onto $L^2(\mathbf{T})$. Define the function **sinus cardinalis** by the usual prescription

$$\operatorname{sinc} z = \begin{cases} \dfrac{\sin z}{z} & \text{for } z \neq 0 \\[2ex] 1 & \text{for } z = 0 \end{cases}$$

In the simplest possible case m=1 the Whittaker-Shannon **sampling theorem** obtains. For a survey see, for instance, Butzer [5] and the informative expository paper by Higgins [10]. Also see the paper [17] which emphasizes the group theoretical point of view.

<u>Theorem 4.</u> Each function $f \in PW(\mathbf{C})$ admits the cardinal series expansion

$$f(z) = \sum_{\mu \in \mathbf{Z}} f(\mu) \operatorname{sinc}(z-\mu)$$

for all $z \in \mathbf{C}$. The convergence of the cardinal interpolation series is uniform on the compact subsets of $\mathbf{C}$.

The fact that each function $f \in PW(\mathbf{C})$ can be recaptured from its values at the integers by the Whittaker-Shannon sampling theorem lies at the foundation of digital signal processing (cf. [23]). For instance, the recently developed CD (=**C**ompact **D**isc) technology forms a very efficient practical realization of this idea. The figures displayed below show the structures of a cross-section and the surface of a CD.

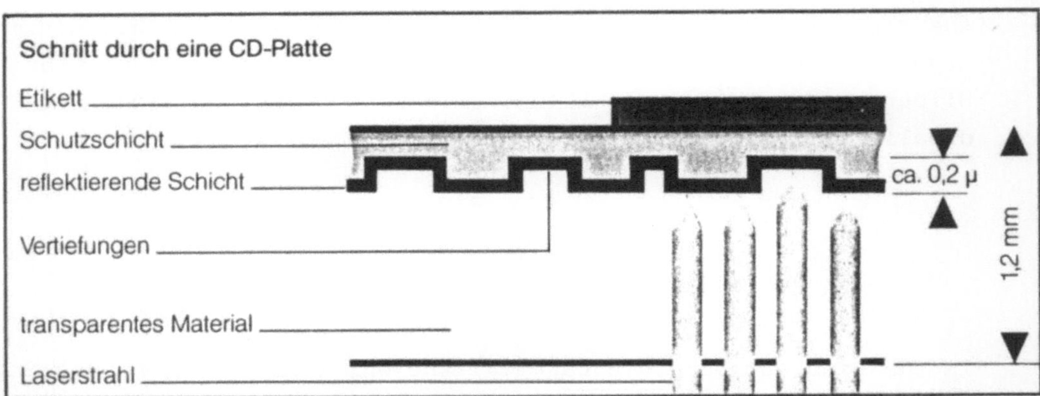

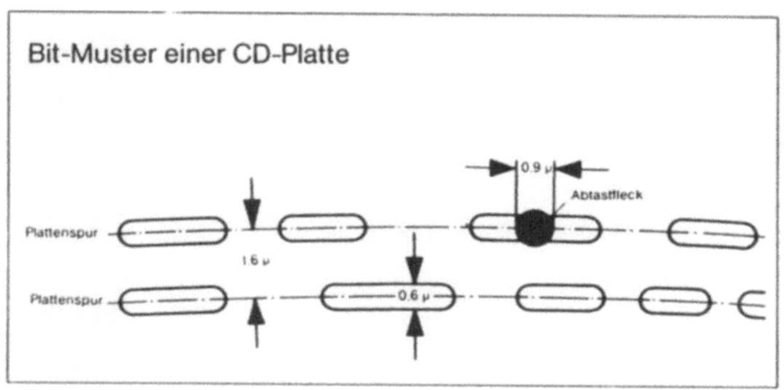

The sequence of digital signals located on the CD are transformed by means of a laser into analog signals.

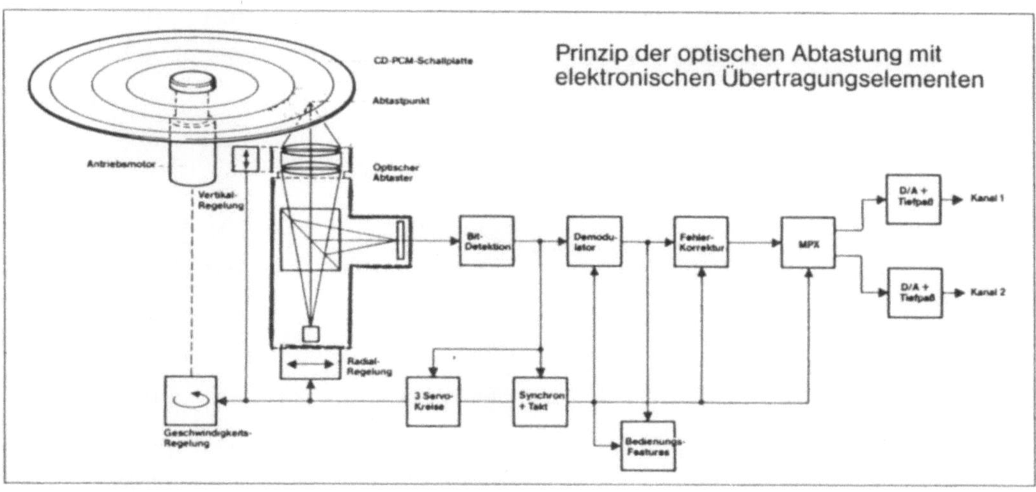

Other examples for the application of the Whittaker-Shannon sampling theorem are the digital typography and the seismic exploration.

6. The Phase Anomaly of Fourier Optics

The evolution of a light wave along a coaxial optical system gives information about the light intensities. It can be described in paraxial approximation with the aid of a quadratic Hamiltonian. The corresponding ray evolution along the optical "time" axis gives rise to a group of operators acting as canonical transformations in the framework of non-relativistic quantum mechanics in two dimensions. We have seen that the phase space of signal theory is the time-frequency plane $\mathbf{R} \oplus \mathbf{R}$ whereas the optical phase space derives from the Fermat principle of "least time" (cf. [16]). In Fourier optics which is the scalar counterpart of linear geometrical optics, an important principle says that the ambiguity surface is **preserved** along the geometrical axis of the optical system. For the sake of simplicity, let us consider the case when the optical system (composed of ideal lenses) is axially symmetric (one degree of freedom). In this case, the phase factor C_u associated with the matrix of the lens operator

$$\begin{pmatrix} 1 & 0 \\ u & 1 \end{pmatrix} \qquad (u \neq 0)$$

(cf. Corollary 1 of Theorem 2 supra) takes the explicit form

$$C_u = e^{-i\frac{\pi}{4}\text{sign}(u)},$$

where $\text{sign}(u) = \frac{u}{|u|}$ for $u \neq 0$. From this and the fact that the ambiguity surface is preserved along the geometrical axis of the optical system we may deduce that the light goes through a phase shift of π when passing through a focus. This phenomenon which is called to be the **phase anomaly** of Fourier optics can be checked by elegant experiments in the setting of microwave optics. The figures below (real and imaginary part) may serve as an illustration of the phase anomaly phenomenon.

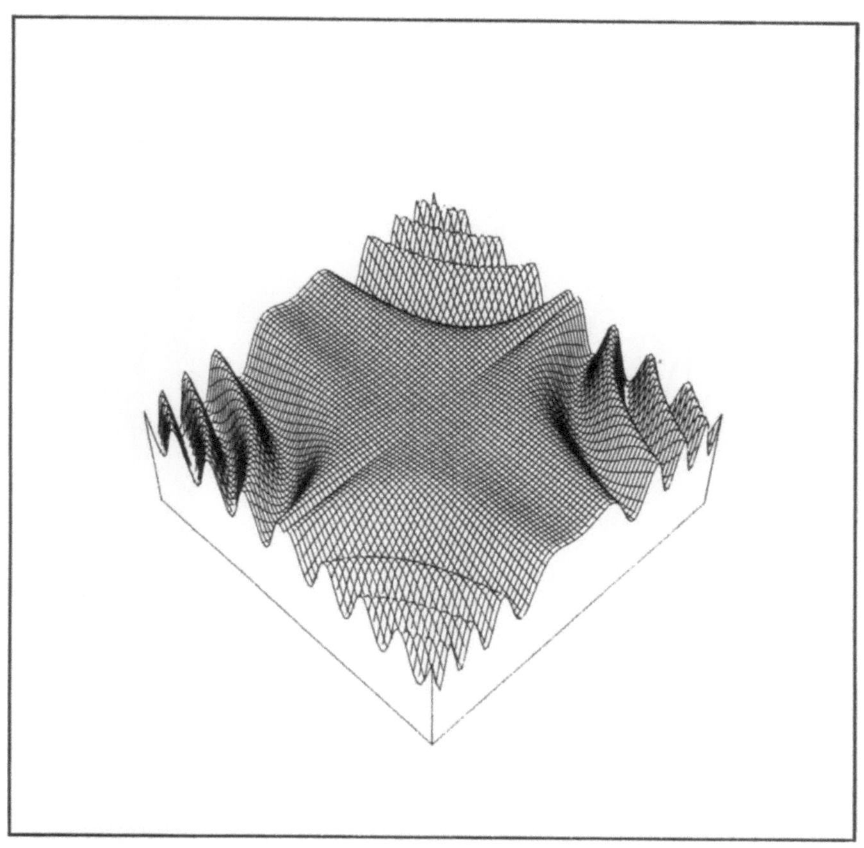

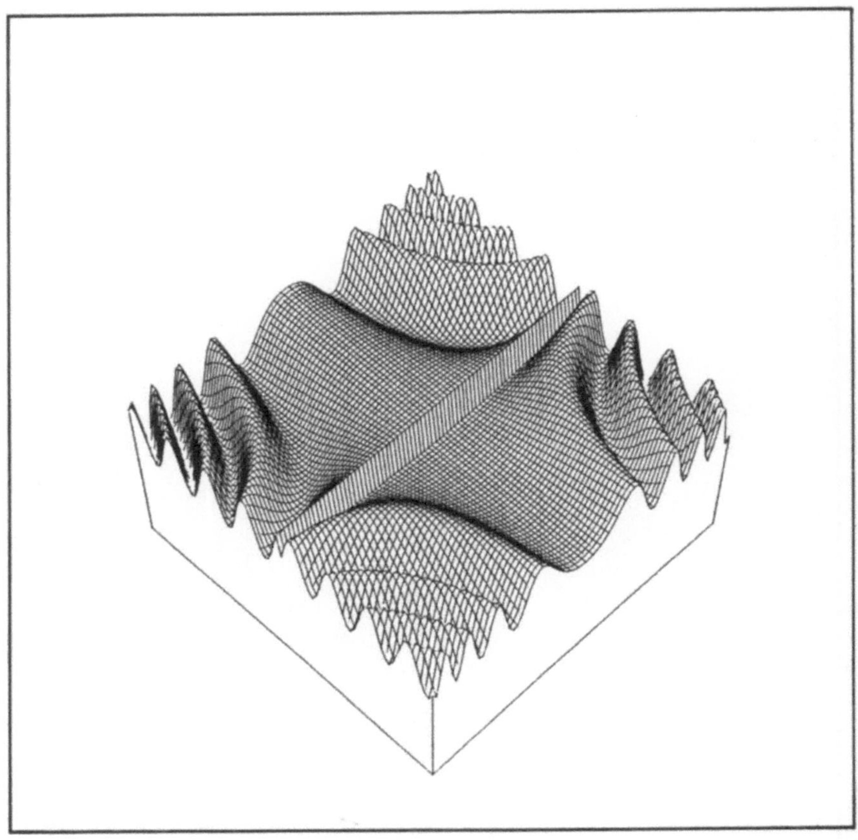

For additional details about the phase discontinuities of
the wave field, see the monograph by Guillemin-Sternberg [9] and
the paper [16]. In fact, it is the contribution of the group-
theoretic Maslov index which properly explains the phase anomaly
of Fourier optics.

7. Summary

Harmonic analysis on the real Heisenberg nilpotent group and the compact Heisenberg nilmanifold plays an important rôle in non-relativistic quantum theory, Fourier optics, and in analog and digital signal processing. In particular, this Lie group stands at the crossroads of quantum mechanics and radar analysis/synthesis. Moreover, within the information theoretic area of applied nilpotent harmonic analysis a "geometric" proof of an identity for Laguerre functions of different orders pops up that deserves mention also from a purely mathematical point of view because of its close relationship to the theory of theta functions.

References

1. Auslander, L.: Lecture notes on nil-theta functions. CBMS Regional Conf. Ser. in Math., no. 34. Providence, R.I.: Amer. Math. Soc. 1977

2. Auslander, L.: A factorization theorem for the Fourier transform of a separable locally compact abelian group. In: Special functions: Group theoretical aspects and applications, pp. 261-269. R.A. Askey, T.H. Koornwinder, and W. Schempp, eds. MIA Series. Dordrecht-Boston-Lancaster: Reidel 1984

3. Balazs, N.L., Jennings, B.K.: Wigner's function and other distribution functions in mock phase space. Phys. Rep. 104 (1984), 347-391

4. Bellman, R.: A brief introduction to theta functions. New York, NY: Holt, Rinehart and Winston 1961

5. Butzer, P.L.: A survey of the Whittaker-Shannon sampling theorem and some of its extensions. J. Math. Res. Expos. (China) 3 (1983), 185-212

6. Gabor, D.: Theory of communication. J. Inst. Elec. Eng. (London) 93 (1946), 429-457

7. Gabor, D.: La théorie des communications et la physique. In: La cybernétique: Théorie du signal et de l'information, pp. 115-149. L. de Broglie, ed. Paris: Éditions de la Revue d'Optique Théorique et Instrumentale 1951

8. Grosjean, C.C.: Note on two identities mentioned by Dr. W. Schempp near the end of the presentation of his paper. Proceedings of the Laguerre symposium at Bar-le-Duc 1984. Lecture Notes in Math. Berlin-Heidelberg-New York-Tokyo: Springer 1985 (to appear)

9. Guillemin, V., Sternberg, S.: Symplectic techniques in physics. Cambridge-London-New York-New Rochelle-Melbourne-Sydney: Cambridge University Press 1984

10. Higgins, J.R.: Five short stories about the cardinal series. Bull. (New Series) Amer. Math. Soc. 1985 (to appear)

11. Hillery, M., O'Connell, R.F., Scully, M.O., Wigner, E.P.: Distribution functions in physics: Fundamentals. Phys. Rep. 106 (1984), 121-167

12. Howe, R.: Quantum mechanics and partial differential equations. J. Funct. Anal. 38 (1980), 188-254

13. Moore, C.C., Wolf, J.A.: Square integrable representations of nilpotent groups. Trans. Amer. Math. Soc. 185 (1973), 445-462

14. Neumann, J. v.: Die Eindeutigkeit der Schrödingerschen Operatoren. Math. Ann. 104 (1931), 570-578

15. Ogden, R.D., Vági, S.: Harmonic analysis of a nilpotent group and function theory on Siegel domains of type II. Advances in Math. 33 (1979), 31-92

16. Raszillier, H., Schempp, W.: Fourier optics from the perspective of the Heisenberg group. Physikalisches Institut der Universität Bonn - HE-84-34, pp. 1-31. Bonn 1984

17. Schempp, W.: Gruppentheoretische Aspekte der Signalübertragung und der kardinalen Interpolationssplines I. Math. Methods Appl. Sci. 5 (1983), 195-215

18. Schempp, W.: Radar ambiguity functions and the linear Schrödinger representation. In: Anniversary volume on approximation theory and functional analysis, pp. 481-491. P.L. Butzer, R.L. Stens, and B. Sz.-Nagy, eds. ISNM, Vol. 65. Basel-Boston-Stuttgart: Birkhäuser 1984

19. Schempp, W.: Radar ambiguity functions, nilpotent harmonic analysis, and holomorphic theta series. In: Special functions: Group theoretical aspects and applications, pp. 217-260. R.A. Askey, T.H. Koornwinder, and W. Schempp, eds. MIA Series. Dordrecht-Boston-Lancaster: Reidel 1984

20. Schempp, W.: Radar ambiguity functions of positive type. In: General inequalities 4, pp. 369-380. W. Walter, ed. ISNM, Vol. 71. Basel-Boston-Stuttgart: Birkhäuser 1984

21. Schempp, W.: Radar reception and nilpotent harmonic analysis VI. C.R. Math. Rep. Acad. Sci. Canada 6 (1984), 179-182

22. Schempp, W.: Radar ambiguity functions, the Heisenberg group, and holomorphic theta series. Proc. Amer. Math. Soc. $\underline{92}$ (1984), 103-110

23. Schempp, W.: On Gabor information cells. In: Multivariate Approximation Theory III. W. Schempp and K. Zeller, eds. ISNM Series. Basel-Boston-Stuttgart: Birkhäuser 1985 (to appear)

24. Schoenberg, I.J.: Cardinal spline interpolation. Regional Conference Series in Applied Math., Vol. $\underline{12}$. Philadelphia, PA: SIAM 1973

25. Souriau, J.M.: Géométrie symplectique et physique mathématique. Gazette des Mathématiciens $\underline{10}$ (1978), 90-130

26. Ville, J.: Théorie et applications de la notion de signal analytique. Câbles et Transmission $\underline{2}$ (1948), 61-74

27. Weil, A.: Sur certains groupes d'opérateurs unitaires. Acta Math. $\underline{111}$ (1964), 143-211. Also in: Collected papers, Vol. III, pp. 1-69. New York-Heidelberg-Berlin: Springer 1980

28. Wigner, E.P.: Quantum-mechanical distribution functions revisited. In: Perspectives in quantum theory, pp. 25-36. W. Yourgrau and A. van der Merwe, eds. New York, NY: Dover 1979

29. Wilcox, C.H.: The synthesis problem for radar ambiguity functions. MRC Tech. Summary Report, no. $\underline{157}$. Madison, WI: The University of Wisconsin 1960

30. Woodward, P.M.: Probability and information theory, with applications to radar. Dedham, MA: Artech House 1980

Lehrstuhl für Mathematik I der Universität Siegen, Hölderlinstraße 3, D-5900 Siegen, Federal Republic of Germany

International Series of
Numerical Mathematics, Vol. 74
© 1985 Birkhäuser Verlag Basel

UNIQUENESS OF BEST L_1-APPROXIMATIONS

OF CONTINUOUS FUNCTIONS

Manfred Sommer

Mathematisch-Geographische Fakultät, Katholische Universität Eichstätt, D-8078 Eichstätt,Federal Republik of Germany

The problem of uniqueness of best approximations of continuous vector-valued functions and continuous functions of one and several variables by finite dimensional subspaces in the weighted L_1-norm is studied. Characterizations of uniqueness of best approximations are obtained. Various examples of subspaces which guarantee uniqueness, including subspaces of polynomials and linear spline functions of two variables, are presented.

0. Introduction

A first result on uniqueness of best approximations in the L_1-norm was obtained by Jackson [6] who showed that if G is the space of all polynomials of degree at most m-1, then every $f \in C[0,1]$ has a unique best L_1-approximation from G. Two more general results were proved by Krein [8] in 1938. He showed that there exists no finite dimensional subspace G of $L_1[0,1]$ (the space of all real-valued Lebesgue integrable functions endowed with the L_1-norm) such that every $f \in L_1[0,1]$ has a unique best L_1-approximation from G (see also Moroney [14]). Therefore

it is reasonable to study the problem of global L_1-uniqueness in
the space $C[a,b]$. Krein's second result states that if G is a
Haar subspace on $(0,1)$, then uniqueness holds for every $f \in C[0,1]$.

In the last years the problem of uniqueness of best
approximations of continuous real-valued functions in the L_1-
norm have been widely investigated, because, unlike the situation
in the uniform norm, it is not necessary for global uniqueness
that the approximating subspace satisfies the Haar property. For
example Galkin [4] and Strauß [21] showed that every subspace
of spline functions with fixed knots (including the Haar subspaces) guarantees uniqueness. Other examples were presented by
Carroll-Braess [1] who proved that for continuously composed
Haar subspaces uniqueness holds and we showed in [17,18] that
every element of a more general class of spline subspaces, including those mentioned above, guarantees uniqueness.

It turned out that all these subspaces have a common
property, the so-called A-property, introduced by DeVore and
Strauß which is sufficient to guarantee uniqueness (see [23]).
Moreover all these subspaces are subspaces of functions of one
variable. Unfortunately this strong restriction is necessary if
we study the problem of uniqueness in the uniform approximation,
because by the wellknown theorem of Mairhuber there exists no
Haar subspace G of functions of several variables with dim $G \geq 2$.

In the L_1-approximation, however, there exist also
unicity subspaces of functions of several variables. For example
Kroó [9 - 11] studied the problem of L_1-uniqueness for continuous
vector-valued functions defined on convex compact subsets of
$\mathbb{R}^n$ ($n \geq 1$) and presented some classes of unicity subspaces. In-
spired by his results we study in this paper the same problem
for more general subsets X of $\mathbb{R}^n$. To do this let X be a compact
subset of the real Euclidean space $\mathbb{R}^n$ ($n \geq 1$) such that $X = \overline{\overset{\circ}{X}}$,
i.e. X is the closure of its interior. By $C^1(X)$ ($1 \geq 1$) we denote
the <u>linear space</u> of <u>continuous vector-valued functions</u>, i.e.

$$C^1(X) = \{f = (f_1, \ldots, f_1): f_i : X \to \mathbb{R} \text{ is continuous,}$$
$$1 \leq i \leq 1\}.$$

Let $W = \{w: X \to \mathbb{R}: w$ is Lebesgue measurable, bounded and positive on $X\}$ be the set of <u>weight</u> <u>functions</u>. Now using the same notations as in [11] we denote, for any $w \in W$, by $C_w^1(X)$ the <u>linear</u> <u>space</u> $C^1(X)$ <u>endowed</u> <u>with</u> <u>the</u> <u>weighted</u> L_1<u>-norm</u>

$$\|f\|_{w,1} = \int_X \|f(x)\|\, w(x)\, dx \quad (f \in C^1(X))$$

where $\|\cdot\|$ denotes the Euclidean norm in $\mathbb{R}^1$. If G is a finite dimensional subspace of $C^1(X)$, then a function $g_o \in G$ is called a <u>best</u> $L_1(w)$ - <u>approximation</u> of $f \in C_w^1(X)$ from G if $\|f - g_o\|_{w,1} \leq \|f - g\|_{w,1}$ for every $g \in G$. The subspace G is called an $L_1(w)$-<u>unicity</u> <u>subspace</u> of $C_w^1(X)$ if every $f \in C_w^1(X)$ has a unique best $L_1(w)$-approximation from G. (If $w \equiv 1$, then the notations will be simplified to best L_1-approximation and L_1-unicity space.) Moreover we denote by $< \cdot, \cdot >$ the inner product in $\mathbb{R}^1$ and we define the function $\operatorname{sgn}_1 f$ by

$$\operatorname{sgn}_1 f = \begin{cases} \dfrac{f}{\|f\|} & \text{on} \quad X \smallsetminus Z(f) \\[2mm] 0 & \quad\quad\ Z(f) \end{cases}$$

where $Z(f) = \{x \in X: f(x) = 0\}$. (In the case when $l = 1$, the index l will be omitted, i.e. $C^1(X) = C(X)$, $C_w^1(X) = C_w(X)$ and $\operatorname{sgn}_1 f = \operatorname{sgn} f$.)

At first we show that three results obtained by Kroó [11] for convex compact subsets X can be generalized, because all arguments in their proofs are still applicable to our general case. The first result characterizes best $L_1(w)$-approximations in the vector-valued case, the second theorem states a characterization of $L_1(w)$-unicity subspaces of $C(X)$ and the third result says that if $G_1, \ldots, G_1$ are A-subspaces of $C(X)$, i.e. if they satisfy the A-property, then the cartesian product $G_1 \times \ldots \times G_1$ is an $L_1(w)$-unicity subspace for every $w \in W$. In particular it follows that every A-subspace G guarantees global $L_1(w)$-uniqueness for every $w \in W$. This generalizes a result of Strauß [23] for real compact intervals $X = [a,b]$. Recently Kroó

[13] gave the converse for $X = [a,b]$. More precisely he proved
that if G is an $L_1(w)$-unicity subspace of $C[a,b]$ for every $w \in W$
satisfying $\inf_{x \in [a,b]} w(x) > 0$, then G satisfies the A-property. At
the same time Pinkus [15] was able to prove this statement under
the weaker hypothesis that G is an $L_1(w)$-unicity subspace for
every continuous $w \in W$. However he had to make minor restrictions
on G. Arguing as in the proof in [13] we are now able to gener-
alize Kroó's result for every compact subset X of $\mathbb{R}^n$ with $X = \bar{\overset{\circ}{X}}$.

Moreover we prove that if X is a subset of the real
line, then every A-subspace is necessarily a weak Chebyshev subspace. In the case when $X = [0,1]$ Pinkus [13] was **even** able to
characterize those subspaces G of $C[0,1]$ which satisfy the A-
property. He showed that every A-subspace is a very spline-like
space similar to those generalized spline spaces which we considered in [17,18].

We are furthermore interested in examples of A-subspaces of $C(X)$. In [19] we defined a subspace of linear spline
functions of two variables which does not satisfy the A-property.
It turns out that this subspace has no basis consisting of
B-splines. Therefore looking for A-subspaces we consider multivariate B-splines and in fact we obtain a class of subspaces of
linear spline functions of two variables which satisfy the
A-property.

Moreover we show that if X is a disconnected subset of
$\mathbb{R}$, then every Haar subspace of $C(X)$ fails to satisfy the A-property. Therefore one could conjecture that there do not exist any
A-subspaces of $C(X)$ where $X \subset \mathbb{R}$ is disconnected. But this is not
true, because using known A-subspaces we define certain classes
of A-subspaces of functions defined on disconnected sets.

Finally we are interested in examples of subspaces G
of $C(X)$ which guarantee global $L_1(w)$-uniqueness for a fixed
weight w. Kroó proved in [9] that if $G = \Pi_{1,m}$, the space of all
polynomials which are linear in the first variable and of degree
at most m in the second one, then G is an $L_1(w)$-unicity subspace
of $C_w(X)$ for $w \equiv 1$ where $X = [0,1] \times [0,1]$. We extend this result

by showing that G is also an $L_1(w)$-unicity subspace for every $w \in W$ with $w(x,y) = w_1(x) \cdot w_2(y)$. But it is easily verified that G does not satisfy the A-property.

1. Some results on $L_1(w)$-unicity subspaces

First of all we give three results obtained by Kroó in [11] for convex and compact subsets X of $\mathbb{R}^n$ with nonempty interior. It turns out that using his arguments the statements remain valid if $X = \overline{\overset{\circ}{X}}$ and $\overset{\circ}{X} \neq \emptyset$.

Theorem 1.1. Let G be a finite dimensional subspace of $C^1(X)$, let $w \in W$ and $f \in C_w^1(X)$. Then the following statements are equivalent:

(1) $0 \in G$ is a best $L_1(w)$-approximation of f from G;
(2)   we have

$$\left| \int_{X \smallsetminus Z(f)} < \text{sgn}_1 f(x), g(x) > w(x)\, dx \right| \leq \int_{Z(f)} \| g(x) \| \, w(x)\, dx$$

for every $g \in G$.

For $l = 1$ this theorem states the wellknown characterization of best L_1-approximations (see e.g. Singer [16, p. 46]).

The next theorem characterizes the $L_1(w)$-unicity subspaces of $C_w^1(X)$ using only inner properties of the approximating subspaces. To do this we define for an arbitrary subset G of $C^1(X)$ a subset G* by

$$G^* = \{g^* \in C^1(X): \text{ there exists a function } g' \in G \text{ such that } |g^*(x)| = |g'(x)| \text{ for every } x \in X\}.$$

Such sets were introduced by Strauß [22] to characterize the L_1-unicity subspaces of $C[a,b]$ where $[a,b]$ is a real compact interval.

<u>Theorem 1.2</u>. Let G be a finite dimensional subspace of $C^1(X)$ and let $w \in W$. Then the following statements are equivalent:

(1) G is an $L_1(w)$-unicity subspace of $C_w^1(X)$;

(2) If for $g^* \in G^*$ the function $O \in G$ is a best $L_1(w)$-approximation of g^* from G, then $g^* \equiv O$.

In the case when $l = 1$ and $X = [a,b]$, this result was obtained by Strauß [22].

In the following we are interested in such subspaces G which are $L_1(w)$-unicity subspaces for every $w \in W$. It turns out that this problem is closely related to the A-property which is defined as follows.

<u>Definition</u>. Let $l = 1$ and let G be a finite dimensional subspace of $C(X)$. We say that G satisfies the A-<u>property</u> (or G is an A-<u>subspace</u> of $C(X)$) if for any $g^* \in G^* \setminus \{O\}$ there exists a function $\tilde{g} \in G \setminus \{O\}$ such that

(1) $\tilde{g} \equiv O$ a.e. on $Z(g^*)$ and

(2) $\tilde{g}(x) \cdot g^*(x) \geq O$ for every $x \in X \setminus Z(g^*)$.

In the case when $X = [a,b]$ and the elements of G have only finitely many separated zeros the A-property was introduced by DeVore and Strauß (see [23]). The above given actual version is due to Kroó [11].

The following result shows that every A-subspace guarantees $L_1(w)$-uniqueness for every $w \in W$.

<u>Theorem 1.3</u>. Let $l \in \mathbb{N}$ and let G_i ($1 \leq i \leq l$) be A-subspaces of $C(X)$. Then for any $w \in W$ the subspace $G = G_1 \times ... \times G_1$ is an $L_1(w)$-unicity subspace of $C_w^1(X)$.

In the case when $l = 1$ and $X = [a,b]$ this statement was verified by Strauß [23].

Now let 1 = 1 and let G be an A-subspace of C(X). Then Theorem 1.3 states that G is an $L_1(w)$-unicity subspace of $C_w(X)$ for every $w \in W$. Kroó was the first to study the problem of whether the converse of this statement is also true. In [11] he obtained the converse for one dimensional subspaces G of C(X) where X is a convex and compact subset of $\mathbb{R}^n$. We also studied this problem on the same subsets X and proved that the converse is true for those finite dimensional subspaces of C(X) which have Chebyshev-rank ≤ 1 (for details see [19]).

Recently Kroó [13] was able to show the converse for all finite dimensional subspaces of C[a,b]. At the same time Pinkus [15] also proved this statement for those subspaces G of C[a,b] for which $\lambda(Z(g)) = \lambda(\overset{\circ}{Z}(g))$ ($g \in G$) where λ denotes the Lebesgue measure and $\overset{\circ}{Z}(g)$ the interior of $Z(g)$, but under the weaker hypothesis that G is an $L_1(w)$-unicity subspace for every <u>continuous</u> $w \in W$.

Now following the lines of the proof in [13] we are able to prove the converse, if $X = \overset{=}{\overset{\circ}{X}} \subset \mathbb{R}^n$.

<u>Theorem 1.4.</u> Let G be a finite dimensional subspace of C(X). Then the following conditions are equivalent:

(1) G is an $L_1(w)$-unicity subspace for every $w \in W$ satisfying
$$\inf_{x \in X} w(x) > 0;$$

(2) G satisfies the A-property.

<u>Proof.</u> It follows directly from Theorem 1.3 that statement (2) implies statement (1).

In order to prove (1) $\Rightarrow$ (2) we follow the lines of the proof in [13]. At first we set $\tilde{W} = \{w \in W: \inf_{x \in X} w(x) > 0\}$. Assume now that G is an $L_1(w)$-unicity subspace of $C_w(X)$ for every $w \in \tilde{W}$. To verify that G is an A-subspace let an arbitrary $g^* \in G^* \setminus \{0\}$ be given. Then by Theorem 1.2, for every $w \in \tilde{W}$ there exists a function $g \in G$ such that

$$(3) \qquad \left| \int_{N(g^*)} \operatorname{sgn} g^*(x) \cdot g(x) w(x) dx \right| > \int_{Z(g^*)} |g(x)| \, w(x) dx$$

where $N(g^*) = X \setminus Z(g^*)$.

Set $\tilde{G} = \{g \in G: g \equiv 0$ a.e. on $Z(g^*)\}$. Since $g^* \in G^* \setminus \{0\}$, there exists a function $g' \in G \setminus \{0\}$ such that $|g^*(x)| = |g'(x)|$ for every $x \in X$. Therefore, $\dim \tilde{G} \geq 1$.

Now we prove that there exists a function $g_o \in \tilde{G}$ such that for every $w \in \tilde{W}$

$$(4) \qquad \int_{N(g^*)} g_o(x) \, \operatorname{sgn} g^*(x) \, w(x) \, dx \neq 0.$$

Assume that in contrary for every $g \in \tilde{G}$ there exists a function $w \in \tilde{W}$ satisfying

$$(5) \qquad \int_{N(g^*)} g(x) \, \operatorname{sgn} g^*(x) \, w(x) \, dx = 0.$$

Let $\{g_1, \ldots, g_k\}$ be a basis of $\tilde{G}$. Then (5) implies that for every $\{b_i\}_{i=1}^{k} \in \mathbb{R}^k$ there exists a $w \in \tilde{W}$ such that

$$(6) \qquad 0 = \int_{N(g^*)} \left(\sum_{i=1}^{k} b_i \, g_i(x) \right) \operatorname{sgn} g^*(x) \, w(x) \, dx = \sum_{i=1}^{k} b_i \int_{N(g^*)} g_i(x) \, \operatorname{sgn} g^*(x) \, w(x) \, dx.$$

Set $A_o = \left\{ \left\{ \int_{N(g^*)} g_i(x) \, \operatorname{sgn} g^*(x) \, w(x) \, dx \right\}_{i=1}^{k} : w \in \tilde{W} \right\}$. Then A_o is a convex subset of $\mathbb{R}^k$.

Now we show that A_o is open. Consider an arbitrary point $C \in A_o$. Then for some $w_C \in \tilde{W}$

$$(7) \qquad C = \{C_i\}_{i=1}^{k} = \left\{ \int_{N(g^*)} g_i(x) \, \operatorname{sgn} g^*(x) \, w_C(x) \, dx \right\}_{i=1}^{k}.$$

By definition of $\tilde{G}$ the functions $g_i \cdot \operatorname{sgn} g^*$, $1 \leq i \leq k$, are linearly independent on $X \setminus Z(g^*)$. Then there exist k distinct points $\tilde{x}_1, \ldots, \tilde{x}_k \in X \setminus Z(g^*)$ such that

$$\det(g_i(\tilde{x}_j) \cdot \operatorname{sgn} g^*(\tilde{x}_j))_{i,j=1}^{k} \neq 0.$$

Since $X = \overline{\overset{\circ}{X}}$, every $x \in \partial X$ (the boundary of X) is the limit of a sequence $(x_m) \subset \overset{\circ}{X}$. Therefore there are k sequences $(x_{im}) \subset \overset{\circ}{X}$, $1 \leq i \leq k$, with $\lim_{m \to \infty} x_{im} = \tilde{x}_i$. Moreover $g_1, \ldots, g_k$, $\operatorname{sgn} g^*$ are continuous at $\tilde{x}_1, \ldots, \tilde{x}_k$ which implies that for every sufficiently large m

$$\det(g_i(x_{jm}) \; \mathrm{sgn}\, g^*(x_{jm}))^k_{i,j=1} \neq 0.$$

Hence the functions $g_i \cdot \mathrm{sgn}\, g^*$, $1 \leq i \leq k$, are linearly independent on $\overset{\circ}{X} \smallsetminus Z(g^*)$. Set $x_j := x_{jm}$, $1 \leq j \leq k$, for such an m. By the above arguments the vectors $1_j = \{g_i(x_j) \; \mathrm{sgn}\, g^*(x_j)\}^k_{i=1}$, $1 \leq j \leq k$, are linearly independent in $\mathbb{R}^k$. Then there exist neighborhoods U_j of x_j, $1 \leq j \leq k$, such that $U_j \subset \overset{\circ}{X} \smallsetminus Z(g^*)$ and the vectors

$$1^*_j = \{\textstyle\int_{U_j} g_i(x) \; \mathrm{sgn}\, g^*(x)\,dx\}^k_{i=1}, \quad 1 \leq j \leq k,$$

are also linearly independent in $\mathbb{R}^k$.

Let $\inf\limits_{x \in X} w_C(x) = \theta > 0$ and define functions $w_j \in \tilde{W}$ by

$$w_j(x) = \begin{cases} 0 & \text{if} \quad x \in X \smallsetminus U_j \\ \theta/2 & \quad\quad\;\; x \in U_j \end{cases}, \quad 1 \leq j \leq k.$$

Then arguing exactly as in [13] we can show that A_o is open. Furthermore by the same kind of arguments we get a contradiction to our assumption that for every $g \in \tilde{G}$ there exists a function $w \in \tilde{W}$ satisfying (5).

Therefore there exists some $g_o \in \tilde{G}$ such that (4) is true for every $w \in \tilde{W}$. Now we show that $g_o \cdot g^*$ does not change sign on $N(g^*)$. Assume that in contrary for some $\tilde{x}_1, \tilde{x}_2 \in N(g^*)$, $(-1)^i g_o(\tilde{x}_i) g^*(\tilde{x}_i) > 0$, $i = 1,2$. Then since $X = \overline{\overset{\circ}{X}}$, there exist two points $x_1, x_2 \in \overset{\circ}{X} \smallsetminus Z(g^*)$ such that $(-1)^i g_o(x_i) g^*(x_i) > 0$ and therefore we find neighborhoods U_i of x_i in X with positive measures ($i = 1,2$) such that $(-1)^i g_o(x) g^*(x) > 0$ for every $x \in U_i$, $i = 1,2$. Now arguing again as in [13] we get a contradiction to our assumption and thus it follows that $\varepsilon g_o(x) g^*(x) \geq 0$ for every $x \in N(g^*)$ where $\varepsilon \in \{-1,1\}$. Since $g_o \in \overset{\circ}{G}$, we have $g_o \in O$ a.e. on $Z(g^*)$ and therefore the A-property of G follows.

2. Examples of A-subspaces and $L_1(w)$-unicity subspaces

As we have mentioned in the introduction, there are many inportant classes of A-subspaces of C[a,b] where [a,b] denotes a real compact interval, for example the subspaces of polynomial spline functions [23], continuously composed Haar subspaces [1] and certain classes of generalized spline functions [17,18]. Recently, a characterization of all A-subspaces of C[a,b] was given in [15]. It turned out that such a subspace is a very spline-like space similar to those generalized spline spaces which were considered in [17,18].

In this section we are interested in examples of A-subspaces of $C(X)$ where $X \subset \mathbb{R}^n$ $(n \geq 2)$ or X is an arbitrary compact subset of the real line. At first we study the second case. (Recall that in any case we set $X = \overline{\mathring{X}}$.)

The following property plays an important role in the uniform approximation (see e.g. [3,7,17,18]). We shall show that it is also closely related to our problem.

__Definition.__ Let $X \subset \mathbb{R}$ and G be an m-dimensional subspace of $C(X)$. Then G is said to be __weak Chebyshev__ if each $g \in G$ has at most m-1 sign changes, i.e. there do not exist points $x_1 < \ldots < x_{m+1}$ in X such that $g(x_i)g(x_{i+1}) < 0$, $i = 1, \ldots, m$.

We need the following equivalent formulation of weak Chebyshev subspaces, due to Deutsch-Nürnberger-Singer [3].

__Theorem 2.1.__ Let $X \subset \mathbb{R}$ and let G be an m-dimensional subspace of $C(X)$. Then the following conditions are equivalent:

(1) G is weak Chebyshev;

(2) Given m-1 points $-\infty = x_0 < x_1 < \ldots < x_{m-1} < x_m = \infty$ with $x_i \in X$, $1 \leq i \leq m-1$, there exists a function $g \in G \smallsetminus \{0\}$ such that $(-1)^i g(x) \geq 0$ for $x \in [x_{i-1}, x_i] \cap X$, $1 \leq i \leq m$.

For the particular case when X = [a,b], this result is due to Jones-Karlovitz [7].

Now we are able to prove an important property of A-spaces.

Theorem 2.2. Let $X \subset \mathbb{R}$ and let G be an m-dimensional A-subspace of C(X). Then G is weak Chebyshev.

<u>Proof.</u> Given m-1 points $-\infty = x_0 < x_1 < \ldots < x_{m-1} < x_m = \infty$ with $x_i \in X$, $1 \leq i \leq m-1$. By Theorem 2.1 it suffices to prove the existence of a function $g \in G \smallsetminus \{0\}$ with $(-1)^i g(x) \geq 0$ for $x \in [x_{i-1}, x_i] \cap X$, $1 \leq i \leq m$.
Since dim G = m, there exists a function $g \in G \smallsetminus \{0\}$ such that $g(x_i) = 0$, $1 \leq i \leq m-1$. Hence we obtain a function $g^* \in G^* \smallsetminus \{0\}$ with $(-1)^i g^*(x) \geq 0$ for $x \in [x_{i-1}, x_i] \cap X$, $1 \leq i \leq m$. From the A-property it therefore follows that there exists a function $\tilde{g} \in G \smallsetminus \{0\}$ such that $(-1)^i \tilde{g}(x) \geq 0$ for $x \in [x_{i-1}, x_i] \cap X$, $1 \leq i \leq m$.
This proves Theorem 2.2.

For X = [a,b] this result was obtained in [19].
Now we study the case when $X \subset \mathbb{R}^n$ ($n \geq 2$) more detailed. A first example of an A-subspace of C(X) was presented by Kroó [11]:

<u>Example 1</u> (<u>Affine-linear functions</u>). Let X be a convex and compact subset of $\mathbb{R}^n$ and let a subspace G of C(X) be defined by

$$G = \{g \in C(X) : g(x_1, \ldots, x_n) = \sum_{i=1}^{n} a_i x_i + a_{n+1}, a_i \in \mathbb{R}\}.$$

Then G satisfies the A-property.

In the following example we consider tensor products of Haar subspaces, in particular subspaces of polynomials of two variables.

<u>Example 2</u> (<u>Polynomials of two variables</u>). Let $X = [0,1]^2 \subset \mathbb{R}^2$ and let for $k, m \in \mathbb{N}$, $k \leq m$, $\tilde{G}_k$ and $\tilde{G}_m$ denote Haar subspaces of C[0,1] of dimension k and m, respectively. Moreover assume that $\tilde{G}_k \subset \tilde{G}_m$. If $\varphi \in C[0,1]$ is an arbitrary increasing

function, then define

$$G = G_{m+k} = \{g(x,y) = \varphi(y) g_k(x) + g_m(x): g_k \in \tilde{G}_k, g_m \in \tilde{G}_m\}.$$

Kroó obtained the following result in [9]:

> G is an L_1-unicity subspace of $C(X)$.

Therefore G guarantees uniqueness of best L_1-approximations for $w \equiv 1$. Recently we proved in [20]:

> G is an $L_1(w)$-unicity subspace of $C_w(X)$ for every weight $w \in W$ satisfying $w(x,y) = w_1(x) w_2(y)$ $((x,y) \in X)$.

As the most important example of G we obtain the set

$$G = \Pi_{1,m} = \{\sum_{i=0}^{1} \sum_{j=0}^{m} a_{ij} x^i y^j: a_{ij} \in \mathbb{R}\}$$

of polynomials of degree one in the first variable and m in the second one. Therefore $\Pi_{1,m}$ is an $L_1(w)$-unicity subspace of $C(X)$ for every $w \in W$ with separated variables.

On the other hand it is easily verified that $G = \Pi_{1,m}$ fails to satisfy the A-property. For example let $g(x,y) = \left(x - \frac{1}{2}\right)\left(y - \frac{1}{2}\right)^m$ $((x,y) \in X)$. Then the function $g*$ defined by

$$g*(x,y) = \begin{cases} -g(x,y) & \text{if } 0 \leq x \leq \frac{1}{2}, \ \frac{1}{2} \leq y \leq 1 \\ g(x,y) & \text{otherwise} \end{cases}$$

belongs to $G*$. Without loss of generality we may assume that m is an odd number. Moreover assume that $g*\tilde{g} \geq 0$ on X for some $\tilde{g} \in G \smallsetminus \{0\}$. Then $\tilde{g}(x,y) = g_1(y) + x g_2(y)$ where g_1 and g_2 are polynomials of degree at most m. Obviously, $\tilde{g}(\frac{1}{2}, \cdot)$ must change the sign at every $y \in (0,\frac{1}{2})$ and therefore $0 = \tilde{g}(\frac{1}{2},y) = g_1(y) + \frac{1}{2} g_2(y)$ $(y \in (0,\frac{1}{2}))$ which implies that $\tilde{g}(x,y) = (x - \frac{1}{2}) g_2(y)$. Let $\tilde{y} \in (\frac{1}{2},1]$ with $g_2(\tilde{y}) \neq 0$ be given. Then it follows that $\tilde{g}(\frac{1}{4},\tilde{y}) = -\frac{1}{4} g_2(\tilde{y})$ and $\tilde{g}(\frac{3}{4},\tilde{y}) = \frac{1}{4} g_2(\tilde{y})$. However we have $g*(\frac{1}{4},\tilde{y}) = -g(\frac{1}{4},\tilde{y}) > 0$ and

$g*\left(\frac{3}{4},\tilde{y}\right) = g\left(\frac{3}{4},\tilde{y}\right) > 0$ which contradicts the assumption that $g*\tilde{g} \geq 0$ on X.

$\underline{\text{Remark}}$. It is still unknown whether the set $\Pi_{k,m}$ $(k,m \in \mathbb{N}, k,m \geq 2)$ of polynomials of degree k in the first varia- ble and m in the second one is an $L_1(w)$-unicity subspace for some weight $w \in W$ or not. A partial solution of this problem was given by Kroó [12] who showed that every rational function has a unique best L_1-appro- ximation from $\Pi_{k,m}$.

Other important classes of subspaces of $C(X)$ where $X \subset \mathbb{R}^2$ are the subspaces of bivariate spline functions. We ob- tained in [19] the following negative result: Let $X = [0,1]^2$ and let the partitions $0 = x_0 < x_1 = \frac{1}{2} < x_2 = 1$ and $0 = y_0 < y_1 = \frac{1}{2} < y_2 = 1$ be given. Denote the rectangular cells of this grid partition by R_{ij}, $i,j = 1,2$. Then the subspace G of linear spline functions defined by

$$G = \{g \in C(X): g(x)\big|_{R_{ij}} = a_{ij} + b_{ij}x + c_{ij}y, \ a_{ij}, b_{ij}, c_{ij} \in \mathbb{R}, \ i,j = 1,2\}$$

fails to satisfy the A-property.

But using similar arguments as in the proof of [9,Theo- rem 3] it is easily verified that G is an L_1-unicity subspace of $C(X)$.

Despite of this negative result we study the bivariate spline functions more detailed. At first it turns out that the preceding spline subspace admits no basis consisting of B-spli- nes (for details of multivariate B-splines see Dahmen-Micchelli [2]). Therefore to look for examples of spline subspaces which satisfy the A-property it seems to be reasonable to prefer those which are spanned by multivariate B-splines. In fact we obtain the following result.

$\underline{\text{Example 3}}$. ($\underline{\text{Bivariate linear spline functions}}$). Let

$R = [a,b] \times [c,d]$ be a compact subset of $\mathbb{R}^2$. Moreover let partitions $x_{-1} < a = x_0 < x_1 < \ldots < x_{r+1} = b < x_{r+2}$ and $y_{-1} < c = y_0 < y_1 < \ldots < y_{s+1} = d < y_{s+2}$ be given. The vertical and horizontal lines $x = x_i$ and $y = y_j$ $(0 \leq i \leq r+1, 0 \leq j \leq s+1)$ divide R into $(r+1)(s+1)$ rectangular cells R_{ij}. Consider a refinement Δ_{rs} of this grid partition by drawing in all diagonals with positive slopes to the rectangulars R_{ij}. This partition is called _uni-diagonal triangulation_. We define a subspace $S_1^0(\Delta_{rs})$ of bivariate linear spline functions by

$$S_1^0(\Delta_{rs}) = \{g \in C(R): g \text{ is linear in every triangular cell of } \Delta_{rs}\}.$$

It turns out that $S_1^0(\Delta_{rs})$ has a basis of B-splines, i.e. a basis consisting of spline functions with minimal support. This fact is one of the reasons why to choose a triangulation of R. The minimal support of a B-spline B_{ij} in $S_1^0(\Delta_{rs})$ is illustrated by the following figure which means that B_{ij} vanishes identically outside the polyedron P_{ij} and has the value one at the grid point (x_i, y_j) $(0 \leq i \leq r+1,\ 0 \leq j \leq s+1)$.

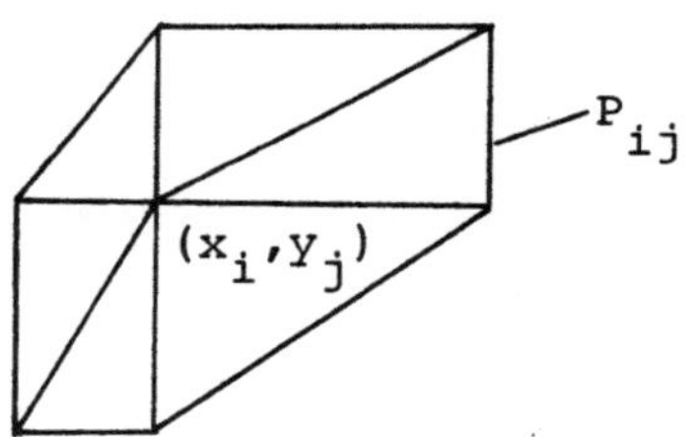

If the grid partition is equidistant, then B_{ij} is a box spline which is called the Courant hat function (for details see Dahmen-Micchelli [2] and the references therein).

Obviously, $S_1^0(\Delta_{rs}) = \text{span } \{B_{ij}|_R\}_{i=0\ j=0}^{r+1\ s+1}$

Now let X be an arbitrary union of some triangular cells of Δ_{rs} and let $I = \{(i,j) \in \{0, \ldots, r+1\} \times \{0, \ldots, s+1\}: (x_i, y_j) \in X\}$. Then we define a subspace G of C(X) by

$$G = S_1^o\,(\Delta_{rs})\Big|_X = \text{span}\ \{B_{ij}\Big|_X\}\ (i,j)\in I\,.$$

We proved in [20]:

G satisfies the A-property.

In particular, if $X = R$ then it follows that $S_1^o(\Delta_{rs})$ is an A-subspace.

Remark. Considering all known examples of A-subspaces one can see that the only A-subspaces where X is not necessarily connected are the subspaces studied in the preceding example. However the following result will show that this observation is not surprizing.

Theorem 2.3. Let X be a disconnected subset of $\mathbb{R}$ with $X = \overline{\overset{o}{X}}$. Then every Haar subspace fails to satisfy the A-property.

The proof of this statement can be found in [20].
Therefore by this statement it seems to be hopeless to find any A-subspace of functions defined on disconnected subsets of $\mathbb{R}$. However we showed in [20] that despite of the negative result in Theorem 2.3 there exist nontrivial A-subspaces of functions defined on disconnected subsets X of $\mathbb{R}^n$.

Theorem 2.4. Let $1 \in \mathbb{N}$ and let $X_1, \ldots, X_1$ be compact, connected and pairwise disjoint subsets of $\mathbb{R}^n$ with $X_i = \overline{\overset{o}{X}_i}$, $1 \leq i \leq 1$. Moreover let $X = \bigcup_{i=1}^{1} X_i$ and let G_i be finite dimensional A-subspaces of $C(X_i)$ with $\dim G_i = m_i$, $1 \leq i \leq 1$. Then the subspace G defined by

$$G = \{g : X \to \mathbb{R} : g\Big|_{X_i} \in G_i,\ 1 \leq i \leq 1\}$$

satisfies the A-property.

Remark. (1) It follows from Theorem 2.2 that if $X_i \subset \mathbb{R}$,

$1 \le i \le 1$, then every subspace G constructed as in Theorem 2.4 is a weak Chebyshev subspace of C(X).

(2) It is easily verified that dim G = $\sum_{i=1}^{1} m_i$ for every subspace G defined in the preceding Theorem.

3. Final Remarks

(1) We want to recall that if G is a Haar subspace of C[a,b], then G satisfies the A-property, but the converse is not true in general. However a partial converse of this statement gave Havinson [5] who showed that if G is an $L_1(w)$-unicity subspace of C[a,b] for every $w \in W$ and no nontrivial element in G vanishes on a subinterval of [a,b], then G is a Haar subspace on (a,b).
(2) The reason why we choose X as the closure of its interior is the following: Every $x \in \partial X$ (the boundary of X) can be obtained as limit of a sequence $(x_m) \subset \overset{\circ}{X}$.
Therefore if $f_1, f_2 \in C(X)$ with $\int_X |f_1(x) - f_2(x)| dx = 0$, then $f_1(x) = f_2(x)$ for every $x \in X$. But if X is a general compact subset of $\mathbb{R}^n$, then we can only conclude that $f_1 = f_2$ a.e. on X. Hence in the general case uniqueness of best $L_1(w)$-approximations would only mean that for any two best approximations $g_1, g_2 \in G$, $g_1 = g_2$ on $X \smallsetminus N$ where $N \subset X$ is a null set.
(3) Two further interesting applications of the A-property were obtained by Kroó in [10,11].
At first let X = [0,1] and f = $(f_1, \ldots, f_1) \in C^1(X)$ such that f_i is continuously differentiable, $1 \le i \le 1$. Moreover let $(n_1, \ldots, n_1) \in \mathbb{N}^1$ and let Π_{n_i} denote the space of polynomials of degree at most n_i, $1 \le i \le 1$. Then f is a curve in $\mathbb{R}^1$ with the length

$$L(f) = \int_o^1 \left(\sum_{i=1}^{1} (f_i'(x))^2 \right)^{1/2} dx = \|f'\|_{w,1} \quad \text{where } w \equiv 1.$$

Then by Theorem 1.3 there exists a unique algebraic curve $p = (p_1, \ldots, p_1) \in \Pi_{n_1} \times \ldots \times \Pi_{n_1}$ such that p(0) = f(0) and L(f-p) is minimal.

For the second application let $\tilde{S}_m(\Delta)$ denote the subspace of complex-valued polynomial spline functions of degree m defined on $[a,b]$ with the fixed simple knots at $\Delta = \{a = x_o < x_1 < \ldots < x_{k+1} = b\}$. It was shown in [11] that by Theorem 1.3 every complex-valued continuous function f defined on $[a,b]$ has a unique best $L_1(w)$-approximation from $\tilde{S}_m(\Delta)$ for every $w \in W$.

4. References

1. M.P. Carroll and D. Braess, On uniqueness of L_1-approximation for certain families of spline functions, J. Approximation Theory 12 (1974), 362-364.

2. W. Dahmen and C.A. Micchelli, Recent progress in multivariate splines, in "Approximation Theory IV" (ed. by C.K. Chui, L.L. Schumaker and J.D. Ward), Academic Press, New York, 27-121 (1983).

3. F. Deutsch, G. Nürnberger and I. Singer, Weak Chebyshev subspaces and alternation, Pacific J. Math. 89 (1980), 9-31.

4. P.V. Galkin, The uniqueness of the element of best mean approximation to a continuous function using splines with fixed knots, Math. Notes 15 (1974), 3-8.

5. S.J. Havinson, On uniqueness of functions of best approximation in the metric of the space L^1, Izv. Akad. Nauk SSSR 22 (1958), 243-270.

6. D. Jackson, Note on a class of polynomials of approximation, Trans. Amer. Math. Soc. 22 (1921), 320-326.

7. R.C. Jones and L.A. Karlovitz, Equioscillation under nonuniquiness in the approximation of continuous functions, J. Approximation Theory 3 (1970), 138-145.

8. M. Krein, The L-problem in abstract linear normed space, in "Some Questions in the Theory of Moments" (ed. by N. Akiezer and M. Krein), Translations of Mathematical Monographs, Vol. 2, Amer. Math. Soc., Providence, R.I. (1962).

9. A. Kroó, Some theorems on unicity of multivariate L_1-approximation, Acta Math. Acad. Sci. Hungar. 40 (1982), 179-189.

10. A. Kroó, Best L_1-approximation of vector valued functions, Acta Math. Acad. Sci. Hungar. 39 (1982), 303-310.

11. A. Kroó, Some theorems on best L_1-approximation of continuous functions, Acta Math. Hungar. (to appear).

12. A. Kroó, On the unicity of best L_1-approximation by polynomials of several variables, Acta. Math. Hungar. 42 (1983), 309-318.

13. A. Kroó, On an L_1-approximation problem, preprint.

14. R.M. Moroney, The Haar problem in L_1, Proc. Amer. Math. Soc. 12 (1961), 793-795.

15. A. Pinkus, Unicity subspaces in L_1-approximation, preprint.

16. I. Singer, Best Approximation in Normed Linear Spaces by Elements of Linear Subspaces, Springer-Verlag, Berlin-Heidelberg-New York (1970).

17. M. Sommer, L_1-approximation by weak Chebyshev spaces, in "Approximation in Theorie und Praxis" (ed. by G. Meinardus), Bibliographisches Institut, Mannheim, 85-102 (1979).

18. M. Sommer, Weak Chebyshev spaces and best L_1-approximation, J. Approximation Theory 39 (1983), 54-71.

19. M. Sommer, Some results on best L_1-approximation of continuous functions, Numer. Funct. Anal. and Optimiz. 6(3) (1983), 253-271.

20. M. Sommer, Examples of unicity subspaces in L_1-approximation, preprint.

21. H. Strauß, L_1-Approximation mit Splinefunktionen, in "Numerische Methoden der Approximationstheorie" (ed. by L. Collatz and G. Meinardus), ISNM 26, Birckhäuser-Verlag, Stuttgart, 151-162 (1975).

22. H. Strauß, Eindeutigkeit in der L_1-Approximation, Math. Z. 176 (1981), 63-74.

23. H. Strauß, Best L_1-approximation, J. Approximation Theory 41 (1984), 297-308.

International Series of
Numerical Mathematics, Vol. 74
© 1985 Birkhäuser Verlag Basel

COMPARISON THEOREMS IN
SPLINE APPROXIMATION

Hans Strauss

INTRODUCTION. Let $C(I)$ be the linear space of all real-valued functions defined on $I = [a,b]$ under the norm $\| \ \|$ and let K be a subset of $C(I)$. Then $E(K,f,\| \ \|) = \inf_{g \in K} \|f - g\|$.

Suppose that Π_{m-1} is the subspace of polynomials of degree $\leq m - 1$ and

$$W_\infty^m = \{f : f^{(m-1)} \text{ absol. continuous, } \|f^{(m)}\|_\infty < \infty \} \ .$$

Let $f, g \in W_\infty^m$ and $|f^{(m)}(x)| \leq g^{(m)}(x)$, $x \in I$ then Bernstein has shown that $E(\Pi_{m-1}, f, \| \ \|_\infty) \leq E(\Pi_{m-1}, g, \| \ \|_\infty)$ (see [9], p.77). In [7,8] these results were extended to regular Haar systems. In particular, the theorems were applied in order to obtain estimates for $E(\Pi_{m-1}, f, \| \ \|_\infty)$ where f are special functions and to determine best approximations in segment approximation. These results were extended to L_p-approximation, $1 < p < \infty$, in [1]. Approximation problems for monotone norms satisfying certain constraints were considered in [6].

In this paper we study comparison theorems for approximation problems with subspaces of spline functions. Moreover, we consider comparison theorems for monosplines which play an important role in numerical integration.

1. SPLINE SUBSPACES WITH FIXED KNOTS. Suppose that G is an n-dimensional subspace of $C(I)$ spanned by $g_1,\ldots,g_n$. Let

$$D\begin{pmatrix} g_1,\ldots,g_n \\ t_1,\ldots,t_n \end{pmatrix} = \det((g_i(t_j))_{i,j=1,\ldots,n})$$

where $a \leq t_1 < t_2 < \ldots < t_n \leq b$. Then the subspace G is called a **weak Chebyshev subspace** iff

$$D\begin{pmatrix} g_1,\ldots,g_n \\ t_1,\ldots,t_n \end{pmatrix} \geq 0$$

for all choices of $a \leq t_1 < \ldots < t_n \leq b$. Let G be a weak Chebyshev subspace. A function $f \in C(I)$ is said to be in the **convexity cone** K_G of G if $\mathrm{span}\{G,f\}$ is a weak Chebyshev subspace. The most important examples are subspaces of spline functions.

Suppose that a partition $\Delta: a = x_0 < x_1 < \ldots < x_r < x_{r+1} = b$ of the interval I is given. Let us denote by

$$S(\Delta,M) = \{s : s(x) = \sum_{i=0}^{m-1} a_i x^i + \sum_{i=1}^{r} \sum_{j=1}^{m_i} b_{ij}(x - x_i)_+^{m-j},$$

$$\sum_{i=1}^{r} m_i = n, \ a_i, b_{ij} \subset \mathbb{R}\}, \ m \geq 2$$

the space of **polynomial spline functions of degree** $\leq m - 1$ **with fixed knots** $\{x_1,\ldots,x_r\}$ **of multiplicity** $M = (m_1,\ldots,m_r)$ where $1 \leq m_i < m$ for $i = 1,\ldots,r$.

A function $g \in W_\infty^m$ is an element of the convexity cone $K_{S(\Delta,M)}$ if

$$\varepsilon_i g^{(m)}(x) \geq 0 \quad \text{a.e.} \quad \text{on } x \in [x_i, x_{i+1}] \tag{1.1}$$

where $\varepsilon_i = (-1)^{m_0 + \ldots m_i + n}$, $m_0 = 0$, for all $i = 0,1,\ldots,r$.

This result can be shown as in Proposition 3.4 in [10]. It follows that $f \in K_{S(M)}$ where $f^{(m)} \geq 0$ if the integers m_i are even for all i.

A comparison theorem for best approximations from weak Chebyshev subspaces is given in [11]. The following theorem is a slight generalization of Corollary 2.1 in [11] for spline subspaces with multiple knots.

A norm $\| \ \|$ is called <u>monotone</u> if $|f(x)| \leq |g(x)|$ for all $x \in I$ implies $\|f\| \leq \|g\|$.

<u>THEOREM 1.</u> Let $S(\Delta,M)$ be given and let $g \in W_\infty^m$ satisfying (1.1). Suppose that $f \in W_\infty^m$ satisfies $|f^{(m)}(x)| \leq |g^{(m)}(x)|$, a.e. for $x \in I$. Then $E(S(\Delta,M),f,\| \ \|) \leq E(S(\Delta,M),g,\| \ \|)$ in any monotone norm.

It follows from the assumptions (1.1) that $g \pm f$ is an element of $K_{S(\Delta,M)}$. Then Theorem 2 in [11] proves the theorem.

2. SPLINE FUNCTIONS WITH FREE KNOTS.

In this section we study comparison theorems for spline approximation with free knots. Suppose that $f \in W_\infty^m$ satisfies $f^{(m)}(x) > 0$, $x \in I$. Then we define

$$M(x) = f(x) - s(x), \qquad s \in S(\Delta,M). \tag{2.1}$$

Such a function can be considered as a <u>Tchebycheffian</u> <u>monospline</u> (T-monospline) (see [13], p. 403). The number of zeros $Z(M)$ of M is counted as in [13], p. 331. Set $\sigma_i = 0$, if m_i is even and $\sigma_i = 1$, if m_i is odd. Then $Z(M) \leq k$, $k = m + \sum_{i=1}^{r} (m + \sigma_i)$. It is also required that x_i is at most an $(m - m_i)$-tuple zero of M.

<u>THEOREM 2.</u> Let f,g be functions of W_∞^m satisfying $0 < g^{(m)}(x) \leq f^{(m)}(x)$, $x \in I$. Suppose that M is a T-monospline of the form (2.1) with a maximal set of zeros. Then there exists a T-monospline N with simple knots

$$N(x) = g(x) - \sum_{i=0}^{m-1} c_i x^i - \sum_{i=1}^{1} d_i (x - z_i)_+^{m-1}$$

285

where $1 = 1/2(k - m)$, c_i, $d_i \subset \mathbb{R}$ and $a < z_1 < \ldots < z_1 < b$ such that

$$|N(x)| \leq |M(x)|, \qquad x \in I.$$

This result is shown in [16] using one-sided L_1-approximation.

If we set $f(x) = g(x) = x^m/m!$ in Theorem 2 then we obtain the general improvement theorem of Karlin in [5]. There is a close relationship between monosplines and quadrature formulae. The result says that in most norms a multiple knot quadrature approximation can always be improved by a simple knot approximation.

In [14, 15] more general comparison theorems for monosplines are given.

Suppose that $S(M) = \bigcup_\Delta S(\Delta,M)$ for all choices of partitions Δ satisfying the properties of Section 1. Let us denote the closure of $S(M)$ by $\overline{S(M)}$.

COROLLARY 3. Let f,g be in W_∞^m satisfying $0 < f^{(m)}(x) \leq g^{(m)}(x)$, $x \in I$. Suppose that s_1 is a best approximation from $\overline{S(M)}$ to g in the L_p-norm, $1 \leq p \leq \infty$ where $M = \{1,\ldots,1\}$. Then there is a function $s_0 \in S(M)$ such that

$$|(f - s_0)(x)| \leq |(g - s_1)(x)|, \quad x \in I.$$

It is well-known that there is a best approximation s_1 from $\overline{S(M)}$ to g in $S(M)$ and that $g - s_1$ has a maximal set of zeros (see [2, 4,5,12]). Then the corollary follows from Theorem 2.

We also obtain that $E(f,S(M),\| \ \|_p) \leq E(g,S(M),\| \ \|_p)$, $1 \leq p \leq \infty$, if $0 \leq f^{(m)}(x) \leq g^{(m)}(x)$, $x \in I$ (see also [3]).

Moreover we are able to prove the following result if the multiplicities of the knots are even.

$\underline{\text{THEOREM 4.}}$ Let $S(M)$ be given where $M = (m_1,\ldots,m_r)$ and m_i is even for all i. Suppose that $f,g \in W_\infty^m$ satisfy $|f^{(m)}(x)| \leq g^{(m)}(x)$, $x \in I$. Then

$$E(S(M),f,\|\ \ \|_p) \leq E(S(M),g,\|\ \ \|_p), \quad 1 \leq p \leq \infty.$$

In the proof of this result Theorem 1 is used. In particular, it is important that g is an element of $K_{S(\Delta,M)}$ for all partitions Δ where the multiplicities of the knots are even.

REFERENCES

1. Augsburger W.: Segmentapproximation in L_p-Räumen. Dissertation, Clausthal, 1967.

2. Braess D.: Chebyshev approximation by spline functions with free knots. Numer. Math. $\underline{17}$ (1971), 357-366.

3. Braess D.: On the degree of approximation by spline functions with free knots. Aequationes Math. $\underline{12}$ (1975), 80 - 81.

4. Karlin S.: On a class of best nonlinear approximation prob- and extended monosplines. In: Studies in spline functions and approximation theory. Karlin S., Micchelli C.A., Pinkus A., Schoenberg I.J. (ed.), 19 - 66. New York, Academic Press, 1976.

5. Karlin S.: A global improvement theorem for polynomial monosplines. In: Studies in spline functions and approximation theory. Karlin S., Micchelli C.A., Pinkus A., Schoenberg I.J. (ed.), 67 - 82. New York, Academic Press, 1976.

6. Kimchi E., Richter-Dyn N.: Restricted range approximation of k-convex functions in monotone norms. SIAM J. Numer. Anal. $\underline{15}$ (1978), 1030 - 1038.

7. Meinardus G.: Über ein Monotonieprinzip bei linearen Approximationen. ZAMM 46 (1966), 227 - 238.

8. Meinardus G.: Zur Segmentapproximation mit Polynomen. ZAMM 46 (1966), 239 - 246.

9. Meinardus G.: Approximation of functions: Theory and numerical methods. Berlin, Springer-Verlag, 1967.

10. Micchelli C.A., Pinkus A.: Moment theory for weak Chebyshev
     systems with applications to monosplines, quadrature formu-
     lae and best one-sided L^1-approximations by spline functions
     with fixed knots. SIAM J. Math. Anal. $\underline{8}$ (1977), 206 - 230.

11. Pinkus A.: Bernstein's comparison theorem and a problem of
     Braess. Aequationes Math. $\underline{23}$ (1981), 98 - 107.

12. Rice J.R.: The approximation of functions, Vol. II, Addison-
     Wesley, Reading, 1969.

13. Schumaker L.L.: Spline functions: Basic theory. New York,
     Wiley-Interscience, 1981.

14. Strauss H.: Comparison theorems for monosplines and best
     one-sided approximation. Numer. Funct. and Optimiz. $\underline{6}$ (1983),
     423 - 445.

15. Strauss H.: Monotonicity of quadrature formulae of Gauss
     type and comparison theorems for monosplines. Numer. Math.
     $\underline{44}$ (1984), 337 - 347.

16. Strauss H.: On Tchebycheffian monosplines and approximation
     by splines with free knots, to appear in: Applicable Anal.

International Series of
Numerical Mathematics, Vol. 74
© 1985 Birkhäuser Verlag Basel

THE DIFFERENTIAL CORRECTION ALGORITHM

G. D. Taylor

1. Introduction

With the advent of high-speed digital computers came an interest in methods of computing algebraic rational uniform approximations. A major incentive for this interest was the need of computer subroutines for the evaluation of the elementary functions [2, 3, 17, 27, 46]. These approximations continue to have significant applications in the approximation of data sets and complicated mathematical functions. In fact, for one-dimensional data sets, the general concensus is that one should use either a piecewise polynomial fit or a rational fit. In view of this and because of interest in the problem itself many researchers have studied the problem of computation of rational fits. At present two algorithms are generally viewed as being the most effective. They are the Remes algorithm [26, 47, 48, 50] and the differential correction algorithm [4, 14, 15].

The first extensive comparative study on this topic was given by Lee and Roberts in 1973 [43] where some eight algorithms were tested on eleven specially chosen functions. The conclusion of this study was that although the linear programming-based algorithms cannot compete with the Remes algorithm in terms of computer time, some of them are considerably more robust. Specifically, the differential correction algorithm properly initialized possesses global convergence properties and successfully handled all the examples where the Remes algorithm failed. In this paper we give a survey of work done on the differential correction algorithm. We begin with a brief description of the rational approximation problem and some theoretical facts.

The usual generalized rational approximation setting that one encounters in computation can be described as follows: Given a finite subset $X \subseteq \mathbb{R}^k$ (Euclidean k-space), a real-valued function f defined on X and two sets of linearly independent functions $\{\phi_1, \ldots, \phi_m\}$, $\{\psi_1, \ldots, \psi_n\}$ defined on X, set

$$\mathbb{R}(X) = \{r = \frac{p}{q}: p(x) = \sum_{j=1}^{m} a_j \phi_j(x), \ q(x) = \sum_{j=1}^{n} b_j \psi_j(x), \ a_j, \ b_j \text{ real}$$

$\forall_{i,j}$ with $q(x) > 0$ $\forall$ $x \in X$ and $\max\limits_{1 \le j \le n} |b_j| = 1\}$.

In this setting $r^* \in \mathfrak{R}(X)$ is said to be a best uniform generalized approximation to f from $\mathfrak{R}(X)$ on X provided $\|f - r^*\|$ = $\inf\limits_{r \in \mathfrak{R}(X)} \|f - r\| = \Delta^*$ where $\|h\| = \max\limits_{x \in X} |h(x)|$. A theory of generalized rational approximation can be found in [16, 44]. Basically, the existence of a best generalized rational approximant is not guaranteed due to both the general form of the approximants and the discrete structure of X, whereas characterization and uniqueness results can be given.

The differential correction algorithm can be applied to this general problem and it consists of two phases. The first phase is an initialization phase which will be discussed later and the second phase is the actual iteration itself. This iteration at the k-th step where one has a current approximation $r_k = p_k/q_k \in \mathfrak{R}(X)$ with error $\Delta_k = \|p - r_k\|$ consists of solving the linear programming problem: Select $p(x) = \sum\limits_{j=1}^{m} a_j \phi_j(x)$ and $q(x) = \sum\limits_{j=1}^{n} b_j \psi_j(x)$ to

$$(1) \quad \begin{cases} \text{minimize } \max\limits_{x \in X} \left\{ \dfrac{|f(x)q(x) - p(x)| - \Delta_k q(x)}{q_k(x)} \right\} \\[2ex] \text{subject to } |b_j| \le 1, \quad j = 1, \ldots, n. \end{cases}$$

Set p_{k+1}, q_{k+1} equal to the solution p, q, respectively, of this problem. This is the form of the differential correction algorithm originally given by Cheney and Loeb [14]. However, in their proof of convergence the additional assumption that there exists $\alpha > 0$ such that $q_k(x) \ge \alpha > 0$ for all k and x was added to this linear programming problem. Because this was not a natural addition they replaced (1) by (2) below in [15] for which a convergence theory was given:

$$(2) \quad \begin{cases} \text{minimize } \max\limits_{x \in X} \{ |f(x)q(x) - p(x)| - \Delta_k q(x) \} \\[2ex] \text{subject to } |b_j| \le 1, \quad j = 1, \ldots, n. \end{cases}$$

In 1974, Barrodale, Powell and Roberts [4] proved that the original differential correction algorithm (1) was convergent without the additional bounds on the denominator being assumed. In fact, they proved that if r_k is not the best approximation then $q_{k+1}(x) > 0 \; \forall \; x \in X$ and $\Delta_k \downarrow \Delta^*$. Furthermore, in the non-degenerate (definition later) one-dimensional case they showed that this algorithm is actually quadratically convergent. (At the time of this result I remember Cheney remarking that he was pleased to see it since users of these codes had found that the original algorithm performed much better. In fact, in [4] both codes were tested and the original algorithm clearly outperformed the second approach (2) which exhibited linear convergence rates.) In the remaining sections of this paper we shall only be referring to (1) as the differential correction algorithm.

At least three FORTRAN production codes of the differential correction algorithm are now available [5, 25, 28, 29]. The third reference was published in 1975 and an improved version is available and was announced in [29]. The code [25] was developed at Bell Laboratories. The third code [5] was published in 1977 and is faster and more robust than the original code given in [28]. The improved version [29] should demonstrate the same robustness as the code [5], although it probably is slightly slower due to the extra features it has.

All of the above codes use Loeb's algorithm for their initialization phase:

$$\text{Select } p_0(x) = \sum_{j=1}^{m} a_j \phi_j(x) \text{ and } q_0(x) = \sum_{j=1}^{n} b_j \psi_j(x) \text{ to be}$$

the normalized solution of the problem (assume $\psi_1 > 0$ on X):

$$(3) \quad \begin{cases} \text{minimize:} \quad \max_{x \in X} |f(x)q(x) - p(x)| \\ \\ \text{with } b_1 = 1 \end{cases}$$

provided this yields a q_0 which does not change sign and is not too small on X; otherwise, take $p_0 \equiv 0$ and $q_0 \equiv \psi_1$. It was found in [43] that this initialization is in general very efficient.

291

The code [5] has a second initialization possibility that attempts to select $\Delta_0 > \Delta^*$ and $q_0 > 0$. It is noted there that this initialization also seems to be quite effective with the efficiency depending primarily upon selecting Δ_0 close to Δ^*. The choice of q_0 does not seem to be very crucial.

For the special case that $X \subset \mathbb{R}^1$ and $\phi_j(x) = x^{j-1}$, $\psi_k(x) = x^{k-1}$ we have the classical rational uniform approximation problem. Here the Remes algorithm can be applied using the alternating characterization of a best uniform rational approximation. For this discussion we shall write (π_n denotes all real algebraic polynomials of degree $\leq$ n)

$$\mathbb{R}_n^m(X) = \{r = p/q: p \in \pi_m, \ q \in \pi_n), \ (p,q)=1 \text{ and } q(x) > 0 \ \forall \ x \in X\}.$$

Uniqueness of best approximations from $\mathbb{R}_n^m(X)$ follows from the following well-known characterization (alternation) theorem.

<u>THEOREM</u>. $r^*(x) = p^*(x)/q^*(x) \in \mathbb{R}_n^m(X)$ is a best approximation to $f(x)$ on X if and only if there exists a set of $N=2+\max(n+\partial p^*, m+\partial q^*)$ points $\{x_i\}_{i=1}^N \subset X$ such that $x_1 < x_2 < \ldots < x_N$, $f(x_i) - r^*(x_i)$ $= -(f(x_{i+1}) - r^*(x_{i+1}))$, $i = 1, \ldots, N - 1$ and $|f(x_i) - r^*(x_i)|$ $= \|f - r^*\|$, $i = 1, \ldots, N$ where ∂p^* and ∂q^* denote the degrees of $p^*(x)$ and $q^*(x)$, respectively. The set of points $\{x_i\}_{i=1}^N$ is said to form an alternating set for f and f is said to be normal if it has a best approximation $r^* = p^*/q^*$ with $\max(n+\partial p^*, m+\partial q^*)$ $= n + m$. (Note that even in this setting existence need not hold when X is a discrete set.)

The possibility of degeneracy (non-normality) of a function is a major difficulty in attempting to generalize the polynomial version of the Remes algorithm to the rational setting. Many of the existing coded versions of the rational Remes algorithm actually assume that the f to be approximated is normal and thereby "ignore" this problem. However, in practice one finds that normal functions that are nearly degenerate may also cause these codes problems.

The Remes algorithm consists in seeking the final alternating set. Assuming for the sake of discussion that the f to be
approximated is normal, then at the beginning of the k-th iteration the Remes algorithm starts with a reference set
$X_k = \{x_1^k < x_2^k < \ldots < x_N^k\} \subseteq X$, $N = m + n + 2$. Next, the algorithm
calculates (if possible) the best rational approximation
$r_k = p_k/q_k$ to f on X_k usually by using Newton's method to solve
the nonlinear system of equations

$$f(x_i^k)q(x_i^k) - p(x_i^k) = (-1)^i \lambda q(x_i^k), \quad i = 1, \ldots, N$$

for the unknowns $a_0, \ldots, a_m$, $b_1, \ldots, b_n$ (the coefficients of p
and q) and λ where $b_0 = 1$ is taken as a normalization. If r_k is
not the best approximation to f on X, the next reference set,
X_{k+1}, is constructed from the extreme points of $f - p_k/q_k$. This
is done by selecting $x_1^{k+1} < \ldots < x_N^{k+1}$ each in X, such that

$$\text{sgn}(f(x_i^{k+1})-p_k(x_i^{k+1})/q_k(x_i^{k+1}))=-\text{sgn}(f(x_{i+1}^{k+1})-p_k(x_{i+1}^{k+1})/q_k(x_{i+1}^{k+1})),$$

$$i = 1, \ldots, N - 1,$$

$$|f(x_i^{k+1})-p_k(x_i^{k+1})/(q_k(x_i^{k+1})| \geq \max_j |f(x_j^k)-p_k(x_j^k)/q_k(x_j^k)|,$$

$$i = 1, \ldots, N$$

and for some η, $1 \leq \eta \leq N$,

$$|f(x_\eta^{k+1}) - p_k(x_\eta^{k+1})/q_k(x_\eta^{k+1})| = \|f - p_k/q_k\| \, .$$

The initial reference set is chosen (normally) to be the points
of X which are closest to the extreme points of the N-th Chebyshev polynomial translated to the smallest closed interval containing X. As noted earlier, this algorithm can fail to converge.
Failure can be caused by the desired solution having a reference
set of less than N points (degeneracy), by not being able to solve
the nonlinear system for an acceptable p_k/q_k (non-existence or
solution with poles on X_k), or by making an improper exchange
due to poles of p_k/q_k off the reference set X_k but between points
of X (this can cause cycling). In spite of these difficulties,
the Remes algorithm remains a very heavily used method. This is
probably due to its extensive dissemination and to a lesser
extent the fact that it tends to be much faster than the

differential correction algorithm (some 19 times faster in the
testing reported in [43]) as well as needing considerably less
storage. A hybrid Remes-differential correction algorithm has
recently been developed which essentially avoids the three types
of failures mentioned above. It will be discussed later [31].
We also wish to note here that the quadratic convergence of the
differential correction algorithm proved in [4] was for the non-
degenerate algebraic rational theory mentioned above. In [36],
quadratic convergence is shown to hold for the generalized
rational setting when it is required that all denominators be
uniformly bounded away from 0. Whether this is true without this
constraint remains open.

Finally, in a recent paper Dunham [24] claimed that in
degenerate cases the differential correction algorithm can be
unstable. However, in a more detailed study Kaufman [32] proved
that if the differential correction algorithm is coded so that it
returns the previous rational approximant whenever $\Delta_{k+1} \geq \Delta_k$
occurs, then the stability of the differential correction algorithm is equivalent to the stability of its linear programming
submodule.

II. Generalizations of the Differential Correction Algorithm

In this section we shall discuss two types of generalizations of the differential correction algorithm. First, we shall
discuss two studies that extend the range of the differential
correction algorithm by using it on appropriate subsets of the
domain of approximation. Then, we shall survey various constrained rational approximation problems for which a modified
differential correction algorithm has actually been run.

One finds in practice that the major cost of the differential correction algorithm in terms of storage and speed is in
its linear programming subroutine with these costs increasing
rapidly with the size of the data set (or domain of function to
be approximated). For the one-dimensional classical rational
approximation problem a hybrid of the Remes and differential

correction algorithms, called Remes-Difcor, has been developed [29, 30, 32]. This algorithm uses the differential correction algorithm only on subsets that are essentially alternating reference sets (small sets) and is no longer subject to the three possible causes of failure for the Remes algorithm. It differs from the Remes algorithm described above in two crucial respects. First, approximation on reference sets are found using the differential correction algorithm rather than solving a nonlinear system of equations. Thus, an approximation with a positive denominator on the reference set is guaranteed even if the system has no solution (say, because of degeneracy). Second, if a g-pole (that is, a point where the denominator is very small in absolute value, or negative) occurs somewhere in X off the reference set, X_k, and $f - p_k/q_k$ changes sign $m + n + 1$ times on the reference set, then the next reference set is expanded to include the point where q_k is smallest. Note that the flexibility achieved by using differential correction on the reference set is essential here, since the new reference set will have $n + m + 3$ points instead of $m + n + 2$. One point will be deleted from this enlarged reference set after p_{k+1}/q_{k+1} is computed. This Remes-Difcor algorithm will have guaranteed convergence regardless of the choice of initial reference set, provided that at some point no more g-poles occur for all succeeding reference sets encountered and best approximations exist on all these reference sets [31]. (It is also shown that in an appropriate setting this algorithm is identical to Remes eventually.) An independent study by Belogus and Liron [9] also developed an algorithm of this character but only with the first feature (use of the differential correction algorithm). Out of seventy test runs comparing the Remes, Remes-Difcor and differential correction algorithms the Remes algorithm failed 13 times, the Remes-Difcor algorithm once (due to apparent non-existence of best approximations on some reference sets) and the differential correction algorithm never [30]. Comparing the speeds of these algorithms for the test cases where Remes ran successfully and the data sets consisted of more than 50 points, one finds that on the average the Remes-Difcor

algorithm is approximately 2.5 times slower than the Remes algorithm and the differential correction algorithm is about 8.5
times slower than the Remes algorithm. Thus, it is claimed that
the Remes-Difcor algorithm inherits the best properties of each
of its parent algorithms and should be seriously considered as
general-purpose library routine in most computation centers.

The second generalization of this flavor is the recent
development of an adaptive differential correction algorithm
[33, 34]. The purpose of this algorithm is to treat data sets
which are too large to allow for the direct use of the differential correction algorithm and where the Remes or Remes-Difcor
algorithms do not apply because of the lack of an alternating
theory. Under the assumption that the differential correction
algorithm calculates a good (not necessarily best) approximation
on each (small) subset to which it is applied, it is proved [33]
that this adaptive algorithm will terminate in a finite number
of steps at a good approximation on the entire data set. The
basic goal of this algorithm is to calculate the desired best
rational fit for the full data set by applying the differential
correction algorithm to a progression of small subsets where
each succeeding subset is selected using information from the
previous subset. Thus, at the k-th step of the algorithm, a
subset X_k of X and a good approximation r_k from $\mathbb{R}(X_k)$ to f on
X_k with error $\Delta_k = \max\{|f(x) - r_k(x)| : x \in X_k\}$ will have been
calculated. Using values of $f - r_k$ and q_k on X_k and $X \sim X_k$ some
points of X_k may be deleted and some points of $X \sim X_k$ are added
to the remaining X_k points to form X_{k+1}. Basically, X_{k+1} will
usually consist entirely of points where $|f - r_k|$ is essentially
larger than or equal to Δ_k or $q_k < \eta$ (some $\eta > 0$ fixed). In
addition, comparing Δ_k with Δ_{k-1} will sometimes also bring
previously deleted points from X_{k-1} back into X_{k+1}. (For precise
details the reader is referred to [33].) For cases where the
domain of the data is rectangular in structure, this code uses a
directional search (based upon a nine-point stencil for the two-
dimensional case) starting from the points of X_k where $|f - r_k|$
is close to Δ_k or $q_k < \eta$ to locate additional points of X to be

added to X_k. For cases of scattered data, a boxing procedure
has been suggested to attempt to treat the set as though it is
rectangular in nature. This code is still an experimental code
and additional testing is necessary. (This is especially the
case for the scattered data problem.)

This adaptive procedure in the few test cases reported
on to date has shown a significant savings in computing time and
storage requirements as compared to the differential correction
algorithm. Furthermore, it can be applied to problems that would
overwhelm the differential correction algorithm; witness a three-
dimensional example given in [34] with a 251 x 126 x 101 grid
(giving 3,194,226 grid points) that would result in a linear
programming problem of nine variables and 6,388,460 constraints
if one were to use the differential correction algorithm. Using
the adaptive differential correction algorithm on this problem,
the desired approximant was found with the differential correction algorithm being applied to no subset consisting of more than
19 points. It was also found that if one combines this method
with a nested mesh refinement approach where the adaptive routine
is applied to each of the nested sets in order (coarse to fine)
additional efficiency can be realized. This particular feature
is included in the code of [34] and is done automatically when
desired.

The other generalization of the differential correction
algorithm that we wish to discuss here is for constrained rational
approximation problems. It was noted in [4] that since the
differential correction algorithm is based upon linear programming, in principal, linear constraints can be readily included
as a part of the algorithm. Here we shall report on situations
where this has actually been done.

The first case to be mentioned here is that of restricted
range approximation [49]. In this theory one has in addition to
X, f and $\mathfrak{R}(X)$ as defined earlier, the existence of two more
finite sets X_u, $X_\ell \subseteq \mathbb{R}^k$ which can be selected completely independ-
dent of each other and X and two real-valued functions u and ℓ

defined on X_u and X_ℓ, respectively, subject to $\ell(x) < u(x)$ for all $x \in X_u \cap X_\ell$. Then one approximates with

$$K(X) = \{r \in \mathfrak{R}(X): \ell(x) \leq r(x),\ x \in X_\ell \text{ and } r(x) \leq u(x),\ x \in X_u\}.$$

The code given in [29] can be used to compute best restricted range rational approximants by simply setting one parameter value and supplying the sets X_u and X_ℓ together with the values of u and ℓ on these sets. In addition, this code also has the capability of computing weighted approximations of the form $W \cdot (f - p/q)$ or $f - W \cdot p/q$. One application of this theory is the calculation of nonnegative approximants for the design of recursive digital filters [22, 23, 35] to be discussed in the next section. The more general restricted range constraints have also been suggested for this design problem [35].

A second constrained problem to which the differential correction algorithm has been applied is rational approximation having restricted denominators. For this problem the approximants are defined in terms of a real-valued nonnegative function ℓ defined on X as

$$K(X) = \{r = p/q: r \in \mathfrak{R}(X) \text{ with } q(x) \geq \ell(x)\ \forall\ x \in X\}.$$

A general theory including a differential correction adaptation for this setting is given in [36]. When $\ell(x) > 0\ \forall\ x \in X$ holds this theory (classical rational case) guarantees the existence of best approximations and avoids some numerical difficulties in both computing and using good approximations. The motivation for studying such a theory was a digital filter study of McCallig [45] where these constraints played an important role.

Another constrained approximation theory problem where the differential correction algorithm has been applied is reciprocal polynomial approximation on $[0, \infty)$ with nonnegative coefficients [37] and nonnegative leading coefficient [38], and uniform reciprocal approximation subject to (general) linear constraints [13]. The above papers develop existence, characterization and uniqueness results for these approximation problems as well as modified differential correction algorithms. In the first two problems, the adaptive differential correction algorithm [33] is extended to this setting. This was done since a major difficulty was to find a large enough finite subset of $[0, \infty)$ to give the $[0, \infty)$ result. The third study modifies the differential correction algorithm and the Remes-Difcor algorithm [30]. Examples of this theory include monotone and bounded coefficient constraints on the approximants. More recently, additional results concerning rational approximation on $[0, \infty]$ have been developed [10, 42]. In [42] the differential correction algorithm was modified to handle the case where ∞ is an extreme point.

The next constrained approximation problem to be mentioned is that of uniform approximation with rational functions having negative poles [39]. This theory appears to be more complicated as it allows for the existence of local best approximations and local uniqueness. Here the Remes-Difcor algorithm was applied to a linearization of the current best approximation candidate in an outer iterative scheme for calculating the best approximation. Since an alternating theory for these linearizations is not known to exist this algorithm possesses no convergence proof although it appears to have converged on all examples considered. The motivation for this theory was a paper by Cody, Meinardus and Varga that studied approximation of e^{-x} on $[0, \infty)$ for the construction of numerical solutions for solving linear systems of ordinary differential equations which arise from semi-discretization of linear parabolic partial differential equations [18]. See [1] and [11] for some recent very interesting results on the approximation of e^{-x} on $[0, \infty)$.

Further uses of the differential correction algorithm include simultaneous approximation of several functions [41] and modifications for approximation by exponential sums [8, 12].

III. Applications to Recursive Digital Filter Design

There are numerous methods for designing digital filters, one of which is uniform approximation theory. In terms of approximation, the one-dimensional filter (magnitude only) problem amounts to approximating a function defined to be 1 on some closed subintervals of $[0, \pi]$, 0 on other disjoint closed subintervals of $[0, \pi]$ and undefined elsewhere, with either a linear combination of cosine functions ($\{\cos kx\}_{k=0}^{N}$ a non-recursive filter) or with nonnegative quotients of these linear combinations (recursive filter) having positive denominators. For the nonrecursive case the Remes algorithm is used under the name of the McClellan-Parks-Rabiner algorithm (who published a coded version of this algorithm for this problem in 1973). For the recursive case the differential correction algorithm has been used as well as other approaches. As mentioned earlier, Dudgeon [22] proposed the use of the differential correction algorithm for this problem in 1974. In this paper the requirement that the approximant be nonnegative was obtained by performing a "shift" to the best unconstrained fit. (An improved approximant could be obtained by solving the constrained problem with the algorithm of [30].) Shortly after this study Dudgeon published a second paper proposing the use of the differential correction algorithm for the design of two-dimensional recursive digital filters [23].

More recently, Corelazzo and Lightner [19, 20] considered a complex form of the digital filter problem. Roughly, this amounts to simultaneously approximating the magnitude of a given complex-valued function (usually step function-like) and an arbitrary linear phase on the unit circle in the complex plane with a rational algebraic function in z having real coefficients. That is, the magnitude of the rational fit should give a good approximation to the magnitude of the function and the phase of the rational function should be close to being linear. The precise theory of such a problem remains to be developed and it may or may not suffer the difficulties inherent in complex rational approximation theory [6]. At present, we are investigating possible modifications to the differential correction algorithm in order to treat this problem. To date, a running code based upon the differential correction algorithm that simultaneously fits the modulus and phase of a given complex-valued function on a finite subset of the unit circle of the complex plane with rational algebraic functions in z with real coefficients has been developed. Initial numerical experiments have appeared to result in convergence. Basically, this code iterates on linearizations of the magnitude and phase approximation problems using a simultaneous version of the differential correction algorithm to solve those linearized problems. A complete study of the theoretical aspects of this problem as well as the above problem where the approximation of the phase is replaced with the requirement that the approximant have essentially a linear phase remains to be done.

Finally, in closing we wish to reference an earlier survey of rational fitting with an emphasis on numerical computation given by Barrodale and Roberts in 1973 [7].

Acknowledgement

This research was supported in part by ONR-N00014-84-0591.

References

1. J. E. Andersson, Approximation of e^{-x} by rational functions with concentrated negative poles, J. Approx. Theory 32 (1981), 85-95.

2. M. Andrews, B. Eisenberg, S. F. McCormick and G. D. Taylor, Evaluation of functions on microcomputers: Rational approximation of the k-th roots, Int. J. Comput. and Math. with Appl. 5 (1979), 163-169.

3. M. Andrews, D. Jaeger, S. F. McCormick and G. D. Taylor, Evaluation of functions on microcomputers: Exp(x), Int. J. Comput. and Math. with Appl. 7 (1981), 503-508.

4. I. Barrodale, M. J. D. Powell and F. D. K. Roberts, The differential correction algorithm for rational ℓ_∞-approximation, SIAM J. Numer. Anal. 9 (1972), 493-504.

5. I. Barrodale, F. D. K. Roberts and K. B. Wilson, An efficient computer implementation of the differential correction algorithm for rational approximation, Manitoba Conf. on Numerical Mathematics and Computing, 1977, U. of Victoria Tec. Rep. DM-115-IR, 1978, pg. 1-21.

6. I. Barrodale, Best approximation of complex-valued data, Proc. 7th Biennial Conf., U. of Dundee, Lecture Notes in Math., Vol. 630, Springer, Berlin, 1978, pgs. 14-22.

7. I. Barrodale and F. D. K. Roberts, Best approximation by rational functions, Proc. 3rd Manitoba Conf. on Numer. Methods, U. of Manitoba, Winnipeg, Manitoba, Canada, 1973, pgs. 3-29.

8. G. G. Belford and J. F. Burkhalter, A differential correction algorithm for exponential curve fitting, Tech. Rep. UIUC-CAC-73-92, Computer Center, U. of Ill., Urbana, 1973.

9. D. Belogus and N. Liron, DCR2: An improved algorithm for ℓ_∞ rational approximation on intervals, Numer. Math. 31 (1978), 17-29.

10. H.-P. Blatt, Rationale Tschebysheff-Approximation über unbeschränkten Intervallen, Numer. Math. 27 (1977), 179-190.

11. P. B. Borwein, Rational approximations with real poles to e^{-x} and x^n, J. Approx. Theory 38 (1983), 279-283.

12. D. Braess, Die Konstruktion der Tschebyscheff-Approximierenden bei der Anpassung mit Exponentialsummen, J. Approx. Theory 3 (1970), 261-273.

13. B. L. Chalmers, E. H. Kaufman, Jr., D. J. Leeming and G. D. Taylor, Uniform reciprocal approximation subject to linear constraints, J. Approx. Theory 41 (1984), 201-216.

14. E. W. Cheney and H. L. Loeb, Two new algorithms for rational approximation, Numer. Math. 3 (1961), 72-75.

15. E. W. Cheney and H. L. Loeb, On rational Chebyshev approximation, Numer. Math. 4 (1962), 124-127.

16. E. W. Cheney and H. L. Loeb, Generalized rational approximation, J. SIAM Numer. Anal. Ser. B, 1 (1964), 11-25.

17. W. J. Cody, Jr., and W. Waite, Software Manual for the Elementary Functions, Prentice-Hall, Englewood Cliffs, NJ, 1980.

18. W. J. Cody, G. Meinardus and R. S. Varga, Chebyshev rational approximations to e^{-x} on [0, ∞) and applications to heat-conduction problems, J. Approx. Theory 2 (1969), 50-65.

19. G. Cortelazzo and M. R. Lightner, Simultaneous design in both magnitude and group-delay of IIR and FIR filters: Problems and results, Proc. IEEE Int. Conf. on Acoust., Sp. and Sig. Proc., April (1983), 201-204.

20. G. Cortelazzo and M. R. Lightner, The use of multiple criterion optimization for frequency domain design of non-causal IIR filters, Proc. IEEE Int. Conf. on Acoust., Sp. and Sig. Proc., ICASSP82, Paris (1982), 1813-1816.

21. S. N. Dua and H. L. Loeb, Further remarks on the differential correction algorithm, SIAM J. Numer. Anal., 10 (1973), 123-126.

22. D. E. Dudgeon, Recursive filter design using differential correction, IEEE Trans. Acoust., Sp. and Sig. Proc. ASSP-22 (1974), 443-448.

23. D. E. Dudgeon, Two-dimensional recursive filter design using differential correction, IEEE Trans. on Acoust., Sp. and Sig. Proc., ASSP-23 (1975), 264-267.

24. C. B. Dunham, Stability of differential correction for rational Chebyshev approximation, SIAM J. Numer. Anal. 17 (1980), 639-640.

25. B. D. Eldredge and D. D. Warner, An implementation of the differential correction, Computer Science Technical Report #48, Bell Laboratories, Murray Hill, NJ, 1976.

26. W. Fraser and J. F. Hart, On the computation of rational approximations to continuous functions, Comm. Assoc. Comput. Mach. 5 (1962), 401-403.

27. J. F. Hart, E. W. Cheney, C. L. Lawson, J. H. Maehly, C. K. Mesztenyi, J. R. Rice, H. C. Thatcher, Jr., and C. Witzgall, Computer Approximations, Wiley, New York, 1968.

28. E. H. Kaufman, Jr. and G. D. Taylor, Uniform rational approximation of functions of several variables, Int. J. Numer. Methods Engrg., 9 (1975), 297-323.

29. E. H. Kaufman, Jr., D. J. Leeming and G. D. Taylor, Uniform rational approximation by differential correction and Remes-differential correction, Int. J. Numer. Methods Engrg. 17 (1981), 1273-1280.

30. E. H. Kaufman, Jr., D. J. Leeming and G. D. Taylor, A combined Remes-differential correction algorithm for rational approximation: experimental results, Comp. and Math. with Appls. 6 (1980), 155-160.

31. E. H. Kaufman, Jr., D. J. Leeming and G. D. Taylor, A combined Remes-differential correction algorithm for rational approximation, Math. Comput. 32 (1978), 233-242.

32. E. H. Kaufman, Jr., The behavior of differential correction in difficult situations, preprint, 16 pg.

33. E. H. Kaufman, Jr., S. F. McCormick and G. D. Taylor, An adaptive differential-correction algorithm, J. Approx. Theory 37 (1983), 197-211.

34. E. H. Kaufman, Jr., S. F. McCormick and G. D. Taylor, Uniform rational approximation on large data sets, Int. J. Numer. Methods Engrg. 18 (1982), 1569-1575.

35. E. H. Kaufman, Jr. and G. D. Taylor, An application of a restricted range version of the differential correction algorithm to the design of digital systems, Int'l. Ser. Numer. Math. 30 (1976), 207-232.

36. E. H. Kaufman, Jr., and G. D. Taylor, Uniform approximation by rational functions having restricted denominators, J. Approx. Theory 32 (1981), 9-26.

37. E. H. Kaufman, Jr., D. J. Leeming and G. D. Taylor, Approximation on $[0, \infty)$ by reciprocals of polynomials with nonnegative coefficients, J. Approx. Theory 40 (1984), 29-44.

38. E. H. Kaufman, Jr., D. J. Leeming and G. D. Taylor, Approximation on subsets of $[0, \infty)$ by reciprocals of polynomials, Approx. Theory IV, L. L. Schumaker, ed., Acad. Press, New York, 1983, pgs. 553-559.

39. E. H. Kaufman, Jr. and G. D. Taylor, Uniform approximation with rational functions having negative poles, J. Approx. Theory 4 (1978), 364-378.

40. E. H. Kaufman, Jr. and G. D. Taylor, Best rational approximations with negative poles to e^{-x} on $[0, \infty)$, Padé and Rational Approximations: Theory and Applications, Acad. Press, NY, 1977, pgs. 413-425.

41. E. H. Kaufman, Jr. and G. D. Taylor, An application of linear programming to rational approximation, Rocky Mtn. J. of Math. 4 (1974), 371-373.

42. E. H. Kaufman, Jr., D. J. Leeming and G. D. Taylor, Uniform rational approximation on subsets of $[0, \infty]$, preprint.

43. C. M. Lee and F. D. K. Roberts, A comparison of algorithms for rational ℓ_∞ approximation, Math. Comp. 27 (1973), 111-121.

44. H. L. Loeb, Approximation by generalized rationals, J. SIAM Numer.Anal. 3 (1966), 34-55.

45. M. T. McCallig, R. Kurth and B. Steel, Recursive digital filters with low coefficient sensitivity, Proc. 1979 Int'l. Conf. on Acoust., Sp. and Sig. Proc., Washington, D.C., April, 1979.

46. S. F. McCormick, D. Pryor and G. D. Taylor, Evaluation of functions in microcomputers: $\ln(x)$, Int. J. Comput. and Math. with Appl. 8 (1982), 389-392.

47. A. Ralston, Rational Chebyshev approximation by Remes algorithm, Numer. Math. 7 (1965), 322-330.

48. J. R. Rice, The Approximation of Functions, Vol. II - Advanced Topics, Addison-Wesley, Reading, MA, 1969.

49. G. D. Taylor, Approximation by functions having restricted ranges III, J. Math. Anal. Appl. 27 (1969), 241-248.

50. H. Werner, Tschebysheff-Approximation in Bereich der rationalen Funktionen bei Vorliegen einer guten Ausgangsnahersung, Arch. Rat. Mech. Anal. 10 (1962), 205-219.

Department of Mathematics
Colorado State University
Fort Collins, Colorado 80523
U.S.A.

International Series of
Numerical Mathematics, Vol. 74
© 1985 Birkhäuser Verlag Basel

Ein mathematisches Modell
für den Reifungsprozess roter Blutkörperchen
bei Neugeborenen

H. WERNER, C. FESSER

Zusammenfassung: In der vorliegenden Arbeit versuchen wir, durch ein mathematisches Modell Meßwerte zu interpretieren. Diese Daten wurden in einigen in der Kinderklinik der Universität Münster unter der Anleitung von Herrn Professor Dr. Schellong angefertigten Dissertationen (vgl. [2], [3]) erhoben. Aus der Lösung des mathematischen Modells können Rückschlüsse für die Medizin gezogen werden. Die zur Beschreibung benutzten Parameter werden hierbei durch optimale Anpassung an die Meßdaten bestimmt.

EINLEITUNG

Ausgangspunkt dieser Arbeit sind empirische Untersuchungen über den Reifungsprozeß roter Blutkörperchen im Säuglingsblut, insbesondere der zeitliche Verlauf der Verteilung auf die Altersgruppen nach der auf Heilmeyer zurückgehenden Einteilung [1] . Das mathematische Modell versucht, den Reifungsprozeß zu simulieren und mit Hilfe einer geringen Anzahl von Parametern das Reifeverhalten genauer zu erfassen und die Verweilzeiten zu schätzen. Dabei stützen wir uns auf Ergebnisse, die bereits in einer früheren Arbeit für einen Teil der Meßkurven abgeleitet worden sind [7] . Es zeigte sich allerdings, daß die Interpretation der für Gruppen sehr junger reifender roter Blutkörperchen (Retikulozyten) gewonnenen Erkenntnisse nicht ohne weiteres auf die älteren Formen der Retikulozyten übertragbar ist. Deshalb wurde die dort angewendete Vorgehensweise modifiziert. Gerade das Reifeverhalten der dort kaum diskutierten älteren Gruppen liefert beispielsweise Erkenntnisse über den zeitlichen Verlauf der Produktion von Retikulozyten im Knochenmark. Die eigentlich vorliegende und dynamisch zu beschreibende Rückkoppelung zwischen Blutvolumen und -produktion wurde ersetzt durch eine geeignete parameterabhängige Form der "Lebenslinien" (Charakteristiken) der Teilchen und ihre dadurch charakterisierte beschleunigte Reifung.
Ziel dieses Manuskriptes ist es, ein Modell aufzustellen, das die medizinischen Messungen hinreichend genau zu reproduzieren ermöglicht. Im ersten Abschnitt werden die zugrunde liegenden medizinischen Fakten erläutert. Abschnitt II beschäftigt sich mit der Aufstellung und der Erläuterung des mathematischen Modells, dessen numerische Behandlung in Abschnitt III beschrieben wird. Abschnitt IV enthält

die durch **Parameteridentifikation** gewonnenen **Resultate** und vergleicht sie mit den bisher bekannten medizinischen Ergebnissen. Abschnitt V schließlich faßt die Ergebnisse zusammen und stellt Modellansätze vor, die im Laufe der Modellentwicklung erprobt, jedoch verworfen wurden, da diese Modelle nicht so gut an die Meßwerte angepaßt werden konnten wie das in Abschnitt II vorgestellte.

Unsere Rechnungen ergeben verhältnismäßig scharfe Aussagen über die Verweildauer reifender roter Blutkörperchen in den einzelnen Altersgruppen, und wir bekommen auch recht genaue Angaben über die Produktion der roten Blutkörperchen im Knochenmark (medulläre Produktion). Bei den von uns getroffenen geometrischen Approximationen der Lebenslinien der einzelnen Teilchen zeigt es sich, daß die Empfindlichkeit des Modells gegen Störungen dieser Linien nicht zu groß ist. Auch dies spricht dafür, daß die hier vorgenommene relativ grobe Approximation dieser glatten Kurven durch Streckenzüge durchaus angemessen ist. Das Ergebnis kann also dahingehend gedeutet werden, daß unsere Berechnungen verhältnismäßig stabil gegenüber der Festlegung der sogenannten Charakteristiken sind. Für die gefundenen quantitativen Aussagen hinsichtlich der Entwicklungszeiten und der Produktion des Knochenmarks darf auf die Zusammenfassung der Ergebnisse im Schlußabschnitt hingewiesen werden.

Frau Professor Dr. med. Kowalewski von der Universitätskinderklinik Bonn danken wir für die Hinweise auf einschlägige Literatur, Herrn Dr. med. Köster darüberhinausgehend für seine Beratung in den medizinischen Fragen.

I. MEDIZINISCHE PHÄNOMENE BEI DER REIFUNG ROTER BLUTKÖRPERCHEN IN DEN ERSTEN LEBENSTAGEN

Die roten Blutkörperchen (Erythrozyten) unterscheiden sich grundsätzlich von anderen Körperzellen durch das Fehlen eines Zellkerns. Wie andere Zellen auch, entsteht ein Erythrozyt durch schrittweise Differenzierung aus multipotenten Stammzellen. Die Entkernung erfolgt während der letzten Stufe dieser Entwicklung: Der Zellkern des Normoblasten wird zerstört, und es entstehen netzartige Strukturen, die stufenweise abgebaut werden. Ein noch nicht vollständig entwickeltes rotes Blutkörperchen, das diese Strukturen aufweist, heißt *Retikulozyt*. Beim erwachsenen Menschen erfolgt ein Großteil der Differenzierung im Knochenmark, ein Teil der Entwicklung des Retikulozyten zum reifen Erythrozyt findet jedoch im Blut statt. Die Retikulozyten werden ins Blut ausgeschwemmt und reifen binnen 48 Stunden zu Erythrozyten heran ([5],S.371). In einigen Arbeiten wird von doppelt so langer Reifezeit gesprochen ([1],S.539). Hingegen verläuft die Erythropoese (d.h. die Produktion der Erythrozyten) beim Säugling gerade in den ersten Lebenstagen grundsätzlich anders als beim Erwachsenen. Die Erythrozyten sind hier von anderer Gestalt, ihr Reifeprozeß dauert etwas länger, sie enthalten ein bindungsaktiveres Hämoglobin und können während des fetalen Lebens und in den ersten Lebenstagen auch außerhalb des Knochenmarks gebildet werden. Diese sogenannte *extramedulläre Erythropoese* findet vor allem in den ersten drei Keimlingsmonaten in Milz und Leber statt. Sie wird dann durch die *medulläre Erythropoese* (Blutbildung im Knochenmark) ersetzt und unterstützt diese in den Tagen vor und nach der Geburt, um dann im postnatalen Leben ganz abzuklingen ([1],S.514). Die extramedullär gebildeten Retikulozyten werden in einem früheren Reifestadium ins Blut ausgeschwemmt als die medullär produzierten. Aufgrund der netzartigen Strukturen der Retikulozyten werden nach Heilmeyer ([1],S.537) vier Reifestadien unterschieden. Da im fetalen Blut alle vier Reifestadien, im postnatalen Blut aber nach einigen Lebenstagen (ca. 1 Woche) nur noch die letzten beiden Stadien zu beobachten sind, nimmt man an, daß die Retikulozyten von Milz und Leber im ersten, vom Knochenmark im dritten Stadium ausgeschieden werden. In der vorliegenden Arbeit wird eine mathematische Simulierung dieser Vorgänge versucht, mit deren Hilfe u.a. Aussagen über die Verweildauer der Retikulozyten in den einzelnen Reifestadien (d.h. die Zeit, in der ein Retikulozyt in einer Reifegruppe bleibt) gemacht werden können.

Um einen möglichen hohen Blutverlust während der Geburt zu kompensieren, wird kurz vor der Geburt die Anzahl der Erythrozyten erhöht. Dies geschieht, wie bereits oben erwähnt, einerseits durch erneute extramedulläre Erythropoese, andererseits durch Beschleunigung des Reifungsvorgangs, d.h. Verkürzung der Reifezeit ([4],S.37). Eine Aufstockung des Stammzellenpools im Knochenmark, um auf diese Weise mehr Erythrozyten zu produzieren, läßt sich nicht beobachten. Trotzdem produziert das Knochenmark in den ersten Lebenstagen vermehrt rote Blutkörperchen. Am ersten Tag nach der Geburt sind 30 - 40 % aller kernhaltigen Zellen im Mark Erythrozyten-Vorgänger, am Ende der ersten Woche jedoch nur noch 8 - 12 %. Es werden also am Anfang anteilig mehr Retikulozyten produziert ([4],S. 20).

Diese rückläufige Entwicklung wird bedingt durch das Ausbleiben des Erythropoetins. Dieses Hormon, das die Erythropoese anregt, läßt sich nach dem ersten Lebenstag bis zum zweiten Lebensmonat nicht mehr im Blut feststellen. Die Produktion der roten Blutkörperchen ist bei der Geburt so hoch, daß in den folgenden Wochen der Körper diese Produktion reduzieren muß, um sich auf einen Normalzustand einzustellen ([4],S.21). Insgesamt ist also nach der Geburt eine Verminderung der anfänglich gesteigerten Erythropoese zu beobachten.

II. Modell

Die Grundlage für das hier angegebene Modell bilden Meßwerte, die in Untersuchungen der Kinderklinik der Universität Münster in den Jahren 1972 und 1973 gewonnen wurden ([2], [3]). Zweimal wurden größere Gruppen von Probanten, einmal über 50, ein zweites Mal 31 untersucht. Bei der hier analysierten zweiten Gruppe wurde der Retikulozytenanteil im Blut in täglichem Abstand innerhalb der ersten beiden Lebenswochen gemessen. Angegeben werden hier die Mittelwerte für die Anzahl $v_1(t)$, $v_2(t)$, $v_3(t)$, $v_4(t)$ der Teilchen (in Promille der roten Blutkörperchen) in den vier Reifestadien nach Heilmeyer in Abhängigkeit vom Alter t des Kindes (Figur 1). Da die Meßdaten eine relativ große Streuung aufweisen, ist es nicht sinnvoll, die Kurven zu glätten. Verwendet werden einfach Polygonzüge. Unter der Prämisse, daß die Retikulozyten nur durch extramedulläre oder durch medulläre Produktion ins Blut eintreten aber nicht während des Reifevorgangs abgebaut werden, versucht das Modell, den Reifeprozeß der Blutkörperchen zu simulieren und somit die Verweildauer der Retikulozyten in den einzelnen Altersgruppen, das genaue Reifungsverhalten und die Knochenmarkproduktion zu schätzen.

2.1 Notation

Die Bezeichnungen stimmen mit den in [7] gewählten überein.

ANZAHL DER RET.
ALTER DES KINDES
MESSDATEN DER GRUPPEN V1,V2,V3 UND V4
V1
V2
V3
V4

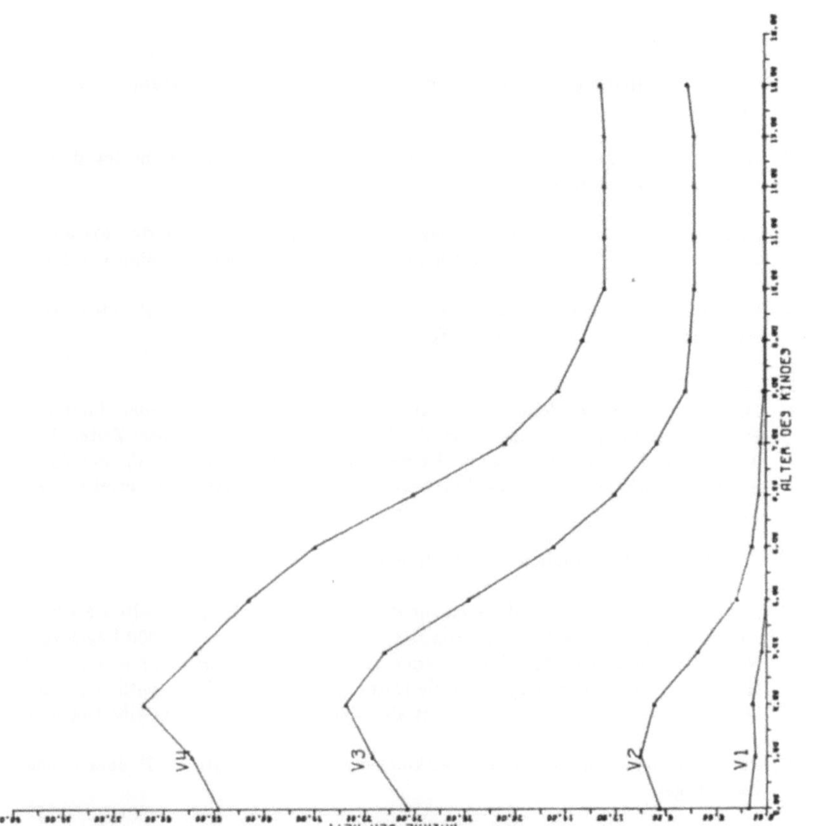

FIGUR 1

t seit der Geburt vergangene Zeit, Alter des Kindes, gemessen in Tagen

T individuelles Alter der Retikulozyten , d.h. deren Reifungszustand, gemessen in Tagen

I_1 := $[0, 14]$, Definitionsintervall für t. Da die Daten nur für $0 \leq t \leq 14$ gegeben sind, wird auch das Modell nur für diesen Bereich diskutiert.

I_2 := $[0, 30]$, Definitionsintervall für T. Aus medizinischen Gründen genügt es, sich für T- Werte auf I_2 zu beschränken, da $T_4 \leq 30$ sicher erfüllt ist. (Dies bedeutet, daß die Retikulozyten im Reifealter von 30 Tagen zum Erythrozyten herangereift sind. Diese Schranken scheinen auch bei Abweichung vom Normalwert von 10 Tagen hinreichend hoch zu liegen.)

I := $I_1 \times I_2$

T_i , $i = 1, 2, 3$; Reifungsalter, in dem die Retikulozyten von der Gruppe v_i zur Gruppe v_{i+1} wechseln

T_4 Reifungsalter, in dem ein Blutkörperchen aus der Gruppe v_4 ausscheidet, d. h. reif ist und zum Erythrozyten wird.

T_0 Reifungsalter, in dem die Retikulozyten der extramedullären Produktion ausgeschwemmt werden. Es wird T_0 := 0 gesetzt (denn nur die Differenzen der T_i sind von Interesse)

T_5 Reifungsalter, in dem die vom Knochenmark produzierten Retikulozyten ins Blut ausgeschwemmt werden. Es gilt: $T_2 < T_5 < T_3$

T_{ij} := T_i - T_j
Die Skalierung dieser Zeiten T_i ist natürlich problematisch, sie werden durch die (mittlere) Reifungszeit für jede Gruppe von Retikulozyten bestimmt. Diese Zeiten können selbst wieder eine Funktion von t sein. Es werden hier die Zeiten T_3 , T_4 und T_5 benutzt, die beobachtet werden, wenn die Erythropoese ihre stationäre Form erreicht hat, etwa für $t \geq 10$.

v_i , $i = 1, 2, 3, 4$; Reifestadien nach Heilmeyer

$v_i(t)$, $i = 1, 2, 3, 4$; $t \geq 0$, Funktionen für die über alle Kinder gemittelten Meßwerte, die die Anzahl von Retikulozyten der Gruppe v_i zur Zeit t bezogen auf 1000 Erythrozyten für ein Normalkind angeben. Meßwerte werden nur an den Gitterwerten $t = 0, 1, \ldots, 14$ erhoben (Figur 1). Medizinisch mag dabei die Entnahme bei einem Kind zeitlich erheblich von der bei einem anderen abweichen. Es liegt also von daher schon eine gewiße Ungenauigkeit vor.

$u(t, T)$ Verteilungsfunktion, Dichte der Retikulozyten im Reifezustand T eines Kindes im Alter von t Tagen

$u_{mod}(t, T)$ die durch das Modell gegebene Verteilungsfunktion

$\tilde{v}_i(t)$, $i = 1, 2, 3, 4$;   Anzahl der Retikulozyten berechnet aus $u_{mod}(t, T)$ in den entsprechenden Altersgruppen bei einem Kind im Alter von t Tagen. (Analogon zu $v_i(t)$)

Aus den gegebenen Daten $v_i(t)$ wird ersichtlich, daß die Anzahl der Retikulozyten eines jeden Reifestadiums nach anfänglichem Anwachsen bis zu einem Normalwert abklingt. Besonders charakteristisch ist

bei den Kurven $v_3(t)$ und $v_4(t)$ das Maximum bei $t = 2$, dem in der folgenden Diskussion noch größere Aufmerksamkeit gewidmet wird. Das in [7] diskutierte Modell gab eine befriedigende Interpretation der Kurven v_1 und v_2. Es wird bei einfacher Übertragung dem beobachteten Verlauf der Kurven $v_3(t)$ und $v_4(t)$ nicht gerecht, denn die Maxima von $v_3(t)$ und $v_4(t)$ werden stark zu großen t hin verschoben. Die plötzliche Akkumulation der Retikulozyten der dritten und vierten Altersgruppe läßt sich nicht durch ein gleichmäßiges Heranreifen der Blutkörperchen beschreiben, vielmehr scheint der Reifeprozeß in den ersten Tagen stark beschleunigt zu werden. Da die Retikulozyten der Gruppen v_1 und v_2 ganz aus dem Blut verschwinden, liegt die Vermutung nahe, daß die extramedulläre Produktion nach wenigen Tagen aufhört und nur noch die medulläre Erythropoese das Blutbild bestimmt. In der Tat ergibt sich nach [7] ein solches Verhalten. In den ersten Tagen vollzieht sich also die Produktion der Retikulozyten an zwei Stellen: in der Leber (als Oberbegriff für die extramedulläre Produktion) und im Knochenmark. Die für die Leberproduktion und die Zeiten T_1 und T_2 gewonnenen Erkenntnisse aus [7] werden hier übernommen, neu diskutiert werden jedoch die Produktion im Knochenmark und die Reifezeiten T_5 , T_3 und T_4 .
Die zur Veranschaulichung benutzten Diagramme haben als Abszisse den Reifezustand T, als Ordinate das Kindesalter t, beides in Tagen gemessen.

2.2 Aufstellung des Modells

Die als Daten gegebenen Kurven $v_i(t)$ lassen sich wie folgt interpretieren:

$$(1) \qquad v_i(t) = \int_{T_{i-1}}^{T_i} u(t, \tau)\, d\tau, \quad t = 0, 1, \ldots, 14$$

Wären die Zeiten T_i $(i = 1, \ldots, 4)$ und die Funktion $u(t, T)$ $((t, T) \in I)$ bekannt, so könnte man $v_i(t)$ berechnen. Hier ist jedoch das Problem genau umgekehrt gestellt. $v_i(t)$ ist bekannt, gesucht sind $u(t, T)$ und T_i . Zu konstruieren ist also ein Modell, das die gegebenen Daten $v_i(t)$ gut reproduziert. In [7] wird ein Modell entwickelt, das die Daten $v_1(t)$ und $v_2(t)$ gut wiedergibt, also gerade die extramedulläre Produktion gut widerspiegelt. Aus diesem Ansatz lassen sich die Zeiten T_1 und T_2 bestimmen, es ergibt sich $T_1 = 2$ und $T_2 = 6$.
Im folgenden sollen die Zeiten T_3 , T_4 und T_5 bestimmt werden, indem anhand des Modells, abhängig von einigen *Modellparametern* modifizierte Kurven $\tilde{v}_3(t)$ und $\tilde{v}_4(t)$ berechnet werden, die den Daten $v_3(t)$ und $v_4(t)$ angepaßt sind. Die Parameter werden so gewählt, daß der relative Fehler

$$\max \left(\frac{\|v_3 - \tilde{v}_3\|_{\mathcal{L}^2(I_1)}}{\|v_3\|_{\mathcal{L}^2(I_1)}} , \frac{\|v_4 - \tilde{v}_4\|_{\mathcal{L}^2(I_1)}}{\|v_4\|_{\mathcal{L}^2(I_1)}} \right)$$

minimiert wird.
Für alle Kurven $v_i(t)$ läßt sich (im Rahmen der Meßgenauigkeit)

$$v_i(t) = const. = v_i(\infty) \text{ für } t \geq 14; i = 1, \ldots, 4,$$

beobachten. Man entnimmt den Meßkurven die Werte

$$v_1(\infty) = 0, \qquad v_2(\infty) = 0, \qquad v_3(\infty) = 6, \qquad v_4(\infty) = 13.$$

Es ist deshalb gerechtfertigt, T_4 aus

$$\frac{T_4 - T_3}{T_3 - T_5} = \frac{v_4(\infty)}{v_3(\infty)}$$

zu bestimmen und

$$(2) \qquad T_4 = T_3 + \frac{13}{6} \cdot (T_3 - T_5) \approx T_3 + 2.17 \cdot (T_3 - T_5)$$

zu setzen. Für die weitere Diskussion des Modells ist folgender Begriff von zentraler Bedeutung:

DEFINITION. *In Anlehnung an die Theorie partieller Differentialgleichungen erster Ordnung heiße $c : I_2 \rightarrow I_1$ Charakteristik der Verteilungsfunktion $u : I \rightarrow \Re$, wenn c stetig ist und $u\,(c\,(T)\,,T) = const \quad \forall T \in I_2 \backslash T_5$ gilt.*

Im weiteren Verlauf der Untersuchung ist noch zusätzlich c monoton in T. Anschaulich gesprochen beschreibt eine Charakteristik eine Kurve, entlang derer ein Blutkörperchen durch alle vier Reifestadien wandert und schließlich zum reifen Erythrozyt wird.

Ist z.B. $c(T) = T + a$, $a \in \Re$, d.h. ist $c(T)$ eine Gerade mit Steigung 1, so altert ein Retikulozyt so schnell wie das Kind . Ist dagegen $\dfrac{dc(T)}{dT} = m < 1$, so reifen die Retikulozyten schneller, die Reifezeiten, bzw. die Verweildauer in den einzelnen Gruppen ist verkürzt. Kennt man die Charakteristiken und die Anfangsverteilungen

$$(3.1) \qquad\qquad\qquad H(t) := u(t,0)$$

$$(3.2) \qquad\qquad\qquad V_{init}(T) := u(0,T)$$

$$(3.3) \qquad\qquad\qquad H_5(t) := u(t,T_5) - \tilde{u}(t,T_5)$$

[wobei $\tilde{u}$ die Verteilung der von der Leber ausgeschütteten und bis zum Reifezustand T_5 entwickelten Blutkörperchen bezeichnen soll und $H_5(t)$ die Produktion im Knochenmark ist] , so läßt sich für $(t,T) \in I$ die Funktion $u(t,T)$ berechnen. Es muß vorausgesetzt werden, daß durch jeden Punkt (t,T) genau eine Charakteristik $c_{t,T}$ gegeben ist. Dann gilt :

$$(4) \qquad u(t,T) = u\,(c_{t,T}(T),T) = \begin{cases} u\,(c_{t,T}(0),0)\ = H\,(c_{t,T}(0)) & t \geq c_{0,0}(T), T < T_5 \\[2ex] u\left(0, c_{t,T}^{-1}(0)\right) = V_{init}\left(c_{t,T}^{-1}(0)\right) & t < c_{0,0}(T), T < T_5 \\[2ex] H\,(c_{t,T}(0))\ \ + H_5\,(c_{t,T}(T_5)) & t \geq c_{0,0}(T), T \geq T_5 \\[2ex] V_{init}\left(c_{t,T}^{-1}(0)\right) + H_5\,(c_{t,T}(T_5)) & t < c_{0,0}(T), T \geq T_5 \end{cases}$$

Deshalb werden im folgenden medizinisch sinnvolle Definitionen für die Charakteristiken c und die Anfangsverteilungen H, V_{init} und H_5 vorgenommen.

Bemerkung: Die Gleichung (4) legt eine Aufspaltung von $u(t,T)$ der Form

$$(5) \qquad\qquad\qquad u(t,T) = u_L(t,T) + u_K(t,T)$$

nahe, wobei u_L die Verteilung der von der Leber, u_K die Verteilung der vom Knochenmark produzierten Blutkörperchen bezeichnet. Aus den oben getroffenen Vereinbarungen folgt dann:

$$u_L(t,T) = 0 \qquad t \geq 14$$

$$u_K(t,T) = 0 \qquad T < T_5$$

2.2.1 Charakteristiken.

Der folgende Abschnitt beschränkt sich auf eine Betrachtung des ersten Quadranten, da $u(t,T)$ nur dort von Interesse ist.

Es handelt sich bei dem hier betrachteten Reifungsprozeß um einen Einschwingvorgang, das Reifungsverhalten stabilisiert sich in den ersten beiden Lebenswochen. Nimmt man nach [7] an, daß die Charakteristiken beim älteren Säugling Geraden mit Steigung 1 sind, so werden durch diese Normierung die Reifezeiten T_i ($i = 1, \ldots 4$) bestimmt. Da der Reifungsprozeß in den ersten Lebenstagen beschleunigt wird, ist es sinnvoll, dort von Charakteristiken mit variabler Steigung < 1 auszugehen.

Die Charakteristiken sollen hier zur Vereinfachung der Rechnung in erster Näherung durch Streckenzüge ersetzt werden. Wir geben zunächst eine geometrische Beschreibung für ihre Konstruktion an.

Zur Festlegung der Charakteristiken definiert man eine Strecke $G_{v,w}(T) = v \cdot \left(1 - \frac{T}{w}\right)$, $0 \leq T \leq w$, abhängig von den *Modellparamteren* v (Schnittpunkt von $G_{v,w}$ mit der t-Achse) und w (Schnittpunkt von $G_{v,w}$ mit der T-Achse) mit $14 \geq v > 0$ und $30 \geq w > 0$. Außerdem wird durch den Punkt $t = v$, $T = 0$ eine Gerade G_v mit der Steigung 1 gelegt. In den oberhalb dieser Geraden liegenden Punkten sollen die Charakteristiken ebenfalls Geraden mit Steigung 1 sein, d.h. schneidet eine Charkteristik die t-Achse im Punkt $\tilde{t} \geq v$, so sei diese Charakteristik von konstanter Steigung 1 (siehe vorn). Auch die unterhalb der Strecke $G_{v,w}$ liegenden Streckenzüge der Charakteristiken in I sollen Steigung 1 besitzen.

Zur Bestimmung des übrigen Richtungsfeldes sei die die T-Achse in w schneidende Charakteristik G_w eine Gerade mit konstanter Steigung $\epsilon < 1$, wobei ϵ als *Modellparameter* gegeben ist. Durch diese Konstruktion ergibt sich ein Punkt (t^*, T^*) als Schnittpunkt von G_w und G_v mit den Koordinaten:

$$T^* \;=\; -\,\frac{v + \epsilon \cdot w}{1 - \epsilon}$$

(6)

$$t^* \;=\; -\,\epsilon \cdot \frac{v + w}{1 - \epsilon}$$

In dem bisher noch nicht diskutierten, oberhalb von $G_{v,w}$ und unterhalb von G_v liegenden Teil von I sollen die Charakteristiken mit den Geraden zusammenfallen, die durch den Punkt (t^*, T^*) gehen.

Damit wird in jedem Punkt von I eine charakteristische Richtung festgelegt, die entweder identisch 1 oder als Steigung der Geraden durch (t, T) und (t^*, T^*) bestimmt ist.

Die hier beschriebene Konstruktion der Charakteristiken ordnet jedem Punkt $(t, T) \in I$ eindeutig eine Charakteristik $c_{t,T}$ zu mit $c_{t,T}(T) = t$. Die Richtung der Charakteristik hängt außerhalb von $G_{v,w}$ stetig von t und T ab.

Es werden also drei verschiedene Gruppen von Streckenzügen unterschieden:

(i)   Die Charakteristiken sind von konstanter Steigung 1, d.h. bei genügend hohem Alter des Kindes ist der Reifevorgang unverkürzt.

(ii) Die Charakteristiken sind Streckenzüge, haben zunächst Steigung 1 und knicken dann auf $G_{v,w}$ ab, d.h. die Retikulozyten werden sukzessive von der Beschleunigung des Reifeprozesses erfaßt.

(iii) Die Charakteristiken sind von konstanter Steigung < 1, wenn sie die T-Achse in $\tilde{T} \geq w$ schneiden. Die Beschleunigung erfaßt die Teilchen schon mit der Geburt, wenn ihr Reifealter bei der Geburt genügend groß ist ($\tilde{T} \geq w$).

Diese Wahl der Charakteristiken erweist sich als ausreichende Näherung für die in Kapitel I dargelegten medizinischen Gegebenheiten

Formal wird also folgender Ansatz gemacht:

Die Charakteristiken c werden durch ihren Schnittpunkt y_c mit der t-Achse paramterisiert. Schneidet eine Charakteristik c_y die T-Achse im Punkt $T = y > 0$, und setzt man die Charakteristik zu $t < 0$ durch eine Gerade der Steigung 1 fort, so wird die t-Achse in $-y$ geschnitten. Es gilt also:

$$c_{\tilde{t}}(T) \quad = T + \tilde{t} \qquad\qquad \forall \tilde{t} \geq v$$

$$c_{-u}(T) \quad = \frac{t^*}{T^* - u} \cdot (T - u) \qquad \forall u \geq w$$

Bezeichnet man für die verbleibenden Charakteristiken mit $a(y)$ die Stelle T_y, in der die Charakteristik c_y ihren Knick aufweist, so ergibt sich $a(y)$ abhängig von den *Modellparametern* v und w wie folgt :

$$(7) \qquad\qquad a(y) = \frac{w}{v + w} \cdot (v - y) \qquad -w \leq y \leq v$$

Die Funktion $m(y)$ schließlich soll die Steigung des Streckenzugs oberhalb von $G_{v,w}$ charakterisieren, d.h.:

$$(8) \qquad\qquad m(y) = \begin{cases} \dfrac{t^*}{T^* + y} & y \leq -w \\[3mm] \dfrac{t^* + a_y \cdot \frac{v}{w} - v}{T^* - a_y} & v > y > -w \\[3mm] 1 & y \geq v \end{cases}$$

Schreibweise: Im folgenden wird die Bezeichnung $m(y), a(y)$ verwendet, wenn Eigenschaften dieser Funktionen erörtert werden, m_y, a_y wenn die Funktionen $c_y(T)$ diskutiert werden.

Die so definierten $a(y)$ und $m(y)$ sind stetig in y, die Charakteristiken sind stückweise Geraden, deren Steigung auf den Teilintervallen monoton steigend in y ist.

Der Parameter ϵ beschreibt, wie stark der Reifeprozeß zu Beginn beschleunigt wird; v gibt die Zeit an, die vergeht, bis diese Beschleunigung auch die "jüngeren" Zellen erfaßt hat; w schätzt das Reifealter, von dem ab Akzeleration bereits zum Zeitpunkt der Geburt einsetzt.

Der Wert y ist gerade der Punkt, an dem man die Anfangsverteilungen auswerten muß, um die Funktion $u_{mod}(t, T)$ zu berechnen, d.h. es gilt:

$$c_y(0) \quad = \quad y \qquad y \geq 0$$

$$c_y(-y) \quad = \quad 0 \qquad y \leq 0$$

Insgesamt werden die Charakeristiken also wie folgt festgelegt (Figur 2) :

$$(9) \qquad\qquad c_y(T) = \begin{cases} T + y & y \geq v \\[3mm] T + y & T \leq a_y,\ -w \leq y < v \\[3mm] m_y \cdot (T - a_y) + y + a_y & T > a_y,\ -w \leq y < v \\[3mm] m_y \cdot (T + y) & y < -w \end{cases}$$

mit a_y wie in (7) und m_y wie in (8).

Diese Kurvenschar $\{c_y(T),\ -30 \leq y \leq 14\}$ erfüllt die oben an die Charakteristiken gestellten Bedingungen, denn es gilt:

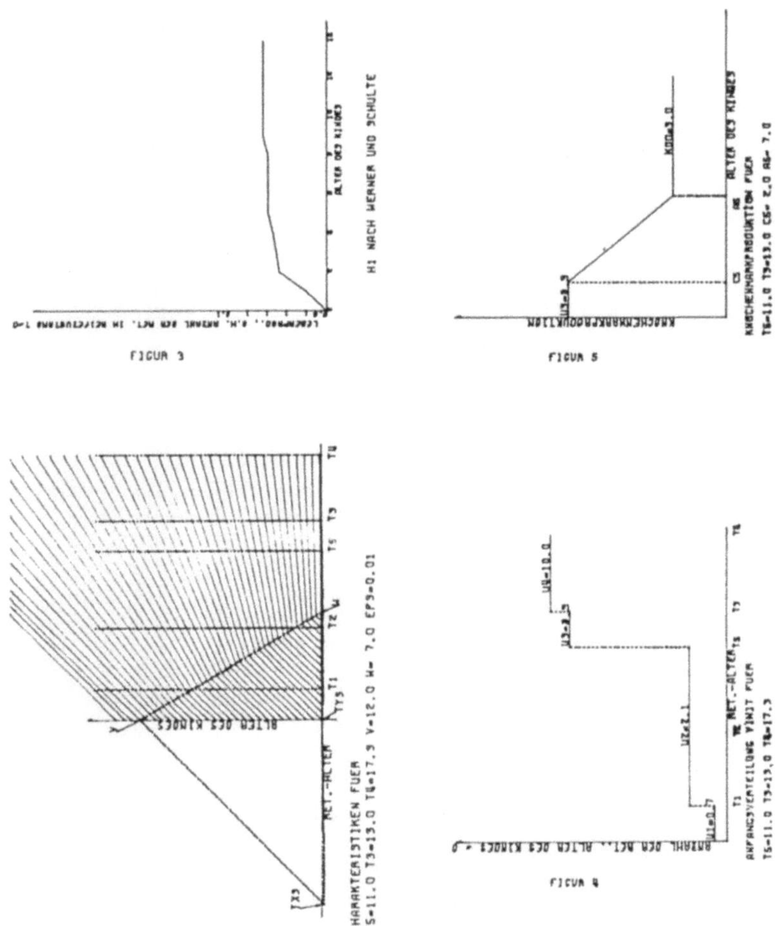

CHARAKTERISTIKEN FUER
TS=11.0 T3=15.0 T6=17.3 V=12.0 W= 7.0 EPS=0.01
FIGUR 2
H1 NACH WERNER UND SCHULTE
FIGUR 3
ANFANGSVERTEILUNG VJW1T FUER
TS=11.0 T3=15.0 T6=17.3
FIGUR 4
KNOCHENMARKPRODUKTION FUER
TS=11.0 T3=13.0 C5= 2.0 A6= 7.0
FIGUR 5

c_y ist stetig $\forall y$ und c_y ist invertierbar, denn da $m_y > 0$ folgt:

$$(10) \qquad c_y^{-1}(t) = \begin{cases} t - y & y \geq v \\[2ex] t - y & t \leq a_y + y \,,\; -w \leq y < v \\[2ex] \dfrac{t - y - a_y}{m_y} + a_y & t > a_y + y \,,\; -w \leq y < v \\[2ex] \dfrac{t}{m_y} - y & y < -w \end{cases}$$

Außerdem gibt es durch jeden Punkt (t, T) mit $t \geq 0, T \geq 0$ genau eine Näherungscharakteristik c_y, d.h. gibt es genau ein $y = y(t, T) \in [-30, 14]$ mit $c_y(T) = t$, das sich aus den Gleichungen (9) leicht bestimmen läßt.

Hiermit sei die Diskussion der Näherungen für die Charakteristiken abgeschlossen.

2.2.2 Anfangsverteilungen.

Nun soll der Modellansatz für die Anfangsverteilungen V_{init} und H, sowie für die Funktion H_5, die die Produktion der Retikulozyten im Knochenmark beschreibt, genauer spezifiziert werden.

2.2.2.1 Festlegung von H

Nach (3.1) gilt: $H(t) = u(t, 0)$ für $t \geq 0$. Es sei wie in [7] gesetzt (Figur 3)[1] :

$$h(t) = \int_0^t u(\tau, 0) \, d\tau \quad , t \geq 0$$

$H(t)$ wird durch eine stückweise, auf den Intervallen

$$I_k := [k \cdot 0.1, (k + 1) \cdot 0.1] \quad , k = 0, 1, \dots$$

konstante Treppenfunktion approximiert, d.h. es wird für $H(t)$ gesetzt :

$$(11) \qquad H(t) = [h((k + 1) \cdot 0.1) - h(k \cdot 0.1)] \cdot 10 \quad \text{für } t \in I_k \quad .$$

Da das Modell in [7] eine gute Approximation der Daten $v_1(t)$ und $v_2(t)$ liefert, und diese die Produktion in der Leber widerspiegeln, liegt die Übernahme der dort entwickelten Funktion $h(t)$ nahe.

2.2.2.2 Festlegung von V_{init}

Es gilt: $V_{init}(T) = u(0, T)$ und

$$v_1(0) \quad = \int_0^2 u(0, \tau) \, d\tau \quad = 1.429 \; ,$$

$$v_2(0) \quad = \int_2^6 u(0, \tau) \, d\tau \quad = 8.5 \quad ,$$

$$v_3(0) \quad = \int_6^{T_3} u(0, \tau) \, d\tau \quad = 28.57 \; ,$$

$$v_4(0) \quad = \int_{T_3}^{T_4} u(0, \tau) \, d\tau \quad = 43.667,$$

---

[1] $H1$ bezeichnet in Figur 3 die Funktion h aus [7]

wobei die Werte : $T_0 := 0$, $T_1 := 2$, $T_2 := 6$ aus [7] übernommen wurden.

Zur Vereinfachung wird auch V_{init} als Treppenfunktion gewählt, und zwar so, daß $u(0,T)$ konstant ist auf den Teilintervallen

$$[0,T_1] \; , \; [T_1,T_2] \; , \; [T_3,T_4] \; .$$

In $[T_2,T_3]$ wird V_{init} spezieller gewählt, da dort in $T = T_5$ die Produktion im Knochenmark beginnt. V_{init} wird wie folgt gesetzt (Figur 4):

$$
(12) \quad
\begin{aligned}
V_{init}(T) &= \frac{v_1(0)}{2} = \frac{1.429}{2} &&=: u_1 \quad 0 \le T \le T_1 \\[2ex]
V_{init}(T) &= \frac{v_2(0)}{T_{21}} = \frac{8.5}{4} &&=: u_2 \quad T_1 \le T \le T_2 \\[2ex]
V_{init}(T) &= \frac{v_4(0)}{T_{43}} = \frac{43.667}{(T_4 - T_3)} &&=: u_4 \quad T_3 \le T \le T_4
\end{aligned}
$$

Für $T \in [T_2, T_3]$ soll gelten:

$$
\begin{aligned}
V_{init}(T) &&= u_2 \quad T_2 \le T \le T_5 \\[1.5ex]
V_{init}(T) &&=: u_3 \quad T_5 \le T \le T_3
\end{aligned}
$$

Da die Produktion im Knochenmark für $T \le T_5$ noch keine Rolle spielt, wird die Anfangsverteilung durch u_2 für $T_2 \le T \le T_5$ konstant fortgesetzt. In T_5 setzt jedoch diese Produktion ein, deshalb können ab dort V_{init} und $u(t,T)$ unterschiedliche Verhaltensweisen zeigen. Der Wert u_3 wird bestimmt aus der Gleichung:

$$v_3(0) = 28.57 \quad \implies \quad u_3 = \frac{28.57 - u_2 \cdot (T_5 - T_2)}{(T_3 - T_5)}$$

Aus diesen Bestimmungsgleichungen lassen sich leicht folgende Relationen für die u_i ableiten:

$$
(13) \quad
\begin{aligned}
&0 < u_1 < u_2 \\[2ex]
&u_2 \le u_3 &&\iff T_3 \le \frac{28.57}{u_2} + 6 &&\approx 17.5 \\[2ex]
&u_3 \le u_4 &&\iff T_5 \ge \frac{28.57 \cdot 2.17 - 43.667}{2.17 \cdot u_2} + 6 &&\approx 9.97 \\[2ex]
&u_3 \ge 0 &&\iff T_5 \le \frac{28.57}{u_2} + 6 &&\approx 17.5
\end{aligned}
$$

Die Bedingung $T_5 \le 17.5$ ist medizinisch sinnvoll, da andernfalls $T_{52} > 11.5$ gelten würde, d.h. die Verweildauer der Retikulozyten im Knochenmark wäre nach allen medizinischen Erfahrungen viel zu groß. Auch die Bedingung an T_3 läßt sich so rechtfertigen.

Beide Bedingungen werden von den unten erhaltenen Ergebnissen erfüllt.

V_{init} wurde also so gewählt, daß immer gilt:

$$\tilde{v}_3(0) = v_3(0) \; , \; \tilde{v}_4(0) = v_4(0)$$

2.2.2.3 Festlegung von H_5

Im Reifestadium $T = T_5$ treten die im Knochenmark produzierten Retikulozyten ins Blut ein. Da die Produktion in der Leber nach einigen Tagen versiegt und dann auch der Reifeprozeß normal, d.h. ohne Beschleunigung verläuft, wurde oben die Steigung der Charakteristiken als konstant und identisch 1, sowie $H(t) = 0$ für $t > T_3 - T_5$ gesetzt. Damit folgt

$$v_3(t) = \int\limits_{T_2}^{T_3} u(t,\tau)\, d\tau = \int\limits_{T_5}^{T_3} u(t,\tau)\, d\tau = \int\limits_{t-(T_3-T_5)}^{t} u(\tau,T_5)\, d\tau$$

für $t > T_3 - T_5$. Unter der Annahme, daß $u\,(t,T_5)$ konstant ist für große t, folgt mit $k_\infty := H_5\,(\infty)$

$$v_3(t) = k_\infty \cdot (T_3 - T_5) \quad \text{für } t \geq a_5\,.$$

Der Parameter a_5, d.h. der Tag, von dem ab die Produktion im Knochenmark konstant ist, ist ein *Modellparameter* und es gelte: $0 \leq a_5 \leq 13$ (d.h. nach spätestens 13 Tagen sollte sich die Produktion auf einen Normalzustand eingestellt haben; diese Annahme wird durch die medizinischen Tatsachen gerechtfertigt). Wegen $v_3(t) \rightarrow 6$ für große t implizieren diese Annahmen:

$$(14) \qquad\qquad k_\infty = \frac{6}{T_3 - T_5} \qquad .$$

Außerdem lehrt die Medizin, daß die Produktion des Erythropoetins erst nach einigen Lebenstagen reduziert wird, d.h. die Reduktion der Retikulozytenproduktion erst mit einiger Verzögerung auftritt. Dies rechtfertigt die Einführung eines weiteren *Modellparameters* c_5, der den Tag bezeichne, von dem ab die Produktion im Knochenmark reduziert wird und dann auf dem Zeitintervall $[c_5, a_5]$ zum konstanten Wert k_∞ abzuklingt. Es gelte: $0 \leq c_5 < a_5$.

Aufgrund dieser Argumente wird H_5 modelliert durch:

$$(15.1) \qquad H_5(t) = \begin{cases} u_3 & 0 \leq t \leq c_5 \\[2mm] \dfrac{k_\infty - u_3}{a_5 - c_5} \cdot (t - c_5) + u_3 & c_5 \leq t \leq a_5 \\[2mm] k_\infty & t \geq a_5 \end{cases}$$

bzw.

$$(15.2) \qquad\qquad H_5(t) = k_\infty \quad \forall t \in I_1\,,\ \text{falls } a_5 = 0\,.$$

H_5 ist also linear auf $[0, c_5]$, $[c_5, a_5]$ und $[a_5, \infty]$ (Figur 5). Durch diese Konstruktion ist sichergestellt, daß H_5 stetig an die Anfangsverteilung V_{init} anschließt ($H_5(0) = u_3$). Dabei gilt:

$$(16) \qquad\qquad u_3 > k_\infty \iff T_5 \; < \; T_2 + \frac{22.57}{u_2} = \; 6 + \frac{22.57}{u_2} \approx 16.6$$

Die Bedingung (16) wird bei optimaler Wahl der Parameter (vgl. Kapitel IV) durch die erhaltenen numerischen Ergebnisse erfüllt.

Bemerkung: $\qquad H_5$ wurde also so gewählt, daß für alle Modellparameter und große t gilt:

$$\tilde{v_3}\,(t) = v_3\,(t) \; , \; \tilde{v_4}\,(t) = v_4\,(t)$$

III. Numerische Behandlung des Modells

Sei im folgenden $P := (T_3, T_5, v, w, a_5, c_5, \epsilon)$. Mit den oben beschriebenen Methoden ist nach Vorgabe der Modellparameter P (die Vorgabe von T_4 ist wegen (2) unnötig) eine durch das Modell bestimmte Verteilungsfunktion $u_{mod}(t, T)$ für $(t, T) \in I$ festgelegt. Es ist nun zu klären, wie gut $u_{mod}(t, T)$ die aus den Meßwerten zu deduzierende Verteilungsfunktion $u(t, T)$ approximiert. Ein Maß für die Abweichung ist gegeben durch

$$(17) \qquad F(P) := \max \left(\frac{\|v_3 - \tilde{v}_3\|_{\mathcal{L}^2(I_1)}}{\|v_3\|_{\mathcal{L}^2(I_1)}}, \frac{\|v_4 - \tilde{v}_4\|_{\mathcal{L}^2(I_1)}}{\|v_4\|_{\mathcal{L}^2(I_1)}} \right)$$

mit

$$\tilde{v}_3(t) = \int\limits_6^{T_3} u_{mod}(t, \tau) \, d\tau, \quad \tilde{v}_4(t) = \int\limits_{T_3}^{T_4} u_{mod}(t, \tau) \, d\tau, \quad t \in I_1$$

analog zu (1). Man sucht, denjenigen Parametersatz P zu finden, für den $F(P)$ minimal wird. Da die aus $u_{mod}(t, T)$ zu berechnenden Größen $\tilde{v}_3(t)$ und $\tilde{v}_4(t)$ Integrale über $u_{mod}(t, T)$ für fest vorgegebene t-Werte in T-Richtung sind, wird zunächst angegeben, wie für beliebige (t, T^1), $(t, T^2) \in I$ und für einen festen Parametersatz P ein solches Integral

$$(18) \qquad \int\limits_{T^1}^{T^2} u_{mod}(t, \tau) \, d\tau$$

numerisch berechnet werden soll.

3.1 Algorithmus

Innerhalb dieser Beschreibung bezeichnet $u(t, T)$ die Modell-Verteilungsfunktion $u_{mod}(t, T)$. Beschrieben wird der Algorithmus, mit dem für beliebige Punkte (t, T^1), $(t, T^2) \in I$ und festen Parametersatz P das Integral (18) und den Fehler $F(P)$ berechnet worden ist, ohne daß die Details ausgeführt werden. Ohne Einschränkung kann $T^1 < T^2$ angenommen werden.
Zugrunde liegt folgende Arbeitshypothese:
Da nach (5) gilt:

$$u(t, T) = u_L(t, T) + u_K(t, T) \ \text{ mit } u_K(t, T) = 0 \ \text{für } T \leq T_5$$

folgt

$$(19) \qquad \int\limits_{T^1}^{T^2} u(t, \tau) \, d\tau = \int\limits_{T^1}^{T^2} u_L(t, \tau) \, d\tau + \int\limits_{\max(T^1, T_5)}^{T^2} u_K(t, \tau) \, d\tau \quad .$$

Die Integrale über u_L und u_K werden getrennt betrachtet.
Weiter wird die Funktion u_L aufgespalten in

$$u_L(t, T) = u_{L^-}(t, T) + u_{L^+}(t, T) \begin{cases} \text{mit } u_{L^-}(t, T) = 0 & \text{für } t < c_0(T) \\ \text{und } u_{L^+}(t, T) = 0 & \text{für } t \geq c_0(T) \end{cases} ,$$

318

wobei c_0 die durch den Punkt $(0,0)$ gegebene Charakteristik bezeichnet. Die Funktion u_{L-} enthält gerade den durch die Anfangsverteilung H gegebenen Anteil der Leberproduktion, u_{L+} den durch die Anfangsverteilung V_{init} gegebenen Anteil von u_L. Der erste Teil des Integrals (19) wird also aufgespalten in:

$$(20) \qquad \int_{T^1}^{T^2} u_L(t,\tau)\,d\tau = \int_{\min\left(T^1,c_0^{-1}(t)\right)}^{\min\left(T^2,c_0^{-1}(t)\right)} u_{L-}(t,\tau)\,d\tau \; + \int_{\max\left(T^1,c_0^{-1}(t)\right)}^{\max\left(T^2,c_0^{-1}(t)\right)} u_{L+}(t,\tau)\,d\tau$$

Das Hauptprogramm erstellt die für die Berechnung des Integrals benötigten Modellgrößen, ein Unterprogramm berechnet dann den Wert $F(P)$ mit Hilfe der iterierten Trapezregel. Die Anfangsverteilung H wird dem Modell in [7] entnommen (hier eingelesen, da vorher bereits berechnet).

Gemäß (2) und (14) werden die Größen T_4 und k_∞ und die Werte der Anfangsverteilung V_{init} nach (13) bestimmt. Zur Berechnung des Fehlers $F(P)$ wird nun der Integrand wie in (19) und (20) aufgespalten.

3.1.1 Berechnung des Integrale über u_{L-}.

Die Berechnung des Integrale über u_{L-} soll mittels der iterierten Trapezregel für vorzugebende Stützstellen erfolgen. Als Funktionswerte werden die durch die Anfangsverteilung H und die Charakteristiken c_y gemäß (4) gegebenen Werte genommen.
Es soll

$$\int_{T^1}^{T^2} u_{L-}(t,\tau)\,d\tau$$

mit $y_{t,T} \geq 0 \;\forall T \in \left[T^1,T^2\right]$ berechnet werden. Da die Funktion $u_{L-}(t,.)$ zunächst nur an diskreten Punkten bekannt ist, sollen zunächst diese Punkte bestimmt werden. Dabei ist nur das Intervall $\left[T^1,T^2\right]$ von Interesse. Man errechnet die zu den Integrationsgrenzen (t,T^i) gehörenden Parameter für die Charakteristiken y_i, $i=1,2$, die nach den oben angestellten Überlegungen eindeutig bestimmt sind. Dann ergeben sich die gesuchten Punkte nach (10) aus:

$$X\left(y^k\right) = c_{y^k}^{-1}(t) \text{ mit } y_2 \leq y^k \leq y_1\,;\, y^k := y^{k-1} + 0.1,\, y^0 := y_2$$

Die Funktion $u(t,T)$ ist bekannt an den Stellen

$$u\left(t,X\left(y^k\right)\right),\, y_2 \leq y^k \leq y_1 \qquad ,$$

und es gilt:

$$u_{L-}\left(t,X\left(y^k\right)\right) = H\left(y^k\right)$$

Sei $n := \lfloor y_1 - y_2 \rfloor \cdot 10$. Dann wird das Integral mit Hilfe der iterierten Trapezregel berechnet durch

$$\int_{T^1}^{T^2} u_{L-}(t,\tau)\,d\tau \approx \sum_{i=0}^{n-1} \frac{X\left(y^i\right) - X\left(y^{i+1}\right)}{2} \cdot \left(H\left(y^i\right) + H\left(y^{i+1}\right)\right) \qquad .$$

3.1.2 Berechnung des Integrale über u_{L+}.

Das Integral über u_{L+} kann einfacher und somit auch genauer bestimmt werden, da die Anfangsverteilung V_{init} stückweise konstant ist. Es ergibt sich, wenn gesichert ist, daß $t < c_0(T)$ erfüllt ist:

$$
\begin{aligned}
\int_{T^1}^{T^2} u_{L+}(t,\tau)\, d\tau \quad &= u_1 \cdot \left(\min\left(c_{-2}^{-1}(t), T^2 \right) - \max\left(c_0^{-1}(t), T^1 \right) \right)_+ \\[2mm]
&+ u_2 \cdot \left(\min\left(c_{-T_6}^{-1}(t), T^2 \right) - \max\left(c_{-2}^{-1}(t), T^1 \right) \right)_+ \\[2mm]
&+ u_3 \cdot \left(\min\left(c_{-T_5}^{-1}(t), T^2 \right) - \max\left(c_{-T_6}^{-1}(t), T^1 \right) \right)_+ \\[2mm]
&+ u_4 \cdot \left(\min\left(c_{-T_4}^{-1}(t), T^2 \right) - \max\left(c_{-T_5}^{-1}(t), T^1 \right) \right)_+
\end{aligned}
\tag{21}
$$

3.1.3 Berechnung des Integrale über u_K.

Die Berechnung des Integrals über u_K wird gemäß folgender Überlegung vorgenommen:
Gegeben sei eine Funktion

$$
g_{\tilde{t},T_5} : [0,14] \to [T_5,30] \quad ,
$$

die bei fest vorgegebenen Werten $\tilde{t}$ und T_5 einem Punkt (t,T_5) den durch $\tilde{t}$ festgelegten Punkt $(\tilde{t},T)$ zuordnet, der auf derselben Charakteristik liegt. Nach Definition der Charakteristiken ist diese Abbildung wohldefiniert und sogar umkehrbar. Dann gilt:

$$
H_5(t) = u_K\left(\tilde{t}, g_{\tilde{t},T_5}(t) \right)
\tag{22}
$$

Mit (22) und $f(T) := g_{\tilde{t},T_5}^{-1}(T)$ folgt dann sofort:

$$
\begin{aligned}
\int_{T^1}^{T^2} u_K(\tilde{t},\tau)\, d\tau \quad &= \int_{f(T^1)}^{f(T^2)} u_K\left(\tilde{t}, g_{\tilde{t},T_5}(\tau) \right) \dot{g}_{\tilde{t},T_5}(\tau)\, d\tau \\[3mm]
&= \int_{f(T^1)}^{f(T^2)} H_5(\tau) \dot{g}_{\tilde{t},T_5}(\tau)\, d\tau \\[3mm]
&= H_5(\tau)\, g_{\tilde{t},T_5}(\tau) \Big|_{f(T^1)}^{f(T^2)} - \int_{f(T^1)}^{f(T^2)} \dot{H}_5(\tau)\, g_{\tilde{t},T_5}(\tau)\, d\tau \\[3mm]
&= T^2 \cdot H_5\left(f(T^2) \right) - T^1 \cdot H_5\left(f(T^1) \right) \\[5mm]
&\quad - \int_{f(T^1)}^{f(T^2)} \dot{H}_5(\tau)\, g_{\tilde{t},T_5}(\tau)\, d\tau
\end{aligned}
\tag{23}
$$

Beziehen sich die für die Integrale benötigten Funktionswerte $H_5(t)$ auf Werte $t \geq a_5$, b.z.w. $t \leq c_5$, so ist $H_5(t)$ dort konstant. Das Integral kann für solche Werte also analog zu (21) ausgewertet werden. Diese Fälle werden mit Hilfe der Werte $c_{y_{a_5},T_5}^{-1}(t)$ und $c_{y_{c_5},T_5}^{-1}(t)$ unmittelbar berechnet.

Der Fall, daß Funktionswerte aus dem linearen Teil von H_5 genommen werden müssen, wird noch weiter unterteilt.

Fall I $\quad T_5 \geq w$

$$g_{\tilde{t},T_5}(t) = \frac{\tilde{t} - t^*}{t - t^*} \cdot (T_5 - T^*) + T^*$$

Fall II $\quad$ Die zu einem Teil des Integrationsweges gehörenden Charakteristiken schneiden $G_{v,w}$ nicht rechts von T_5, sie knicken also vor T_5 nicht ab. Dies ist erfüllt, wenn der Integrationsweg links von $c_{y_{G_5},T_5}^{-1}(t)$ liegt, wobei $G_5 := G_{v,w}(T_5)$

$$g_{\tilde{t},T_5}(t) = \frac{\tilde{t} - t^*}{t - t^*} \cdot (T_5 - T^*) + T^*$$

Fall III $\quad$ Ein Teil des Integrationsweges liegt unterhalb von $G_{v,w}$.

$$g_{\tilde{t},T_5}(t) = \tilde{t} + T_5 - t$$

und das Integral kann wegen der Linearität von $g_{\tilde{t},T_5}$ sehr einfach berechnet werden.

Fall IV $\quad$ Ein Teil des Integrationsweges liegt oberhalb von $G_{v,w}$ und rechts der Charakteristik durch G_5.

$$g_{\tilde{t},T_5}(t) = T^* + \frac{w \cdot (t^* - \tilde{t})}{v} \cdot \left(1 + \frac{T^*\left(1 + \frac{v}{w}\right) + t^*\left(1 + \frac{w}{v}\right) - w - v}{t + w - T_5 - t^*\left(1 + \frac{w}{v}\right)}\right)$$

3.2 Testreihe zum Parameteridentifikationsproblem

Nachdem nun festgelegt ist, wie die Werte $\tilde{v}_3(t)$ und $\tilde{v}_4(t)$, sowie die Funktion $F(P)$ für festen Parametersatz P berechnet werden, kann als nächstes das Auffinden der "besten" Modellparameter diskutiert werden. Bei der gewählten Form des Modells und der erwarteten Genauigkeit beschränken wir uns darauf, die Berechnung von $\tilde{v}_3(t)$ und $\tilde{v}_4(t)$ für ganzzahlige t durchzuführen, da auch die Daten nur an diesen Punkten gemessen worden sind.

Um die Suche nach dem besten Parametersatz P, d.h. nach dem Parametersatz, für den das Fehlerfunktional minimal wird, zu vereinfachen, werden die Parameter unterteilt in drei Gruppen:

(i) $\quad P_T := (T_5, T_3)$

Diese beiden Parameter bestimmen die Verweildauer der Retikulozyten in den einzelnen Altersgruppen. Sie beeinflussen nicht nur die Anfangsverteilung V_{init}, sondern sind auch insofern die Parameter mit der größten Bedeutung, als sie die Länge des Integrationsweges bestimmen. Aufgrund dessen liegt die Vermutung nahe, daß schon leichte Änderungen der Parameter P_T den Fehler stark verändern. Es ist also zu erwarten, daß diese Reifezeiten durch das Modell gut bestimmt werden.

Um das Fehlerminimum bezüglich P_T zu finden, läßt man diese Parameter die ganzen Zahlen durchlaufen mit $T_3 > T_5 > 6$, solange $T_4 \leq 30$ erfüllt ist (vgl. Kapitel *II*). Eine genauere Bestimmung der optimalen Parameter P_T als für ganzzahlige Werte scheint unter den gestellten Anforderungen nicht sinnvoll: Es werden Schätzwerte für die Reifezeiten ermittelt, und die Parameter sind ohnehin nur tageweise gegeben.

(ii) $P_G := (v, w, \epsilon)$

Diese Parameter bestimmen die "Geometrie" des Reifeprozesses, d.h. sie legen die Charakteristiken fest. Die Rechnungen zeigen, daß $F(P)$ monoton mit ϵ anwächst, deshalb wird im folgenden $\epsilon \equiv 0.01$ (d.h. ϵ sehr klein) und $P_G = (v, w, 0.01)$ oder kürzer $P_G = (v, w)$ gesetzt. Dies ist auch insofern keine große Einschränkung, als die Steigung der Charakteristiken stärker von der Lage der Punkte v und w abhängt als von ϵ. Da ein leichtes Drehen der Geraden $G_{v,w}$ den Verlauf der Charakteristiken nur schwach verändert, vergrößert P_G den Fehler in einer größeren Umgebung des Fehlerminimums kaum.

Für die Optimierung durchlaufen v und w die ganzen Zahlen mit $0 < v \leq 14$ und $0 < w \leq 30$.

(iii) $P_K := (c_5, a_5)$

Durch P_K die Produktion im Knochenmark festgelegt. Ähnlich wie bei P_T wird auch hier eine große Einflußnahme auf den Fehler und somit eine scharfe Bestimmung dieser Größen erwartet, da je nach Lage der Punkte c_5 und a_5 unterschiedlich lang über unterschiedlich große Funktionswerte integriert wird.

Für die Optimierung durchlaufen c_5 und a_5 die ganzen Zahlen mit $0 \leq c_5 < a_5 < 14$.

IV. DISKUSSION DER NUMERISCHEN RESULTATE

Bevor die durch die oben beschriebenen Testläufe gefundenen Fehlerminima diskutiert werden, wird zunächst analysiert, welchen Einfluß die einzelnen Parametersätze P_T, P_G und P_K auf die resultierenden Werte $\tilde{v}_3(t)$ und $\tilde{v}_4(t)$, $t = 0, \ldots, 14$, haben. Die im Programm errechneten Werte $\tilde{v}_3(t)$ und $\tilde{v}_4(t)$ werden zur graphischen Veranschaulichung der Resultate linear interpoliert. Durch die spezielle Wahl von V_{init} und H_5 gilt für die praktisch interessanten Fälle :

$$v_i(0) = \tilde{v}_i(0), \quad v_i(14) = \tilde{v}_i(14), \quad i = 3, 4$$

Die Modellkurven werden in dem Intervall $[1, 13]$ mehr oder weniger stark von den Daten abweichen. Von besonderem Interesse ist hierbei das Maximum von v_3 und v_4 bei $t = 2$, dessen Wiedergabe besondere Anforderungen an die Modellierung stellt und die Parameter stark einengt, denn schon bei geringer Abweichung vom optimalen Wert treten starke Veränderungen dieser Extrema und damit hohe Werte des Fehlerfunktionals auf.

4.1 Abhängigkeit von den Parametern T_5 und T_3

Da sich das Modell nicht für eine analytische Analyse eignet, wird die Diskussion auf numerische Betrachtungen gestützt. Der Einfluß der Zeiten T_5 und T_3 bei festen Parametersätzen P_G und P_K wird durch folgenden Test besonders deutlich:

Für verschiedene jeweils fest vorzugebende Parametersätze P_G und P_K werden T_5 und T_3 in folgender Weise variiert:

(i) $T_5 = const.$, $T_{35} = 1, \ldots, 5$
 Die Funktionen $\tilde{v}_3(t)$ und $\tilde{v}_4(t)$ wachsen monoton mit T_{35} stark an.

(ii) $T_3 = const.$, $T_5 = 7, \ldots, T_3 - 1$
 Die Funktionen $\tilde{v}_3(t)$ und $\tilde{v}_4(t)$ fallen monoton mit T_5 stark ab.

(iii) $T_{35} = const.$, $T_5 = 7, \ldots, 16$
 Die Funktion $\tilde{v}_3(t)$ bleibt im wesentlichen unverändert, hingegen fällt $\tilde{v}_4(t)$ wieder monoton mit wachsendem T_5 stark ab.

Es zeigt sich, daß sich die Kurven abhängig von den Zeiten T_5 und T_3 für alle Parametersätze analog verhalten (siehe Figuren 6,7,8)[1] .
Das Anwachsen in (i) läßt sich mit dem Vergrößern des Integrationsweges erklären. Analog dazu wird der Weg in (ii) verkürzt. In (iii) bleibt der Integrationsweg wegen $T_{35} = const.$ unverändert. Die Variation über T_5 spielt nur mittelbar über V_{init} eine Rolle. Da die Werte von $\tilde{v}_3(t)$ sich im wesentlichen auf H, u_1, u_2 und die Produktion im Knochenmark beziehen und diese Größen bei konstantem T_{35} unverändert bleiben, wird $\tilde{v}_3(t)$ von T_5 allein kaum beeinflußt. Für $\tilde{v}_4(t)$ dagegen macht sich die Variation über T_5 durch V_{init} bemerkbar.
Insgesamt ergibt sich also der bereits in 3.2 erwartete starke Einfluß der Parameter P_T. Der Fehler varriiert über 80 Prozentpunkte.
Bei Durchlaufen der in Kapitel III beschriebenen Testreihe ergibt sich, daß Fehlerwerte ≤ 0.05 nur angenommen werden für $P_T = (11, 10)$ und $P_T = (13, 11)$, die Fehlerschranke 0.04 wird sogar nur für $P_T = (13, 11)$ unterschritten, d.h.:

$$P_T \neq (13, 11) \qquad \implies \quad F(P_T, P_G, P_K) > 0.04 \quad \forall P_G, P_K$$

$$P_T \neq (11, 10) \quad \wedge \quad P_T \neq (13, 11) \qquad \implies \quad F(P_T, P_G, P_K) > 0.05 \quad \forall P_G, P_K$$

Dies zeigt, wie scharf die optimalen Reifungszeiten durch das Modell bestimmt sind.

4.2 Abhängigkeit von den Parametern v und w

Der Einfluß der Parameter v und w soll analog zu 4.1 verdeutlicht werden. Für konstante P_T und P_K werden v und w variiert. Die Kurven $\tilde{v}_3(t)$ und $\tilde{v}_4(t)$ zeigen in Abhängigkeit von P_G für feste Werte von P_T und P_K ein ähnliches Verhalten wie bei der Variation von P_T (siehe Figuren 9,10). [1] Folgende Beobachtungen können bei einem Vergleich der Kurven angestellt werden:

(i) $w = const.$, $v = 1, \ldots, 14$
 Beide Kurven steigen mit wachsendem v leicht an. Die Funktion $\tilde{v}_4(t)$ fällt an der Stelle $t = 1$, sodaß ein Knick nach unten entstehen kann. Der Verlauf der Kurven wird für große t nur wenig verändert.

---

[1]Die Kurven wurden berechnet für die Parameterwerte $T_5 = 11$, $T_{35} = 2$, $v = 12$, $w = 7$, $a_5 = 7$, $c_5 = 2$ und $\epsilon = 0.01$. Verändert wurde nur der angegebene Parameterwert. Die durch Punkte gekennzeichneten Werte sind die gemessenen, die unmarkierten die durch das Modell errechneten Zahlen.

(ii) $v = const.$, $w = 1, \ldots, 30$

Beide Kurven steigen mit wachsendem w leicht an und werden etwas nach rechts verschoben. Ihr qualitativer Verlauf bleibt unverändert.

Auch hier werden die in 3.2 ausgesprochenen Erwartungen erfüllt. Der Fehler variiert um ca. 20 Prozentpunkte, die beobachteten Veränderungen sind also verhältnismäßig gering. Beim Anwachsen von v und w haben die Charakteristiken lange eine große Steigung, die für große T hohen Werte $V_{init}(T)$ und für kleine t hohen Werte $H_5(t)$ wandern also langsam ab. Dies erklärt das Ansteigen der Kurven. Ist dagegen w zu groß, so wird das Nachfließen der hohen Werte von $H_5(t)$ zu lange herausgezögert, dies erklärt das Abknicken von $\tilde{v}_4(t)$.
Die Parameter P_G beeinflußen den Fehler nur schwach. Dies ist sehr vorteilhaft, da die medizinisch relevanten Werte P_T und P_K unabhängig von P_G (der Form der Charakteristiken) recht genau bestimmt werden können.

4.3 Abhängigkeit von den Parametern c_5 und a_5

Die Vorgehensweise ist der in 4.1 und 4.2 ähnlich. Für konstante P_T und P_G wird P_K variiert. Wieder verhalten sich die Kurven $\tilde{v}_3(t)$ und $\tilde{v}_4(t)$ abhängig von den Parametern P_K für fest vorgegebene P_T und P_G in ähnlicher Weise (siehe Figuren 11,12).[1] Es ergibt sich:

(i) $c_5 = const.$, $a_5 = c_5 + 1, \ldots, 14$
Beide Kurven wachsen monoton mit a_5, es werden jedoch nur die Werte für große t beeinflußt.

(ii) $a_5 = const.$, $c_5 = 1, \ldots, a_5 - 1$
Beide Kurven wachsen monoton mit c_5, es werden jedoch nur die Werte für kleine t beeinflußt.

Die in 3.2 angestellten Überlegungen können übernommen werden. Die Fehlerwerte variieren stark (um ca. 40 Prozentpunkte). Der Einfluß der Parameter auf die unterschiedlichen Stellen der Kurven läßt sich durch die Lage der Punkte a_5 und c_5 (vgl. Figur 2), die Art der Beeinflußung durch ihr Einwirken auf die Größe der Werte von H_5 erklären.
Es ergibt sich, daß für alle Parametersätze P, die "vernünftige" Fehlerwerte liefern (d.h. der Ungleichung $F(P) \leq 0.05$ genügen) $c_5 = 2, 3, a_5 = 6, 7$ gilt, genauer:

$$P_K \neq (3,7) \quad \wedge \quad P_K \neq (2,7) \quad \wedge \quad P_K \neq (2,6) \quad \Longrightarrow \quad F(P_T, P_G, P_K) > 0.05 \quad \forall P_T, P_G$$

Diese Parameter sind folglich besonders scharf bestimmt, erlauben also eine genauere medizinische Interpretation.

4.4 Ergebnisse der Testreihe

Die in den vorhergehenden Abschnitten beschriebenen Änderungen werden nun in dem Testprogramm überlagert. Um die auftretenden Fehlerminima genauer zu veranschaulichen, werden die Fehlerwerte

---

[1] Die Kurven wurden berechnet für die Parameterwerte $T_5 = 11$, $T_{35} = 2$, $v = 12$, $w = 7$, $a_5 = 7$, $c_5 = 2$ und $\epsilon = 0.01$. Verändert wurde nur der angegebene Parameterwert. Die durch Punkte gekennzeichneten Werte sind die gemessenen, die unmarkierten die durch das Modell errechneten Zahlen.

$F(P)$ in Diagramme eingetragen und Höhenlinien der Interpolierenden von F diskutiert. Dabei wird jeweils ein Parameterwert konstant gelassen. Diskutiert werden also die Höhenlinien folgender Funktionen:

$$F_{P_G,P_K}(T_3,T_5) \quad := F(T_3,T_5,P_G,P_K) \quad ,P_G,P_K \text{ fest}, \quad 6 < T_5 < T_3 < \frac{30+2.17\cdot T_5}{3.17}$$

$$F_{P_T,P_K}(v,w) \quad := F(P_T,v,w,P_K) \quad ,P_T,P_K \text{ fest}, \quad \begin{aligned} &0 < v \le 14 \\ &0 < w \le 30 \end{aligned}$$

$$F_{P_T,P_G}(c_5,a_5) \quad := F(P_T,P_G,c_5,a_5) \quad ,P_T,P_G \text{ fest}, \quad 0 \le c_5 < a_5 \le 14$$

Von Interesse sind im folgenden vor allem die Parametersätze P mit $F(P) \le 0.05$.

4.4.1 Diskussion der Höhenlinien von F_{P_G,P_K}.

Es ergeben sich zwei Fehlerminima bei (11,10) und (13,11). Das Fehlertal bei (13,11) tritt zwar nur bei wenigen Parametersätzen P_G und P_K auf, ist aber wesentlich tiefer als bei (11,10), also besonders scharf bestimmt. Figur 13[1] stellt einen Ausschnitt der Höhenlinien um die Fehlertäler herum dar. Der Verlauf der Höhenlinien legt die Vermutung nahe, daß F_{P_G,P_K} auf Geraden mit $T_{35} = const.$ nahezu konstant ist. Deshalb wurde in einer weiteren Figur T_{35} gegen T_5 abgetragen. Man sieht hier noch deutlicher, daß das Fehlertal in (10,11) nur eine Erweiterung des Fehlertales um (11,13) ist.

4.4.2 Diskussion der Höhenlinien von F_{P_T,P_K}.

Im Gegensatz zu 4.4.1 ergeben sich keine eindeutig bestimmten Fehlerminima (Figur 14). Das Fehlertal ist vergleichsweise breit und scheint sich entlang Funktionen $v \cdot w = const.$ zu erstrecken. Ein in diesem Sinne leichtes Drehen der Geraden $G_{v,w}$ hat also kaum einen Einfluß auf das Reifeverhalten. Insgesamt ergibt sich, daß der Fehler sehr klein wird für $w = 7,8$, $v \ge w$. Dies beschreibt einen anfänglich durchaus normalen, dann rasch beschleunigten Reifeprozeß.

4.4.2 Diskussion der Höhenlinien von F_{P_T,P_G}.

Wie in 4.4.1 läßt sich für alle Parametersätze ein eindeutig bestimmtes Fehlerminimum finden, und zwar für $c_5 = 2,3$ und $a_5 = 6,7$ (Figur 15). Da die Höhenlinien fast konzentrisch um das Minimum herum verlaufen, läßt seine Schärfe nichts zu wünschen übrig. Die Produktion im Knochenmark klingt also nach einer Verzögerung von 2 Tagen innerhalb von 5 Tagen auf den Normalwert ab. Dies scheint sinnvoll zu sein, da so ein möglicher hoher Blutverlust bei der Geburt rascher kompensiert werden kann als bei ständig gleich bleibender Produktion.

---

[1] Die in den Figuren 13,14 und 15 angegebenen Fehlerwerte wurden, um die Lesbarkeit zu erhöhen, mit 100 multipliziert

V. Schlussbemerkung

Abschließend werden die in Kapitel IV diskutierten Ergebnisse zusammengefaßt und eine Reihe der im Laufe der Entwicklung des vorliegenden Modells erstellten, aber verworfenen Modellansätze vorgestellt.

5.1 Zusammenfassung der Ergebnisse

Für alle drei Gruppen von Tests ergibt sich, daß für folgende Parameterwerte P der Fehler $F(P)$ minimal wird (d.h. die Fehlerschranke 0.04 unterschreitet):

T_5	T_3	T_4	v	w	c_5	a_5	F
11	13	17.3	7	8	2	7	0.03775
11	13	17.3	8	8	2	7	0.03651
11	13	17.3	9	8	2	7	0.03785
11	13	17.3	11	7	2	7	0.03516
11	13	17.3	12	7	2	7	0.03591
11	13	17.3	13	7	2	7	0.03678
11	13	17.3	14	7	2	7	0.03776

Gerade die medizinisch interessanten Werte P_T und P_K sind also gut bestimmt.
Diese Ergebnisse decken sich mit den im ersten Kapitel erläuterten medizinischen Tatsachen. Dort wurde ausgeführt, daß die vom Knochenmark ins periphere Blut ausgeschwemmten Retikulozyten innerhalb von 2-4 Tagen reifen. Da dieser Wert beim erwachsenen Menschen gemessen wurde, die Reifezeit der frühkindlichen Retikulozyten aber verlängert ist, kann das hier erhaltene Ergebnis $T_{45} \approx 6$ als gute Näherung angesehen werden. Es entspricht einer Verdoppelung der Reifezeit gegenüber der beim Erwachsenen. Die Produktion im Knochenmark wird nach der Geburt reduziert und ist am Ende der ersten Lebenswoche konstant. Die Parameter P_K beschreiben eine nach zwei Tagen abklingende Knochenmarkproduktion, die nach Ende der ersten Lebenswoche ihren Normalwert erreicht hat. Die Parameter P_G beschreiben eine anfänglich stark verkürzte (beachte: $w < v$), jedoch bald normalisierte Reifezeit.
Obwohl das Modell stark vereinfachend ist, werden durch einige medizinisch sinnvolle Annahmen also medizinisch recht plausible Ergebnisse erzielt.

5.2 Weitere Modellansätze

Außer dem in den vorhergehenden Kapiteln diskutierten Modell wurden noch weitere Modellansätze durchgerechnet, die jedoch nicht zu solch guten Approximationen von v_3 und v_4 führten. Einige davon seien im folgenden angegeben.
Zu Beginn der Modellierung wurde durchweg die Produktion im Knochenmark als konstant angesetzt, d.h. die bei Werner und Schulte [7] für v_1 und v_2 gemachten Annahmen auf v_3 und v_4 ausgedehnt. In unserer Terminologie hieße dies:
Es wurde $c_5 = 0$ und $a_5 = 0$ gesetzt. Außerdem wurde der Effekt eines anfänglich verlangsamten Reifeverhaltens vernachlässigt, es wurde also $v = 0$ und damit $G_{v,w} \equiv 0$ gesetzt.
Unter diesen Annahmen erwies es sich als unmöglich, das Maximum der Kurve v_4 zu reproduzieren, was mit dem zu frühen Abwandern der hohen Werte von V_{init} durch $v = 0$ und dem fehlenden anfänglichen Impuls von H_5 durch $c_5 = a_5 = 0$ erklärt werden kann.

Als Charakteristiken wurden teils Geradenscharen, teils Wurzeln aus linearen Funktionen (die z.B. für $t = 0$ (bei der Geburt) die Steigung 1 haben) untersucht, wobei die Geraden nicht nur zu einfacheren Rechnungen führen, sondern den Wurzelfunktionen insofern überlegen waren, als bei letzteren die Retikulozyten gerade in den ersten Lebenstagen zu rasch reiften und dort $\tilde{v}_3$ und $\tilde{v}_4$ sogar fielen.

Das gleiche Verhalten zeigt sich bei Vernachlässigung des Parameters c_5 (gleichbedeutend mit $c_5 = 0$), d.h. wenn angenommen wird, daß die Produktion im Knochenmark unmittelbar nach der Geburt und nicht erst mit einer gewissen Verzögerung abklingt.

Das hier vorliegende Modell entstand aus seinen Vorgängern durch Hinzufügen weiterer *Modellparameter* aufgrund immer genauerer Analysen des medizinischen Sachverhaltes. Die abgleleiteten Ergebnisse sollten von Medizinern verifiziert werden.

LITERATUR

[1] L. Heilmeyer, "Blut und Blutkrankheiten", Springer, Berlin, 1968.

[2] H.J. Herrmann, "Der Verlauf der Retikulozytenwerte bei Neugeborenen in den ersten acht Lebenstagen", Dissertation, Universitäts-Kinderklinik Münster, 1972.

[3] G. Herrmann, "Die Reifungsstufen der Retikulozyten bei Neugeborenen in den ersten acht Lebenstagen", Dissertation, Universitäts-Kinderklinik Münster, 1972.

[4] F.Oski / L.Naiman, "Hematologic Problems in the Newborn", W.B. Saunders Company, Philadelphia, 1982.

[5] Piper, "Innere Medizin", Springer, Berlin, 1982.

[6] I. Schulte, "Ein mathematisches Modell zur Produktion der roten Blutkörperchen bei Neugeborenen", Staatsexamensarbeit, 1983.

[7] H. Werner / I. Schulte, *Ein mathematisches Modell für die Blutbildung bei Neugeborenen*, ZAMM **63** (1983), 391–393.

[8] H. Werner, *Mathematische Modelle in der Medizin*, in G. Meinardus, "Approximation in Theorie und Praxis– ein Symposionbericht", BI, Mannheim / Wien / Zürich, 1979, pp.251–267.

H. Werner, Institut für Angewandte Mathematik, Wegelerstr. 6, D-5300 Bonn

C. Fesser, Institut für Angewandte Mathematik, Wegelerstr. 6, D-5300 Bonn

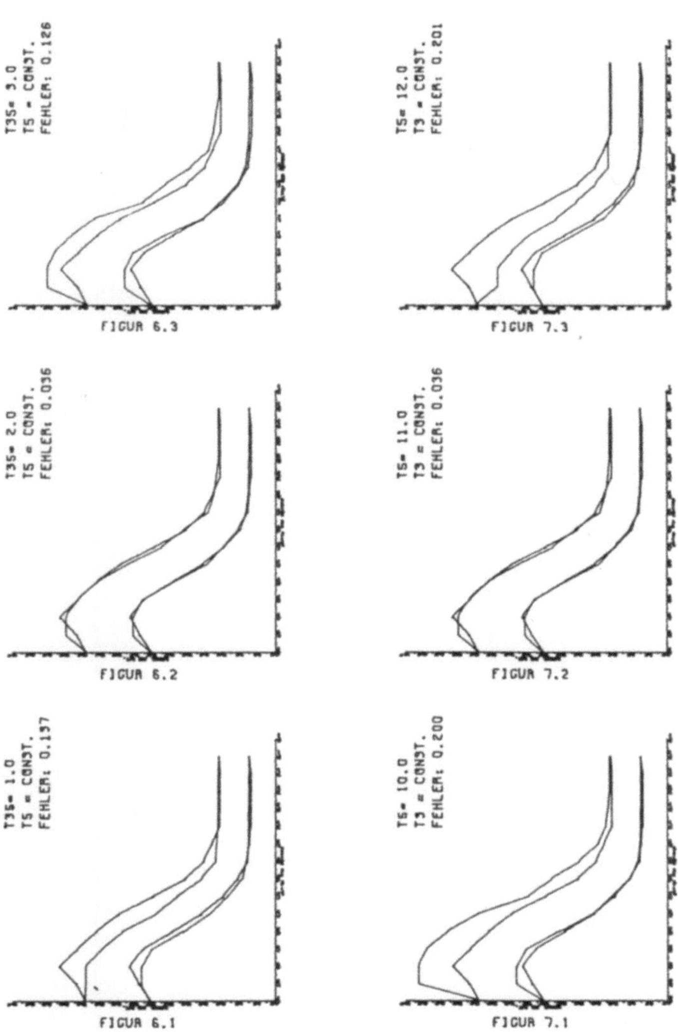
T3S= 3.0
T5 = CONST.
FEHLER: 0.126
FIGUR 6.3
T5= 12.0
T3 = CONST.
FEHLER: 0.201
FIGUR 7.3
T3S= 2.0
T5 = CONST.
FEHLER: 0.036
FIGUR 6.2
T5= 11.0
T3 = CONST.
FEHLER: 0.036
FIGUR 7.2
T3S= 1.0
T5 = CONST.
FEHLER: 0.137
FIGUR 6.1
T5= 10.0
T3 = CONST.
FEHLER: 0.200
FIGUR 7.1

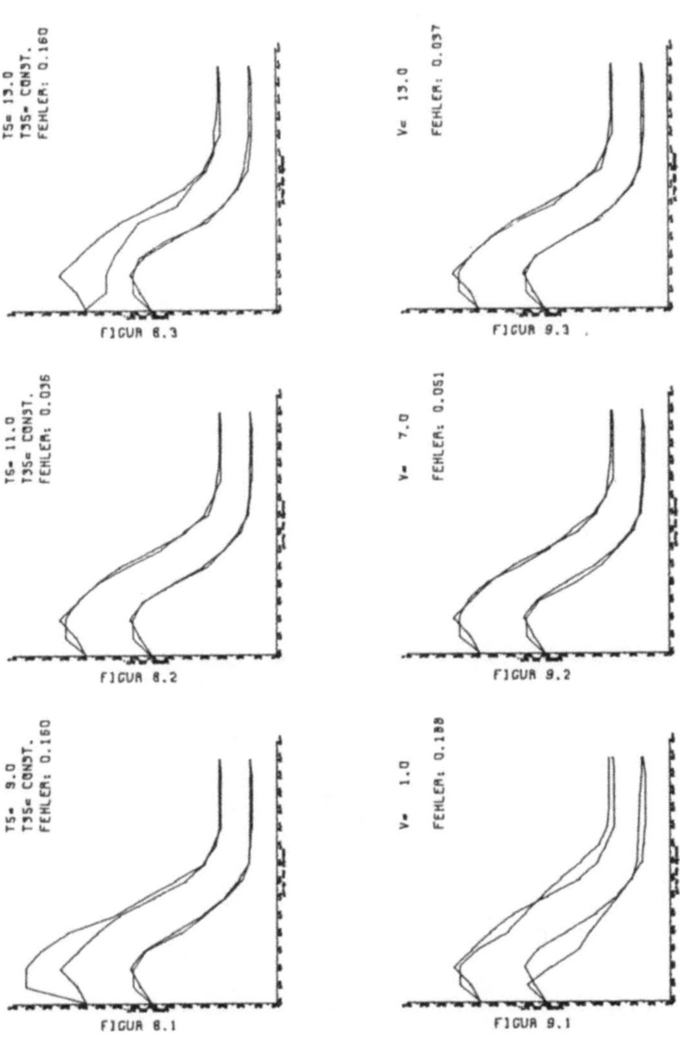

TS= 13.0
T35= CONST.
FEHLER: 0.160
FIGUR 8.3
V= 13.0
FEHLER: 0.037
FIGUR 9.3
TS= 11.0
T35= CONST.
FEHLER: 0.036
FIGUR 8.2
V= 7.0
FEHLER: 0.051
FIGUR 9.2
TS= 9.0
T35= CONST.
FEHLER: 0.160
FIGUR 8.1
V= 1.0
FEHLER: 0.188
FIGUR 9.1

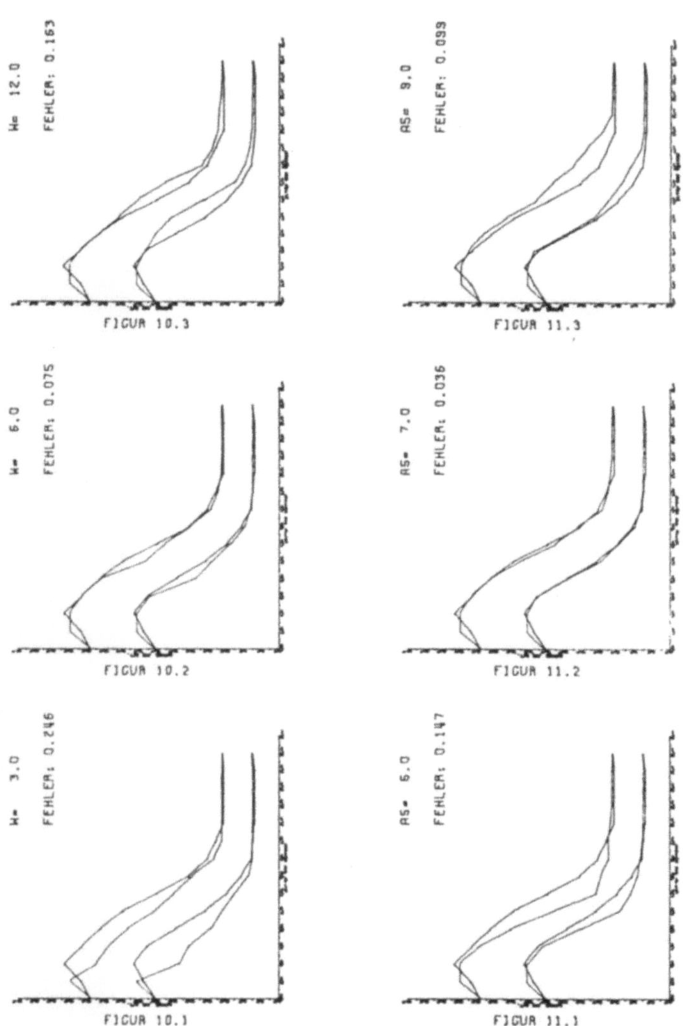
H= 12.0
FEHLER: 0.163
FIGUR 10.3
AS= 9.0
FEHLER: 0.099
FIGUR 11.3
H= 6.0
FEHLER: 0.075
FIGUR 10.2
AS= 7.0
FEHLER: 0.036
FIGUR 11.2
H= 3.0
FEHLER: 0.246
FIGUR 10.1
AS= 6.0
FEHLER: 0.147
FIGUR 11.1

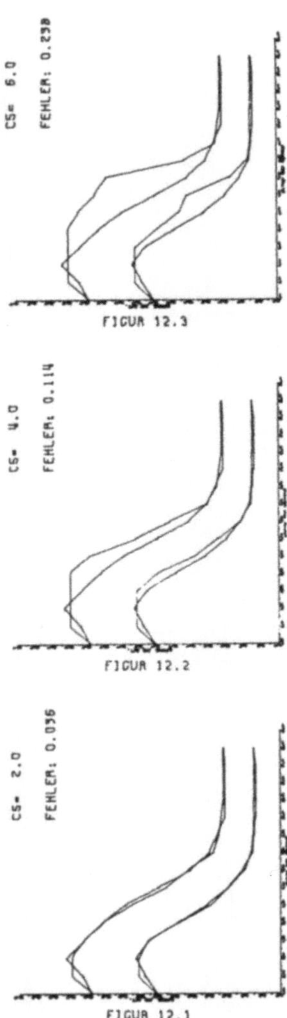
CS= 6.0
FEHLER: 0.238
FIGUR 12.3
CS= 4.0
FEHLER: 0.114
FIGUR 12.2
CS= 2.0
FEHLER: 0.056
FIGUR 12.1

HOEHENLINIEN FUER PT.MINIMUM=0.0352
V=11.0 W= 7.0 AS= 7.0 CS= 2.0
T5=14
T5=13
T5=12
T5=11
T5=10
T5= 9
T3=18
T3=17
T3=16
T3=15
T3=14
T3=13
T3=12
T3=11
T3=10

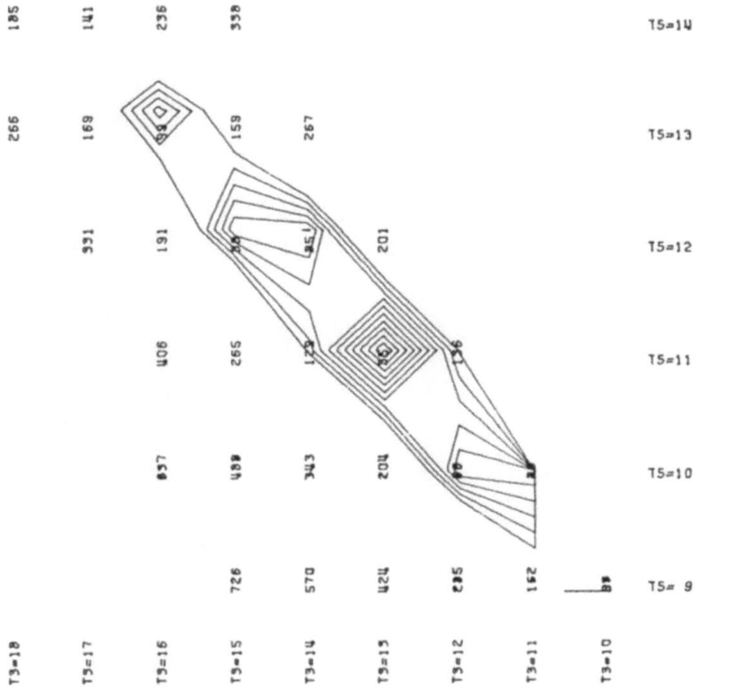

FIGUR 13

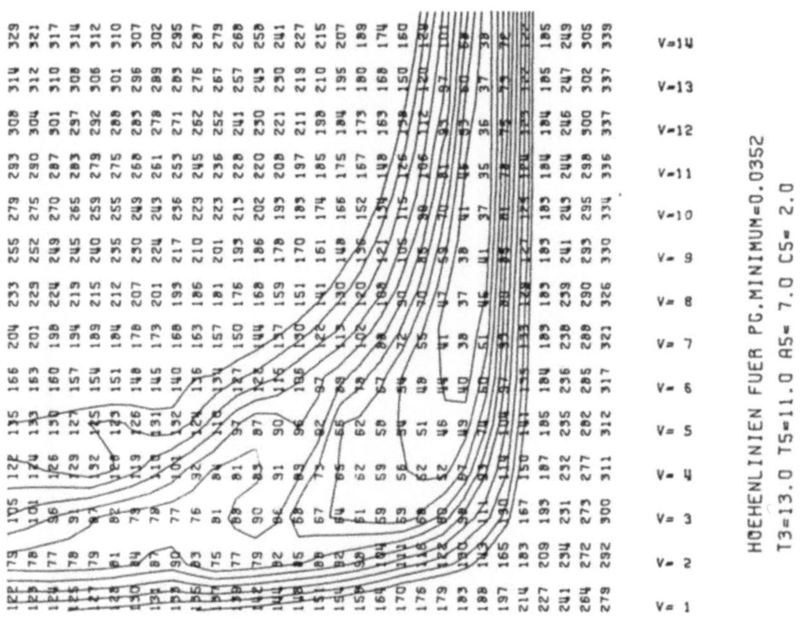

HOEHENLINIEN FUER PG.MINIMUM=0.0352
T3=13.0 T5=11.0 AS= 7.0 CS= 2.0

FIGUR 14

HOEHENLINIEN FUER PK.MINIMUM=0.0359
T3=13.0 T5=11.0 V=12.0 W= 7.0
CS=13:   623
CS=12:   593 606
CS=11:   541 582 578
CS=10:   480 505 526 544
CS= 9:   410 439 464 486 506
CS= 8:   328 361 390 417 442 465
CS= 7:   238 272 306 338 368 395 420
CS= 6:   146 174 210 248 283 316 347 374
CS= 5:   128 98 114 151 191 231 267 300 329
CS= 4:   185 127 71 50 96 100 183 221 256 288
CS= 3:   265 209 147 93 58 63 88 109 183 220 252
CS= 2:   360 302 244 185 125 78 63 86 122 159 193 225
CS= 1: 432 386 331 276 220 166 121 86 97 118 146 177 206
       A5=1 A5=2 A5=3 A5=4 A5=5 A5=6 A5=7 A5=8 A5=9 A5=10 A5=11 A5=12 A5=13

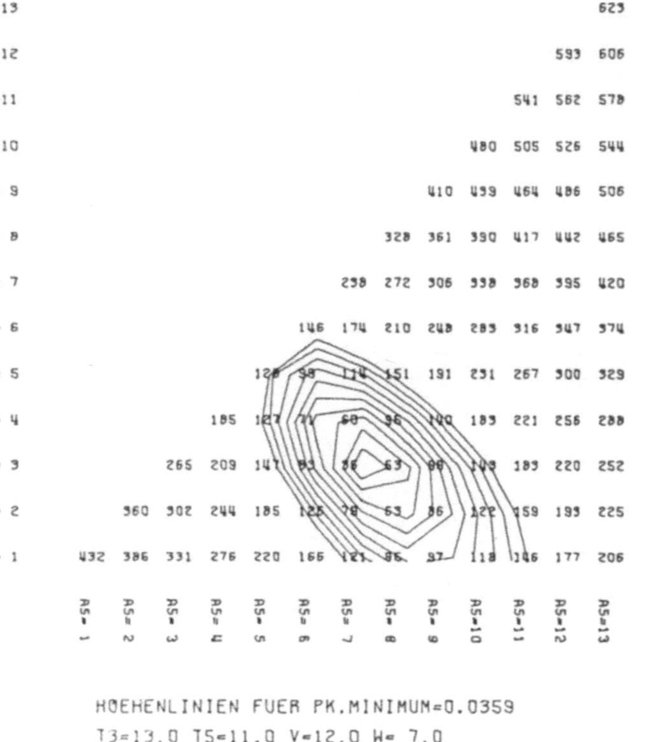

FIGUR 15

International Series of
Numerical Mathematics, Vol. 74
© 1985 Birkhäuser Verlag Basel

ON THE P-STABILITY OF ONE-STEP COLLOCATION FOR DELAY DIFFERENTIAL EQUATIONS*

Marino Zennaro

Istituto di Matematica

Università degli Studi

I-34100 Trieste(Italy)

Abstract. In this paper we present a stability analysis of the one-step collocation method at Gaussian points for DDEs, based on the test equation

$$y'(t)=ay(t)+by(t-\tau), \quad t>0$$
$$y(t)=\phi(t) \quad \text{for} \quad -\tau \le t \le 0$$

where a and b are complex coefficients and τ is any positive constant delay. Convergence and superconvergence results for this method were recently proved. We show that it is P-stable, that is it yields an approximate solution u which, as y itself does, asymptotically vanishes if $|b|<-\text{Re}(a)$.

1. INTRODUCTION. This paper concerns stability properties of numerical methods for the solution of DDEs of the form

$$y'(t)=f(t,y(t),y(t-\tau)), \quad t>t_0$$
$$(1)$$
$$y(t)=\phi(t) \quad \text{for} \quad t_0-\tau \le t \le t_0$$

where y, ϕ and f are n-vectors. The initial function ϕ is supposed to be continuous. The delay τ is any positive real number.

Recently the literature on numerical methods for the approximate solution of (1) has been growing more and more. Consequently some papers concerning their stability properties have appeared. There are many concepts of stability for DDEs, based

____

*Work supported by the Italian CNR, Project INFORMATICA, SOFMAT.

on different test equations (e.g. [1],[2],[5],[7],[8],[11]).

We use the definition of Barwell [2] of P-stability, based on the test equation

$$(2) \quad \begin{array}{l} y'(t)=ay(t)+by(t-\tau), \ t>0 \\ y(t)=\phi(t) \quad \text{for} \ -\tau \leq t \leq 0 \end{array}$$

where a and b are complex numbers, $\tau>0$ and ϕ is continuous and complex valued.

It is known that the solution y vanishes for $t \to +\infty$, all delays τ and all initial functions ϕ, if

$$(3) \quad |b|<-\text{Re}(a).$$

Barwell's definition is the following.

DEFINITION 1. A numerical method for DDEs is said to be P-stable if, under condition (3), the numerical solution of (2) vanishes for $t \to +\infty$ for all step-lengths h satisfying

$$(4) \quad h=\tau/m, \quad \text{m positive integer.}$$

Thus a numerical method is P-stable if it preserves the asymptotic behaviour of the solution y of (2) under the restriction (4) on the step-length.

If we drop the restriction (4), then we get the definition of GP-stability, which is obviously stronger (see still Barwell [2]).

It is obvious that the GP-stability implies the P-stability and that the P-stability implies the A-stability (when the method is applied to an ODE).

In this paper we prove the P-stability of the one-step collocation method at r Gaussian points. This method, wellknown for the solution of ODEs (remark that it is A-stable), has been recently investigated by Bellen [3] for application to DDEs.

In general we assume a variable delay $\tau=\tau(t) \geq \rho>0$ in (1). We have the so called breaking points $\xi_0=t_0<\xi_1<\ldots<\xi_j<\ldots,\ \xi_{j+1}$

solution of $\sigma(\xi)=\xi_j$, $\sigma(\xi):=\xi-\tau(\xi)$, where the solution y has discontinuities in its derivatives, caused by the delayed argument.

By putting $I_j:=[\xi_{j-1},\xi_j]$, Bellen [3] defines a <u>constrained mesh</u> Δ satisfying the following conditions:

(i)   - it includes the breaking points;

(ii) - it is arbitrarily chosen in I_1;

(iii) - it is such that $\Delta\cap I_j=\sigma^{-1}(\Delta\cap I_{j-1})$ for $j\geqq2$.

Remark that, if the delay τ is constant, then a uniform mesh satisfying (4) is a constrained mesh as above.

By choosing such kinds of meshes, he retaines the superconvergence order 2r at the nodes, although he uses the collocation solution to approximate the retarded part, which is known to be uniformly accurate of order r+1 only.

These results were later extended by Vermiglio [12] to another implicit Runge-Kutta method and, still later, generalized to all Runge-Kutta processes by Zennaro [13].

The main disadvantage in choosing a constrained mesh is that one cannot use a step-size control technique. Moreover, for non constant delays $\tau=\tau(t)$, the precomputation of the mesh (imposed by condition (iii)) can be expensive. On the other hand neither extra evaluations of the function are needed for the approximation of the retarded part (like in Bellen-Zennaro [4]), nor heavy interpolation formulas involving several mesh points are required (like in Oberle-Pesh [9]).

The results of this paper suggest that the choice of a constrained mesh is the best one for stability reasons, at least for the one-step collocation method at r Gaussian points when the delay τ is constant. In fact, as previously said, this method is P-stable; on the contrary it seems not to be GP-stable. For the case r=1 this last statement follows from some results of Jackiewicz [7].

We present some numerical experiments (for the case $r=1$) which completely confirm our theory. Nay, the difference between constrained and free meshes is so great that, for certain equations, the step-size control technique turns out to be by far worse than the use of a simple constant step-length.

It will be the subject of further papers to extend our analysis of stability to all Runge-Kutta processes applied to DDEs in the way proposed in [13] (considering also variable delays).

Our conjecture is that the use of constrained meshes is quite appropriate for solving certain classes of DDEs which, in some sense, present particular stiffness features.

2. PROOF OF THE P-STABILITY. In this section we prove the P-stability of the one-step collocation method at r Gaussian points described in [3].

In applying the method to (2) under the restriction (4), we easily get, for $m:=\tau/h$ and $i \geq m$:

$$(5) \quad \begin{aligned} \underline{k}_{i+1} &= a(y_i\underline{u} + hA\underline{k}_{i+1}) + b(y_{i-m}\underline{u} + hA\underline{k}_{i-m+1}) \\ y_{i+1} &= y_i + h\underline{w}^T\underline{k}_{i+1} \end{aligned}$$

where $\underline{u} := (1,1,\ldots,1)^T$ is the unitary r-vector, $\underline{w}$ is the r-vector of the weights and A is the $r \times r$-matrix of the coefficients of the equivalent r-level implicit Runge-Kutta process. The values y_i and $\underline{k}_i$, $i=0,\ldots,m$, are determined by the initial function ϕ.

By putting $\underline{Y}_i := (y_i, h\underline{k}_i^T)^T$, which is an $(r+1)$-vector, $\alpha := ah$, $\beta := bh$ and $\eta := \underline{w}^T(I-\alpha A)^{-1}\underline{u}$, we can write (5) in the following compact form, provided that the complex number α is such that the matrix $I-\alpha A$ is non singular:

$$(6) \quad \underline{Y}_{i+1} = L\underline{Y}_i + M\underline{Y}_{i-m+1} + N\underline{Y}_{i-m}, \quad \underline{Y}_i \text{ given for } i=0,\ldots,m,$$

where L, M and N are $(r+1) \times (r+1)$-matrices and are defined as follows:

$$L := \left\| \begin{array}{c|c} 1+\alpha\eta & 0 \,.\,.\, 0 \\ \hline \alpha(I-\alpha A)^{-1}\underline{u} & 0 \end{array} \right\| \qquad M := \left\| \begin{array}{c|c} 0 & \beta\underline{w}^{\mathbf{T}}(I-\alpha A)^{-1}A \\ \hline 0 & \\ \vdots & \beta(I-\alpha A)^{-1}A \\ 0 & \end{array} \right\|$$

$$N := \left\| \begin{array}{c|c} \beta\eta & 0 \,.\,.\, 0 \\ \hline \beta(I-\alpha A)^{-1}\underline{u} & 0 \end{array} \right\|$$

LEMMA 2. If a complex number γ is such that $\mathrm{Re}(\gamma)<0$, then the matrix $I-\gamma A$ is non singular and, moreover,

$$|1+\gamma\underline{w}^{\mathbf{T}}(I-\gamma A)^{-1}\underline{u}|<1.$$

Proof. These are wellknown results regarding the one-step collocation method at Gaussian points for ODEs (see, for example, Prothero-Robinson [10]). Q.E.D.

Our aim reduces to prove the following theorem.

THEOREM 3. The recurrence relation (6) is such that $\lim\limits_{i\to\infty} \underline{Y}_i=\underline{0}$ if α and β satisfy

$$(7) \qquad\qquad |\beta|<-\mathrm{Re}(\alpha).$$

Proof. Note that (7) implies $\mathrm{Re}(\alpha)<0$ and hence, by Lemma 2, the recurrence relation (6) is well defined.

We have to prove that all the roots of the characteristic equation

$$(8) \qquad\qquad \det(\lambda^{m+1}I-\lambda^{m}L-\lambda M-N)=0$$

are inside the unit circle (see, for example, Gohberg-Lancaster-Rodman [6]).

To this aim, let us suppose the existence of a solution λ of (8) such that $|\lambda|\geqq 1$. Then there exists an $(r+1)$-vector $\underline{x}^*\neq\underline{0}$ such that

$$(9) \qquad\qquad \lambda^{m+1}\underline{x}^*-\lambda^{m}L\underline{x}^*-\lambda M\underline{x}^*-N\underline{x}^*=\underline{0}.$$

By putting $\underline{x}^*:=(\rho,\underline{x}^{\mathbf{T}})^{\mathbf{T}}$, $\underline{x}$ suitable r-vector, we can split

(9) into two parts:

(9a)
$$\lambda^{m+1}\rho - \lambda^m(1+\alpha\eta)\rho - \lambda\beta\underline{w}^{T}(I-\alpha A)^{-1}A\underline{x} - \beta\eta\rho = 0$$

(9b)
$$\lambda^{m+1}\underline{x} - \lambda\beta(I-\alpha A)^{-1}A\underline{x} - \rho(\lambda^m\alpha+\beta)(I-\alpha A)^{-1}\underline{u} = \underline{0}.$$

If $\rho=0$, then we get $\underline{x}\neq\underline{0}$; moreover (9b) yields

$$(I-\gamma A)\underline{x}=\underline{0}, \quad \gamma:=\alpha+\beta/\lambda^m.$$

On the other hand assumption (7) and $|\lambda|\geqq 1$ easily imply $Re(\gamma)<0$. Thus, by Lemma 2, we have an absurdity, for then it must be $\rho\neq 0$.

In this case, by simple calculations, (9a) and (9b) yield

$$\lambda=1+\gamma\underline{w}^{T}(I-\gamma A)^{-1}\underline{u}, \quad \gamma:=\alpha+\beta/\lambda^m.$$

Since $Re(\gamma)<0$, this leads to a contradiction. In fact, by Lemma 2, we get $|\lambda|<1$. Q.E.D.

3. NUMERICAL EXPERIMENTS. In Section 1 we said that the method we are considering seems not to be GP-stable. Now we prove this fact for $r=1$.

In this simple case the nodal and the uniform order of accuracy are both equal to 2. Moreover the collocation polynomial is nothing but the linear interpolant of nodal values.

We follow the approach of Jackiewicz [7] and drop the restriction (4) on the step-length. The simplicity of the method allows us to write the following scalar recurrence relation for the nodal values y_i, $i\geqq m$:

$$(1-\alpha/2)y_{i+1}=(1+\alpha/2)y_i+\beta u y_{i-m+1}+\beta(1-u)y_{i-m}$$

where m is the integer part of $\tau/h-1/2$ and $u:=\tau/h-1/2-m$ belongs to $[0,1)$.

The GP-stability region of the method is

$$S:=\{(\alpha,\beta)\in\mathbb{C}\times\mathbb{C} \mid \lim_{i\to\infty} y_i=0 \text{ for all } m\geqq 1 \text{ and } u\in[0,1)\}.$$

As in [7], we restrict ourselves to consider a and b real numbers. By following the same arguments, we can prove that

$$S\cap\mathbb{R}\times\mathbb{R}=\{(\alpha,\beta)\in\mathbb{R}\times\mathbb{R} \mid \min[2,-\alpha]>|\beta|\}.$$

This is the same GP-stability region of the trapezoidal rule (θ-method for $\theta=1/2$). Moreover it is clear that the method is not GP-stable.

We have tested our results on the following example:

$$y'(t)=-50y(t)+40y(t-1), \quad t>0$$

(10)

$$y(t)=\exp(-t) \quad \text{for} \quad -1\leq t\leq 0.$$

The coefficients of (10) satisfy (3), but they require a step-length $h<1/20=0.05$ in order that the couple (α,β) lies in the GP-stability region S of the method. On the other hand, by the P-stability of the method itself, any step-length $h=1/m$, m positive integer, is allowed. These theoretical considerations fit in with the numerical experiments very well.

In Table 1 one can see that the difference between a step-length satisfying (4) and another which does not is very surprising. The comparisons are made at the point $t_f=10$, with respect to the reference solution $y(t_f)=0.131413237\ldots$ (this value was computed by means of a higher order method. However, in principle, the solution y of (10) can be computed analytically).

Near the error obtained with a constant step-length $h=1/m$ (which satisfies (4)), we have written the error obtained by using a mesh of the following type: in each interval $I_j=[j-1,j]$, $j=1,\ldots,10$, the uniform mesh of step-length $h=1/m$ is shifted with respect to the mesh of the preceding interval I_{j-1} by a quantity $jh/11$; moreover it includes the breaking points of the solution, i.e. the integers $j=0,1,\ldots,10$.

This latter mesh is not a constrained mesh, since it heavily violates condition (iii) (see Section 1). On the other hand, since for $r=1$ the order of uniform approximation of the collocation solution is equal to the nodal order, and since the mesh includes the breaking points, there is no loss in the resulting order of the method for DDEs (see Bellen-Zennaro [4]). Nay, since each step-length h_i of the latter mesh is $\leq h=1/m$, one could expect that the results obtained in the second way are better.

We therefore conclude that the great discrepancy is due to the different properties of stability of the two methods. This is suggested also by the weakening of the difference as the step-length becomes smaller.

Remark that a comparison with a uniform mesh which does not satisfy (4) would not be correct, since it does not include the breaking points of the solution, causing a loss in the resulting order of the method.

m	h=1/m	error (constrained mesh)	error (free mesh)
2	0.500	0.18E+00	0.17E+01
3	0.333	0.57E-01	0.85E+01
4	0.250	0.54E-02	0.34E+01
5	0.200	0.90E-02	0.51E+00
10	0.100	0.10E-02	0.85E-01
20	0.050	0.47E-04	0.21E-03
50	0.020	0.75E-05	0.32E-04
100	0.010	0.19E-05	0.79E-05
200	0.005	0.47E-06	0.20E-05

Table 1.

The situation in Table 1 makes the step-size control technique a not practical tool. In fact Table 2 shows that it yields errors the magnitudes of which are much larger than those obtained by constant step-lengths satisfying (4). On the other hand the proportionality of the global error to the given tolerance per unit step and to the square of the average step-length is almost perfect, showing that the local error is well computed. Note that

there are many rejected steps. We have used the local error estimate proposed by Zennaro [14] .

We want to remark that we took care in order that the breaking points of the solution were included in the mesh; so no loss
of accuracy is due to the non smoothness of the problem.

tolerance	total number of steps	rejected steps	average step-length	error
10^{-1}	235	18	0.04255	0.17E-02
10^{-2}	649	27	0.01541	0.23E-03
10^{-3}	2071	34	0.00483	0.26E-04
10^{-4}	6556	39	0.00153	0.29E-05

Table 2.

It seems clear that when a DDE such as (1) presents particular stiffness features as well as the linear equation (10), the
most appropriate way to apply the one-step collocation method at
Gaussian points is that of using a constrained mesh (that is the
method presented in [3]). We conjecture that this situation extends to many other Runge-Kutta methods, also for equations
with variable delay.

REFERENCES

1. A.N.Al-Mutib, Stability properties of numerical methods for
   solving delay differential equations, J.Comput.Appl.Math. 10
   (1984), 71-79.
2. V.K.Barwell, Special stability problems for functional differential equations, BIT 15(1975), 130-135.
3. A.Bellen, One-step collocation for delay differential equations, J.Comput.Appl.Math. 10(1984), 275-283.
4. A.Bellen, M.Zennaro, Numerical solution of delay differential
   equations by uniform corrections to an implicit Runge-Kutta

method, preprint.

5. C.W.Cryer, Highly stable multistep methods for retarded differential equations, SIAM J.Numer.Anal. 11(1974), 788-797.

6. I.Gohberg, P.Lancaster, L.Rodman, Matrix polynomials, Academic Press, New York, 1982.

7. Z.Jackiewicz, Asymptotic stability analysis of θ-methods for functional differential equations, Numer.Math. 43(1984), 389-396.

8. Z.Jackiewicz, V.L.Bakke, Stability analysis of linear multistep methods for delay differential equations, preprint.

9. H.J.Oberle, H.J.Pesh, Numerical treatment of delay differential equations by Hermite interpolation, Numer.Math. 37(1981), 235-255.

10. A.Prothero, A.Robinson, On the stability and accuracy of one-step methods for solving stiff systems of ordinary differential equations, Math.Comp. 28(1974), 145-162.

11. P.J. van der Houven, B.P.Sommeijer, Stability in linear multistep methods for pure delay equations, J.Comput.Appl.Math. 10(1984), 55-63.

12. R.Vermiglio, A one-step subregion method for delay differential equations, preprint.

13. M.Zennaro, Natural continuous extensions of Runge-Kutta methods, preprint.

14. M.Zennaro, One-step collocation: uniform superconvergence, predictor-corrector method, local error estimate, SIAM J.Numer.Anal. (in press).

International Series of
Numerical Mathematics, Vol. 74
© 1985 Birkhäuser Verlag Basel

SOME OPEN PROBLEMS

D. BRAESS, Rational Approximation.

Let $E_{nn}(f)$ denote the sup-norm of the error of the best approximation to f by rational functions of degree (n,n) on the unit interval.

Due to the Lip 1 Theorem one has

$$E_{nn}(f) = o(n^{-1})$$

for any Lip 1 function f. In this context, the following question arises. Given a sequence (ε_n) with $\varepsilon_n > 0$ and $\lim \varepsilon_n = 0$, does there exist a Lip 1 function f with

$$E_{nn}(f) \geq \frac{\varepsilon_n}{n} \tag{$*$}$$

for all $n \geq 1$? It seems to be known only how to construct a Lip 1 function such that $(*)$ holds for infinitely many n's: Let N be a subset of natural numbers such that $\sum_{k \in N} \varepsilon_{3^k} < \infty$. Note that $(*)$ holds for $k \in N$, if f is defined by

$$f(x) = \sum_{k \in N} \varepsilon_{3^k} 3^{-(k+1)} T_{3^{k+1}}(x) \quad .$$

L. COLLATZ, Periodic Solutions for Hyperbolic and Parabolic Delay
Equations.

Ordinary differential equations (D.E.) with delay-term have
been investigated since long time extensively in the linear
and in the nonlinear case, compare f.i. workshop [83], but
it may be perhaps desirable, to develop corresponding theo-
ries for special types of partial D.E., compare Hoffmann-
Sprekels [83]; for ordinary D.E. one has results for initial
value problems, for asymptotic behaviour and many other questions, as f.i. possible periodic solutions (compare Bellen-
Zennaro [83], an der Heiden [83] a.o.). Periodic solutions
are known also for nonlinear partial D.E. usually without a
delay-term; but nonlinear retarded partial D.E. of the type

$$u_t(x_j,t) = \Phi(Lu(x_j,t)) + ku(x_j,t-\tau)$$

or $\qquad\qquad\qquad\qquad\qquad\qquad\qquad\qquad\qquad\qquad\qquad\qquad$ (1)

$$u_{tt}(x_j,t) = \Phi(Lu(x_j,t)) + ku(x_j,t-\tau)$$

could be of interest for the applications. Here u is a function of $x_1,\ldots,x_n$, Lu a given (linear) differential operator,
f.i. the Laplacean operator Δu, Φ a given nonlinear function,
k and τ given constants; time-dependent functions $\tau(t)$ may
also occur. Then one can ask for periodic solutions.
New phenomena occur even in the simple linear case with one
independent real coordinate x.

I. Hyperbolic delay-equation

A string of length L may be fixed at the ends $x = 0$ and $x = L$;
we look for periodic displacements $u(x,t)$:

$$u(x,t) = \sin(ax) \cdot \sin(bt) \quad . \qquad\qquad\qquad (2)$$

Here may be $a = m\frac{\pi}{L}$ (integer m, number of inner knots m-1) and we suppose a constant time-delay τ.

We have $u(x,t-p\frac{\pi}{b}) = (-1)^p u(x,t) = \rho u(x,t)$ for integer p with $\rho = \pm 1$.

$u(x,t)$ satisfies the retarded hyperbolic equation

$$u_{tt}(x,t) = c^2 u_{xx}(x,t) + ku(x,t-\tau) \tag{3}$$

for

$$b^2 = c^2 a^2 - k\rho \quad , \quad \tau = p\frac{\pi}{b} \tag{4}$$

or

$$b^2 = m^2 \gamma - k\rho \quad \text{with} \quad \gamma = \frac{c^2 \pi^2}{L^2} . \tag{5}$$

For a given string L, c and therefore γ are fixed positive constants.

<u>Case 1</u>: We choose m,k,ρ and calculate b and τ from (5),(4); for $k\rho > m^2\gamma$ no oscillations are possible ($b^2 < 0$), and for $k\rho < m^2\gamma$ the frequency (which is proportional to b) of the oscillations depends on the delay.

<u>Case 2</u>: We choose b (as large as we wish) and satisfy (5) by suitable k,m.

Of course one can consider more general cases.

II. Parabolic delay-equation

We consider analogeously the heat conduction equation with delay-term in one dimension f.i. a beam, a pipe, a ring,

$$u_t(x,t) = cu_{xx}(x,t) + ku(x,t-\tau) \tag{6}$$

and we ask for periodic solutions $u(x,t)$ with prescribed period $\frac{2\pi}{a}$ with respect to x and unknown period $\frac{2\pi}{b}$ with respect to t; we can use the same function u as in (2), and u satisfies the parabolic delay-equation (6), if

$$k \cos b\tau = ca^2 \quad , \quad k \sin b\tau = -b \tag{7}$$

or

$$k^2 = c^2 a^4 + b^2 \quad , \quad \tan b\tau = \frac{-b}{ca^2} \tag{8}$$

hold. c and a are given constants.

<u>Case 1</u>: We prescribe k and get b and then τ from (8).

<u>Case 2</u>: We choose b (prescribed frequency) and get τ and then k from (8).

<u>Result</u>: We wave equation and the heat-conduction equation with one space-dimension (a finite interval on the real axis) admits periodic (harmonic solutions with arbitrarely prescribed frequencies, if one is adding to the equation a suitable delay-term. Again here one can consider more general cases, and one would be interested also here in nonlinear problems.

<u>References</u>.

Bellen, A. and Zennaro, H. [83], Maximum principles for periodic solutions of linear delay differential equations, in Workshop [83], 19-24.

an der Heiden, U. [83], Periodic, aperiodic and stochastic behavior ... modeling biologican and economical processes, in Workshop [83], 91-108.

Hoffmann, K.-H. and Sprekels, J. [83], Automatic Delay-Control in a Two Phase Stefan-Problem, in Workshop [83], 119-135.

Workshop Oberwolfach [83], Differential-Difference Equations, ed. Collatz, Meinardus, Wetterling, Birkhäuser 1983, 196 p.

R.S. VARGA, Rational and Polynomial Approximation.

1. Let $E_n(|x|) := E_n(|x| ; [-1,+1],\infty)$ be the best uniform approximation of $|x|$, by polynomials from Π_n, on $[-1,+1]$.

S. Bernstein (1914) has shown that there exists a $\beta > 0$ for which

$$\lim_{n \to \infty} 2n \, E_{2n}(|x|) = \beta \ .$$

Is there another equivalent definition of β which can be based on confluent hypergeometric, or elliptic functions?

2. Let $\Pi_{m,n}^r$ and $\Pi_{m,n}^c$ be respectively the sets of rational functions with real (and with complex) coefficients. It is known (A.A. Gončar, K.N. Lungu, E.B. Saff/R.S. Varga) that if

$$\gamma_{m,n}^I := \inf \left\{ \frac{E_{m,n}^c(f;[-1,+1],\infty)}{E_{m,n}^r(f;[-1,+1],\infty)} : f \in C^r[-1,+1] \text{ with } f \notin \Pi_{m,n}^r \right\}$$

then $\gamma_{m,n}^I$ can be less than unity. Recently, Gutknecht/Trefethen has shown, somewhat surprisingly, that

$$\gamma_{m,n}^I = 0 \quad \text{for all} \quad n \geq m+3 \ ,$$

and they <u>conjecture</u> that

$$\gamma_{m,n}^I = 0 \quad \text{iff} \quad n \geq m+3 \ .$$

This open conjecture is an interesting problem.

H.O. WALTHER, A Uniqueness Problem for a Nonlinear Differential Delay Equation.

The class of autonomous retarded functional differential equations which is relatively best understood today is given by

$$\dot{x}(t) = \alpha f(x(t-1)) \qquad\qquad (\alpha f)$$

with $f : R \to R$ satisfying $xf(x) < 0$ for $x \neq 0$, $f(0) = 0$. If positive parameters α are considered then (αf) represents a

simple case of delayed negative feedback: A deviation
$x(t-1) \lessgtr 0$ from the equilibrium solution $t \to 0$ is followed by
a move $\dot{x}(t) \gtrless 0$ in the opposite direction. The dynamics of
(αf) depend in a very subtile way on the graph of f, and many
of the bifurcation phenomena which are of interest in dissi-
pative O.D.E.s may be found in this small class of F.D.E.s.
Understanding (αf) will also help to get a better feeling
for the equations

$$\dot{x}(t) = p(x_t) - d(x_t) \tag{p,d}$$

where the functionals p,d describe autocatalytic production
and destruction respectively. p and d are defined on
$C([-1,0],R)$ and $x_t \in C$ is given by $x_t(a) = x(t+a)$ for $[t-1,t]$
in the domain of x - as usual in F.D.E.s [3]. The interplay
of autocatalytic production and destruction is common to
many control processes in living systems [4].
A special and also historically important case is Hutchinson's equation for delayed logistic growth of a single spe-
cies

$$\dot{n}(t) = rn(t)[1 - \frac{n(t-\tau)}{K}] , \qquad r,\tau,K \text{ positive} \tag{n}$$

[5]. The positive solutions (which are the biologically
meaningful ones) correspond to the set of all solutions of
(αf) with $\alpha = r\tau$, $f = f_H$, $f_H(x) = 1-e^x$, via $x(t) = \log \frac{n(\tau t)}{K}$.
It is well known that for every $\alpha > \pi/2$ equation (αf_H) has a
periodic solution x with $x(-1) = 0$, $\dot{x} > 0$ on $[-1,0)$, $\dot{x} < 0$ on
an interval $(0, z_1+1)$ with $x(z_1) = 0$, $\dot{x} > 0$ on (z_1+1, z_2), and
$x(t) = x(t+z_2+1)$ for all real t. Numerical results strongly
suggest that x has a stable and attractive orbit in C, and

that for $\alpha < \pi/2$ no periodic solutions exist.

<u>Problem</u>: Prove uniqueness and stability properties of x!
The tools might be available, compare [6,8]. Related results
are contained in [15,11,12,9]. Solving this problem will be
instructive for the investigation of a larger set of equations: f_H is not an <u>odd</u> function - one can give reasons that
most of the nonlinearities f related to applications are far
from being odd - most of the more detailed results on bifurcation of periodic solutions were obtained for odd functions
f only.

Another promising, possibly harder <u>problem</u> is to show for <u>all</u>
equations (αf) that the set of initial conditions $\phi \in C$ which
define slowly oscillating solutions - i.e. solutions x with
$|z-z'| > 1$ for every pair of zeros $z \neq z'$ in some unbounded
interval $[t_x,\infty)$ - is open and dense. For a partial result,
see [13]. The first statement of the conjecture is in [6].

<u>Suggestions for numerical analysis</u>: It is desirable to im-
prove and develop algorithms for the computation of bifurcation diagrams for slowly oscillating solutions of (αf)
[2,10]. In particular stability properties (Floquet multipliers) of these periodic solutions should be studied, compare [14,1].

<u>References</u>

1   S. Chapin: Periodic solutions of some nonlinear differential delay equations. Ph.D. thesis, Michigan State U.
1983.

2   K.P. Hadeler: Effective computation of periodic orbits
and bifurcation diagrams in delay equations. Numer. Math.
34, 457-467 (1980).

3   J.K. Hale: Theory of functional differential equations.
     New York-Heidelberg-Berlin: Springer 1977.

4   U. an der Heiden, M.C. Mackey: The dynamics of produc-
     tion and destruction: Analytic insight into complex be-
     havior. J. Math. Biology 16, 75-101 (1982).

5   G.E. Hutchinson: Circular causal systems in ecology.
     Annals of the New York Acad. of Sci. 50, 221-246 (1948).

6   J.L. Kaplan, J.A. Yorke: On the stability of a periodic
     solution of a differential delay equation. SIAM J. Math.
     Analysis 6, 268-282 (1975).

7   R.D. Nussbaum: A global bifurcation theorem with appli-
     cations to functional differential equations. J. Func-
     tional Analysis 19, 319-339 (1975).

8   R.D. Nussbaum: Uniqueness and nonuniqueness for periodic
     solutions of $x'(t) = -g(x(t-1))$. J. Differential Equations
     34, 25-54 (1979).

9   R.D. Nussbaum: Asymptotic analysis of functional diffe-
     rential equations and solutions of long period. Arch.
     Rat. Mech. Analysis 81, 373-397 (1983).

10   D. Saupe: Beschleunigte PL-Kontinuitätsmethoden und perio-
     dische Lösungen parametrisierter Differentialgleichungen
     mit Zeitverzögerung. Ph.D. thesis, Bremen 1982.

11   H.O. Walther: Stability for attractivity regions of auto-
     nomous functional differential equations. Manuscripta
     math. 15, 339-363 (1975).

12   H.O. Walther: A theorem on the amplitudes of periodic
     solutions of differential delay equations with applica-
     tions to bifurcation. J. Differential Equations 29,
     396-404 (1978).

13   H.O. Walther: Density of slowly oscillating solutions of
     $\dot{x}(t) = -f(x(t-1))$. J. Math. Analysis Appl. 79, 127-140
     (1981).

14   H.O. Walther: Bifurcation from periodic solutions in
     functional differential equations. Math. Z. 182, 269-289
     (1983).

15   E.M. Wright: A nonlinear differential-difference equation.
     J. Reine Angew. Math. 194, 66-87 (1955).